T0142319

Lecture Notes in Physics

Springer
Berlin
Heidelberg
New York
Barcelona
Hong Kong
London
Milan
Paris
Tokyo

Physics and Astronomy

ONLINE LIBRARY

http://www.springer.de/phys/

Editorial Policy

The series *Lecture Notes in Physics* (LNP), founded in 1969, reports new developments in physics research and teaching -- quickly, informally but with a high quality. Manuscripts to be considered for publication are topical volumes consisting of a limited number of contributions, carefully edited and closely related to each other. Each contribution should contain at least partly original and previously unpublished material, be written in a clear, pedagogical style and aimed at a broader readership, especially graduate students and nonspecialist researchers wishing to familiarize themselves with the topic concerned. For this reason, traditional proceedings cannot be considered for this series though volumes to appear in this series are often based on material presented at conferences, workshops and schools (in exceptional cases the original papers and/or those not included in the printed book may be added on an accompanying CD ROM, together with the abstracts of posters and other material suitable for publication, e.g. large tables, colour pictures, program codes, etc.).

Acceptance

A project can only be accepted tentatively for publication, by both the editorial board and the publisher, following thorough examination of the material submitted. The book proposal sent to the publisher should consist at least of a preliminary table of contents outlining the structure of the book together with abstracts of all contributions to be included.

Final acceptance is issued by the series editor in charge, in consultation with the publisher, only after receiving the complete manuscript. Final acceptance, possibly requiring minor corrections, usually follows the tentative acceptance unless the final manuscript differs significantly from expectations (project outline). In particular, the series editors are entitled to reject individual contributions if they do not meet the high quality standards of this series. The final manuscript must be camera-ready, and should include both an informative introduction and a sufficiently detailed subject index.

Contractual Aspects

Publication in LNP is free of charge. There is no formal contract, no royalties are paid, and no bulk orders are required, although special discounts are offered in this case. The volume editors receive jointly 30 free copies for their personal use and are entitled, as are the contributing authors, to purchase Springer books at a reduced rate. The publisher secures the copyright for each volume. As a rule, no reprints of individual contributions can be supplied.

Manuscript Submission

The manuscript in its final and approved version must be submitted in camera-ready form. The corresponding electronic source files are also required for the production process, in particular the online version. Technical assistance in compiling the final manuscript can be provided by the publisher's production editor(s), especially with regard to the publisher's own Latex macro package which has been specially designed for this series.

Online Version/ LNP Homepage

LNP homepage (list of available titles, aims and scope, editorial contacts etc.):
http://www.springer.de/phys/books/lnpp/
LNP online (abstracts, full-texts, subscriptions etc.):
http://link.springer.de/series/lnpp/

G. Reiter J.-U. Sommer (Eds.)

Polymer Crystallization

Observations, Concepts and Interpretations

 Springer

Editors

Günter Reiter
CNRS-UHA
Institut de Chimie des Surfaces
et Interfaces
15, rue Jean Starcky, BP 2488
68057 Mulhouse Cedex, France

Jens-Uwe Sommer
CNRS-UHA
Institut de Chimie des Surfaces
et Interfaces
15, rue Jean Starcky, BP 2488
68057 Mulhouse Cedex, France

Cover Picture: (see contribution "Crystallization of Polymers in Thin Films" by Günter Reiter, Gilles Castelein and Jens-Uwe Sommer in this volume)

Library of Congress Cataloging-in-Publication Data

Polymer crystallisation: observations, concepts, and illustrations / G. Reiter, J.-U. Sommer (eds.).
   p. cm. -- (Lecture notes in physics, ISSN 0075-8450 ; 606)
   Includes bibliographical references.
   ISBN 3-540-44342-8 (acid-free paper)
   1. Crystaline polymers. 2. Polymers. 3. Crystallisation. I. Reiter, G. (Günther), 1960-
II. Sommer, J.-U. (Jens-Uwe), 1963- III. Series.

QD382.C78P65 2003
547'.7--dc21

2002045271

ISSN 0075-8450
ISBN 3-540-44342-8 Springer-Verlag Berlin Heidelberg New York

Springer-Verlag Berlin Heidelberg New York
a member of BertelsmannSpringer Science+Business Media GmbH

http://www.springer.de

© Springer-Verlag Berlin Heidelberg 2003
Printed in Germany

Typesetting: Camera-ready by the authors/editor
Camera-data conversion by Steingraeber Satztechnik GmbH Heidelberg
Cover design: *design & production*, Heidelberg

Printed on acid-free paper
54/3141/du - 5 4 3 2 1 0

Preface

The classical view on polymer crystallization basically focused on the explanation of a few macroscopically observable parameters like the thickness of the resulting lamellar structure and the corresponding growth rates. However, the emerging paradigm for the description of chain crystals is too simple and cannot account for the complex non-equilibrium processes responsible for structure formation on various levels, ranging from the nanometer up to the millimeter scale. This complexity detected by several novel experimental results led to a renewed interest in this "old" topic of polymer crystallization. These new findings concern the early stages of the crystallization process, crystal formation in confined geometries like ultra-thin films and the competition between (micro)phase separation and crystallization in copolymers and blends. In particular, high spatial resolution techniques such as atomic force microscopy provided deeper insight into the molecular organization of crystallizable polymers. Computer simulations based on microscopic processes were used to improve our understanding of how polymer crystals are nucleated and how they grow. New ideas emerged about possible multistage pathways which are followed during the formation of polymer lamellae. The importance and the consequences of the non-equilibrium character of polymer crystals got significantly more attention. Links and analogies to growth phenomena and pattern formation in general are being developed. However, these ideas are still subject of intensive and controversial discussions. As a result of these discussions, a number of novel experiments and computer simulations have been designed with the aim to discriminate between the underlying basic assumptions. Obviously, this present situation needs a common platform, which is the aim of this volume of "Lecture Notes in Physics".

We tried to assemble a collection of novel experimental results and theoretical concepts reflecting the state-of-the-art in polymer crystallization. This comprises phenomena at the onset of crystallization, kinetically controlled growth and subsequent relaxation, responsible for the formation of complex structures with order on several length-scales. Although the content of this volume is already rather substantial, we do not want to pretend that all aspects and new results currently under consideration are included. Nevertheless, it provides a broad overview of the ongoing research on the subject of polymer crystallization to readers with a general background in polymer physics as well as to experts in this field.

Although this book cannot give definite answers to many questions still controversially discussed, it may stimulate new works and attempts shedding more light on these problems. The new results also demand profound revisions of the existing theoretical models and ask for a new conceptual understanding of polymer crystallization as a non-equilibrium process.

We hope that such a collective endeavour will help to advance our understanding of polymer crystallization!

Mulhouse   *Jens-Uwe Sommer*
August 2002   *Günter Reiter*

Table of Contents

List of Contributors

Mahmoud Al-Hussein
Albert-Ludwigs-Universität,
Fakultät für Physik,
Hermann-Herder-Str. 3
79104 Freiburg, Germany

Hans-Georg Braun
Max Bergmann Center for Biomaterials and
Institute of Polymer Research
Dresden,
Microstructure Group,
01069 Dresden, Germany

Gilles Castelein
Institut de Chimie des Surfaces et
Interfaces CNRS
et Université de Haute Alsace,
Mulhouse, 15 rue Jean Starcky,
68057 Mulhouse, France

Wim H. de Jeu
FOM-Institute for Atomic and
Molecular Physics,
Kruislaan 407,
1098 SJ Amsterdam, The Netherlands

Gaetano Di Marco
C.N.R.,
Istituto per i Processi Chimico–Fisici,
sez. Messina,
Via La Farina 237,
98123 Messina, Italy

Tiberio A. Ezquerra Sanz
Instituto de Estructura de la Materia,
C.S.I.C.,
Serrano 119,
Madrid 28006, Spain

Jamie K. Hobbs
H.H. Wills Physics Laboratory,
Tyndall Avenue,
Bristol BS81TL. UK

Dimitri A. Ivanov
Laboratoire de Physique des
Polymères,
Université Libre de Bruxelles,
CP223, Boulevard du Triomphe,
B-1050
Brussels, Belgium

Bernd-J. Jungnickel
Deutsches Kunststoff-Institut,
Schlossgartenstr. 6,
64289 Darmstadt, Germany

Zebene Kiflie
Università di Palermo,
Dipartimento di Ingegneria Chimica,
Viale delle Scienze,
90128 Palermo, Italy

Tsunehisa Kimura
Tokyo Metropolitan University,
Department of Applied Chemistry,
1-1 Minami-ohsawa, Hachoiji,
Tokyo 192–0397, Japan

Sergei N. Magonov
Digital Instruments/Veeco Metrology
Group,
112 Robin Hill Rd.,
Santa Barbara, CA 93117, U.S.A.

Evelyn Meyer
Max Bergmann Center
for Biomaterials and
Institute of Polymer Research
Dresden,
Microstructure Group,
D-01069 Dresden, Germany

Hendrik Meyer
Institut Charles Sadron, CNRS,
6, rue Boussingault,
67083 Strasbourg, France

Alaa Mohammed
University of Rostock,
Department of Physics,
18051 Rostock, Germany

Aurora Nogales Ruiz
University of Reading,
J.J. Thomson Physics Laboratory,
Witheknights,
Reading RG6 6AF, U.K.

Norimasa Okui
Tokyo Institute of Technology,
Department of Organic and Polymeric
Materials,
International Research Center of
Macromolecular Science,
Ookayama, Meguroku, Tokyo, Japan

Gerrit W.M. Peters
Eindhoven University of Technology,
Materials Technology
(www.mate.tue.nl),
Dutch Polymer Institute,
P.O. Box 513,
5600 MB Eindhoven, The Netherlands

Stefano Piccarolo
Università di Palermo,
Dipartimento di Ingegneria Chimica,
Viale delle Scienze,
90128 Palermo, Italy

Marco Pieruccini
C.N.R.,
Istituto per i Processi Chimico–Fisici,
sez. Messina,
Via La Farina 237,
98123 Messina, Italy

Sanjay Rastogi
Department of Chemical Engineer-
ing/Dutch Polymer Institute,
P.O.Box 513,
5600 MB Eindhoven, The Netherlands

Günter Reiter
Institut de Chimie des Surfaces et
Interfaces CNRS
et Université de Haute Alsace,
Mulhouse, 15 rue Jean Starcky,
68057 Mulhouse, France

Jens Rieger
BASF Aktiengesellschaft,
Polymer Physics,
67056 Ludwigshafen, Germany

Christoph Schick
University of Rostock,
Department of Physics,
18051 Rostock, Germany

Jens-Uwe Sommer
Institut de Chimie des Surfaces et
Interfaces CNRS
et Université de Haute Alsace,
Mulhouse, 15 rue Jean Starcky,
68057 Mulhouse, France

Gert Strobl
Albert-Ludwigs-Universität,
Fakultät für Physik,
Hermann-Herder-Str. 3
79104 Freiburg, Germany

Joan Josep Suñol
Universitat de Girona,
Grup de Recerca en Materials i
Termodinàmica,
17071 Girona, Spain

Susumu Umemoto
Tokyo Institute of Technology,
Department of Organic and Polymeric
Materials,
International Research Center of
Macromolecular Science,
Ookayama, Meguroku, Tokyo, Japan

Elena Vassileva
Universität Kaiserslautern,
Institut für Verbundwerkstoffe GmbH,
Erwin Schrödingerstrasse Geb.58,
67663 Kaiserslautern, Germany

Mingtai Wang
Max Bergmann Center
for Biomaterials and
Institute of Polymer Research
Dresden,
Microstructure Group,
01069 Dresden, Germany

Andreas Wurm
University of Rostock,
Department of Physics,
18051 Rostock, Germany

1 Introduction

Jens-Uwe Sommer and Günter Reiter

Institut de Chimie des Surfaces et Interfaces CNRS et Université Haute Alsace
Mulhouse, 15 rue Jean Starcky, F–68057 Mulhouse, France

The discovery of folded chain crystals by Andrew Keller in 1957 has confronted the community of polymer scientists with the fact that one of the most important forms of polymer matter is not in thermodynamic equilibrium. The basic units of a crystallized polymer substance are thin quasi two-dimensional lamellae separated by amorphous material. The thickness of these lamellae is usually much smaller than the length of fully extended chains but only the latter generally corresponds to an equilibrium thermodynamic state. A huge number of experiments has been carried out on well reproducible non-equilibrium states of such comparatively thin lamellae. As a consequence of the fact that non-equilibrium states depend of their history, a number of standard preparation methods have been classified such as rapid quenching, "cold crystallization" by annealing from the glassy state or "self-seeding" by re-cooling after melting. Standard techniques for the observation of polymer crystallization are optical microscopy (observation of spherulitic super-structures and their growth rates), electron microscopy (observation of the morphology of the lamellar structure), scattering techniques (crystallinity, lamellar thickness and lamellar separation) and calorimetric methods such as differential scanning calorimetry (DSC) for measuring thermodynamic properties of the crystallization and melting behavior. These experiments have provided, among many other observations, some quit general empirical results for the lamellar thickness, growth rate and apparent melting points as a function of the crystallization temperature.

The first attempts to understand the state of crystallized polymers based on molecular properties used ad-hoc arguments about the growth of polymer lamellae from disordered chain molecules. The central assumption of these models was that the once formed polymer lamellae are quasi-stable (or completely frozen-in), so that all interesting processes take place exclusively at the growth front. The small thickness of the lamellae was explained by either assuming a secondary nucleation barrier (which increases with thickness) or by assuming a self-organized stretching process of chain parts at the growth front (which imposes a kinetic barrier increasing with lamellar thickness). These classical models clearly focused on the growth phenomenon of chain crystals. Many other problems, such as primary nucleation or anomalies at the melting point of crystalline polymers are still lacking consistent and well accepted explanations.

In recent years, the experimental studies of crystalline polymers made substantial and qualitative advances due to newly developed methods (such as atomic force microscopy) or significant improvement of previously used techniques (such as simultaneous wide- and small angle X-ray scattering). Moreover, the interest in nano-meter sized structures, thin polymer films and copolymers has opened new windows for the observation of crystallization processes. Viewed in the light of these new developments, it becomes more and more clear that the crystalline state of polymer matter is much more complex then previously anticipated and that many other problems such as nucleation, morphological reorganization and non-equilibrium annealing processes must be taken into account for a consistent model of polymer crystallization. An overview of over some of the new developments and results in the general context of crystal formation is given in Chap. 2.

Deviations from the simplified picture of a one-step growth processes leading to quasi-stable lamellae which melt at their thermodynamic stability limit have been observed very early. In the first place we note the detection of relaxation and reorganization phenomena taking place after growth. The classical example for which reorganization and different morphological phases have been observed is polyethylene (PE) crystallized under high pressure. Here, a particularly mobile hexagonal phase (PE crystallizes in a orthorhombic phase under ambient conditions) has been be found which allows considerable thickening of the lamellae after growth. However, it can be shown that also under low pressure conditions the hexagonal phase is obtained, but as a meta-stable intermediate phase which undergoes a spontaneous change into the orthorhombic phase. This led to an ad hoc interpretation in terms of a generalized state diagram. A discussion of this phenomenon is given in Chap. 3 where novel experiments and their interpretation are presented. Reorganization phenomena of the non-equilibrium lamellae are also observed in other systems like polypropylene (PP) or polystyrene (PS). Noting that there the effect takes place on a smaller scale, in Chap. 4 it is shown that annealing of crystalline samples below the melting-point results in a change of state of the lamellae (detected by an increase of the averaged lamellar thickness). Explanations of the different phenomena observed here can hardly be obtained using a simple one-step process for polymer crystallization. On the other hand a multi-stage model proposed by G. Strobl can been applied to interprete the behavior obtained in these experiments. In this model intermediate meso-phases are introduced which are prone to reorganize in later stages of the formation of lamellae. A novel method to detect such intermediate or precursor phases in the formation of polymer crystals is the use of an external magnetic field. Usually, the applied field strength is much too weak to orient the chains in the amorphous state. Therefore, measuring simultaneously the anisotropy of the sample (birefringence) and the degree of crystallinity (for instance as obtained by scattering techniques) one can identify pre-oriented structures in the induction period before a crystalline fraction is observable in scattering experiments as discussed in Chap. 5.

Considering the complexity of polymer crystallization, the question arises of what kind of theoretical models can help to understand the experimental findings. Here, in the first place computer simulations are suitable tools for exploring general aspects of the formation of chain crystals. Two strategies can be embarked on in the search for gaining more insight into the processes of polymer crystallization. First, sufficiently simplified chain models can be simulated at high under-coolings using Molecular Dynamics methods. Using comparatively short chains the formation of chain crystals can be followed right from the onset. In Chap. 10 it is shown that even without attractive intermolecular forces large under-coolings drive dense polymer systems into stretched or folded states of crystalline order. Here, energy is gained by slightly stiffening the segments. Spontaneous (homogeneous) nucleation takes place at high under-coolings and large hysteresis effects are observed when temperature cycles are applied. For longer chains, the empirically observed relation between the lamellar thickness and under-cooling is obtained also for these systems. Alternatively, simulations can be used to investigate the behavior of much more coarse-grained and generalized models which treat growth and reorganization events on a more abstract level. Here, one asks the question of what parameters are indispensable to understand crystallization processes of chain molecules and to study the impact of these model parameter on large length scales which are experimental accessible. Then, experiments can be designed to test the predictions of such a model directly. In Chap. 9 this approach has been chosen for polymer crystallization in thin films. There, diffusion processes strongly control the morphology of the obtained crystalline structure resulting in finger-like growth patterns. The latter depend significantly on the temperature and the degree of folding of the individual chains. It is possible to compare directly the simulations results and predictions to experiments on polyethyleneoxide monolayers as presented in Chap. 8. Most interestingly, the obtained growth structures show reorganization processes and significant morphological changes during annealing. It can be shown that not the originally grown structure melts but only its descendants in the course of morphlogical evolution. This is in marked contrast to classical models where the structure obtained during growth is directly responsible for the limiting thermal stability, i.e. melting.

One of the greatest advances in the study of polymer crystallization is related to real space observation of polymer crystals down to the nano-meter scale. This has recently been made possible by the Atomic Force Microscopy (AFM) technique. In Chap. 7 the AFM-techniques is introduced and applied to several systems of varying degree of complexity. It is shown that such real space information obtained by AFM helps to remove some ambiguity inherent to information obtained in reciprocal space as obtained by scattering techniques. Thus, the combination of complementary techniques is sometimes necessary. In Chap. 6 first in-situ observations of the growth of melting of individual polymer lamellae are presented. It is found that growth of individual lamellae apparently occurs sporadically in space and time and that the growth front of the lamellae is significantly more irregular than predicted in secondary nucleation models. The strong influence of diffusion processes even at low under-coolings is indicated as well by these experiments. The heating of individual lamellae shows that parts of the lamellae melt at a given temperature but melting is then stopped and only continued at higher temperatures. This indicates that either parts of the crystal have very different stabilities or, parallel to melting, reorganization and improvement of lamellae takes place.

Thin polymer films are increasingly attractive both for applications and for experimental studies. The combination of AFM with scattering methods allows to gain new insights into the chain ordering. When polymers are confined in a monolayer it has been shown that spontaneous nucleation events are not observable. However, if nucleation sites are provided to such systems a clear-cut growth-only process is started from these nucleation sites which results in the formation of single, flat-on polymer lamellae. Since area density during crystallization is strongly changed (upright chain stems in the crystalline phase in constrast to rather flattened adsorbed polymers in the molten state) diffusion processes take over the control of the resulting patterns. The morphology of the growth patterns depend strongly on temperature due to different degrees of folding of the chains. Therefore, the obtained morphologies "magnify" the underlying ordering processes. In Chap. 8 such experiments are described in detail and the resulting structures have been observed using AFM. In contrast to investigations in the bulk, now the fine-structure of individual lamellae (as also demonstrated in Chaps. 6 and 7) can be measured and analyzed after morphological changes. As a result of these experiments, it can be shown again that individual lamellae undergo considerable reorganization in the crystalline state and show a non-homogeneous surface topography. When annealed these lamellae do not melt directly but undergo various morphological transformations as predicted in the simulations of Chap. 9. The growth of dendritic 2D lamellar structures due to polymer film crystallization on patterned surfaces is investigated in Chap. 13.

Another area of recent intensive studies concerns the crystallization in block-copolymer meso-phases. Here, the crystallizable chains are restricted and partially separated in (soft-)ordered phases having characteristic length scales which are comparable to the size of the crystalline lamellae. Generally, the driving force for crystallization is much higher than that for soft-order mechanisms such as demixing and phase separation. Nevertheless, the pre-existing structure will influence the nucleation behavior and the kinetics of crystallization considerably. In Chap. 11 nearly symmetric copolymers are investigated which were spin-cast as a thin film onto a solid substrate. Here, the use of X-ray reflectometry allows to indentify the order of chains within the stack of lamellae forming the film of alternatingly crystalline anf amorphous layers. In particular, the temperature dependence of such order can be detected with high precision. In Chap. 8 highly asymmetric copolymers have been studied where the much shorter block provides the crystallizable phase. This phase consists now of isolated spheres having a diameter of about 12 nm only. Spontaneous i.e homogeneous nucleation takes place at high under-coolings in each nano-sphere individually. On the way to melting the individual spheres show remarkable reorganization processes resulting in a different thermal stability although all spheres can be regarded as equivalent and have exactly the same thermal history. Also mixing a crystallizing polymer with another polymer can yield new insights into the crystallization process. Here, crystallization kinetics competes either with demixing kinetics or thermodynamic forces which can yield different super-molecular structures. This does not only complicate the problem of homopolymer crystallization but also enhances effects which are difficult to observe in pure systems. An example is a liquid blend located in the miscibility region of the thermodynamic phase diagram for the given temperature and composition. When crystallization of one component starts, diffusion and adsorption processes at the growth front change the local composition of the liquid part. This can drive the system in the immiscible state, leading to droplet formation in the liquid part, and, in turn, also to visible changes in the morphology of the crystalline structures. Various aspects of crystallization in polymer blends are discussed in Chap. 12.

Thermodynamic analysis, in particular using Differential Scanning Calorimetry (DSC) belongs to the standard tools for the investigation of crystallization and melting processes in polymers. The area of a DSC melting-peak, for instance, gives an alternative method to define the crystallinity of the sample. In Chap.16 it is shown how several model parameters of the polymer crystallization process can be obtained using DSC. In crystallization from the melt usually an amorphous part remains which separates crystalline lamellae from each other so that amorphous-crystalline interfaces are always present. Therefore, the term "semicrystalline" is frequently used in this context. The physical reasons for the existence of such a two-phase structure have been already discussed for a long time. One obvious argument is that during crystallization also a structural phase separation between stretched chains parts (stems) on one side and loops and entanglements on the other side takes place and that the highly entangled parts are unable to form a crystalline phase. Also statistical arguments have been discussed were an equilibrium between loops in the amorphous phase and the crystalline phase has been anticipated. However, it is very likely that also for this two-phase problem, kinetic arguments are predominant. In Chap. 14, it is shown that a simple two-phase model (crystal/amorph) is insufficient to reflect the thermodynamic observations made by using temperature modulated DSC. A third, so called rigid amorphous phase, which separates the crystalline phase from the mobile amorphous phase has to be introduced as a consequence of analyzing the glass transition of the amorphous phase. A Combination of scattering techniques and dielectric spectroscopy is used in Chap. 15 to follow the evolution of both crystalline and amorphous region during isothermal crystallization.

Another general and outstanding problem is of nucleation of polymer crystals. Usually, heterogeneous nucleation results in large spherulitic structures after the growth process. For chain molecules, homogeneous nucleation is particularly difficult since chains must spontaneously align to each other. Only high supercoolings can force homogeneous nucleation as is demonstrated in Chap. 8. Also computer simulations can be designed to exclude heterogeneous nucleation effects (by using cyclic boundary conditions) as demonstrated in Chap. 10. On the other hand, conformational changes forced by external constraints influences the nucleation rate considerably such as stretching of chains, either in polymer networks or by shearing. The latter is of particular interest also for applications since there crystallization usually takes place under processing conditions such as extrusion. In Chap. 17, the impact of nucleation and growth processes on flow equations are considered in detail. There, relations are provided between the number of nuclei and rheological relaxations times. The nucleation points can be either modeled as physical cross-links or alternatively as increase of the number of branches of the chains, both leading to an increase of the measurable relaxation times. The relation between physical cross-linking and nucleation of polymer crystals is also the subject of Chap. 18, where fast cooling rates are used to freeze-in the sample before crystalline growth will play a significant role.

The growth of polymer lamellae after nucleation belongs to the classical topics of research in polymer crystallization. However, lamellar growth is not only influenced by temperature, pressure and nucleation effects but also by chain length and composition effects. In Chap. 19, an extensive study of the influence of the chain length on various characteristic properties of polymer crystallization is presented using data from many different polymers. In Chap. 20 the change of the properties of the crystalline lamellae due to the effect of small molecules (impurities) added to the system is studied.

2 Polymer Crystallization Viewed in the General Context of Particle Formation and Crystallization

Jens Rieger

BASF Aktiengesellschaft, Polymer Physics, 67056 Ludwigshafen, Germany
jens.rieger@basf-ag.de

Abstract. Aim of the contribution is to establish a connection between the advanced knowledge about liquid to solid transitions in nanoparticulate dispersions to crystallization phenomena in polymeric systems. General and new aspects of solid phase formation will be discussed in a chapter on classical nucleation theory and on precursor structures. The question of whether particle formation and crystallization are to be treated as spinodal or binodal phase transition will be addressed. The role which computer simulations play for a deeper understanding of structure formation will be exemplified. Nucleating agents will be considered briefly. The importance of metastable states in colloid science as well as in semicrystalline polymers will be discussed.

2.1 Introduction

Crystallization is an interesting case of phase transition which determines the final properties of many technologically relevant and scientifically exciting systems: semicrystalline polymers – the topic of this book –, metals, ceramics, pigments, pharmaceuticals, minerals, and biological structures such as bone and teeth. Though the physical mechanisms of formation should be the same in all cases it is interesting to note that each of the above cited material groups is considered by subgroups of the scientific community that are acting more or less independently. This fragmentation of the scientific field unfortunately prohibited a cross-fertilization that should be strived for in order to enhance progress. The present contribution is an attempt to bring different branches together. It originates from a review on nanoparticle formation in water [1]. At first sight dispersions and semicrystalline systems do not have much in common. But looking closer the relationships become obvious: in both cases a liquid-solid transition is involved; the dimensions of the resulting structures lie in the range from 1 to 1000 nm; and textbooks tell us that both processes are due to nucleation and growth. A closer look at this last assumption – with regard to recent literature – reveals that the processes during solid state formation in the case of nanoparticle formation and polymer crystallization are by far more complex than usually assumed.

It is the aim of this contribution to establish a connection between the advanced knowledge about liquid to solid transitions in nanoparticulate systems to crystallization phenomena in polymeric systems. We will proceed as follows: general and new aspects of solid phase formation will be discussed in a chapter on classical nucleation theory and on precursor structures. Then, the question of whether particle formation and crystallization are to be treated as spinodal or binodal phase transition will be addressed. The role which computer simulations play for a deeper understanding of structure formation will be exemplified. Nucleating agents will be considered briefly. We will conclude with a discussion of the importance of metastable states in colloid science as well as in semicrystalline polymers.

Considering the vast amount of literature that has been published on polymer crystallization over the past decades it is obvious that the present review can not be comprehensive. But it is hoped that at least some essential points will be met and that some ideas from the physics of particle formation find their way into the domain of polymer physics. To set the stage the reader is referred to a number of books on the topic discussed here [2-6].

2.2   Classical Nucleation Theory

Textbook knowledge might be summarized as follows [7,8]. One starts with a system consisting of one or more components that initially exist as one phase. By changing, e.g., temperature or pressure the free energy changes such that a phase separated state is energetically favorable. In the case of polymers the phase separation considered here includes crystallization either from a polymer solution or from the polymer melt. Usually it is assumed that phase separation proceeds via nucleation and growth from a metastable state (the case of instantaneous, i.e. spinodal, phase separation will be treated below). Nuclei of the new (crystalline) phase appear and redissolve due to statistical fluctuations; the bulk enthalpy of the nuclei is favoring the growth whereas the surface energy term of opposite sign favors a redissolution. Only nuclei above a critical size will grow. The corresponding nucleation rate, i.e. the number of nuclei formed per time unit, can be derived using a classical Arrhenius approach relating the rate to supersaturation. This approach still dominates the field of crystallization though many deficiencies have been recognized [7,9]. Here we want to mention only three points of criticism: (i) It is questionable whether it is justified to use the bulk free energy of the crystalline phase for the nuclei because the structure might differ from that of grown crystals. (ii) Secondly and relating again to the smallness of the nuclei it is not at all evident that the surface structure is as well defined as in the case of larger crystals. Thus it must be asked what the appropriate value for the surface energy is. (iii) From a fundamental point of view it is to be asked what structure is identified as a nucleus in the thermodynamic sense [10]. An almost countless number of publications deal with these and related questions in the field of crystallization of atomic, ionic, or low molecular weight systems.

Interestingly, comparatively little has been published with respect to the formation of primary nuclei for crystallization in the field of high molecular weight (polymeric) systems, but see, e.g., [6,11]. On the one hand, this is surprising since the number of nucleation centers affects some of the application-related properties of semicrystalline polymers such as, e.g., transparency. On the other hand, this is understandable for several reasons: (i) the basic relationship between the number of primary nuclei – as detected, e.g., as the centers of spherulites – and the degree of supercooling is well established for most conventional polymers [12]. (ii) Historically, much effort has been put into the understanding of crystal growth and the development of morphology rather than into the first step of nucleation, which is notoriously difficult to monitor. (Meanwhile this field has gained renewed impetus due to new techniques such as atomic force microscopy. This technique allows for the direct visualization of the nucleation core of spherulites [13,14].) (iii) Nucleating agents are often used in order to pre-determine the number of nucleation centers. (iv) Shear, which is applied during processing of polymer melts, affects the nucleation behavior in a fundamental way.

Apart from these annotations it must be recognized that the concept of nucleation and growth played a pivotal role in the development of crystallization theories for polymers with respect to crystal growth [4,6]. The deposition of polymer chains onto growing surfaces was treated using the above concept by considering the free energy of the resulting clusters as the basis for the rate equations describing the growth of the crystal. The following considerations comprise the original concept of homogeneous nucleation as well as the latter case of what might be termed secondary or pseudo-heterogeneous nucleation, i.e. deposition of clusters onto existing crystalline surfaces of the same material.

2.3 Precursors

One of the basic ingredients of classical nucleation theories is the assumption that the system is homogeneous before the actual nucleation and growth processes set in. There are a number of experiments in the field of ionic and low molecular weight systems that show that this assumption is not always justified. It is found that precursor structures appear with dimensions below the size of the critical nucleus. These precursors are not meant to be transient aggregates, which are due to statistical fluctuations, but more long-living species, which are not considered in classical nucleation theories. Examples for such experiments are reviewed in two monographs [15,16] and a review [17]. Especially particle formation processes involving multi-valent cations – as, e.g., hydrolysis and condensation of metal alkoxides – are of impressing complexity [18,19]. Another interesting observation fits into the range of phenomena discussed here: based on findings from crystallization experiments with chiral molecules Botsaris et al. postulates that cluster/nuclei form preferentially near the surfaces of existing crystals [20].

Precursor structures are considered in the field of polymer crystallization within the context of memory effects and in the model of crystal growth via pre-structuring (Strobl's block model for polymer crystal growth[1]. By the term memory effects we denote the observation that a polymer sample may be completely molten (according to calorimetric or rheological data) and still keeps

---

[1] See also the contribution by Al-Hussein and Strobl on page 48 in this book.

partial memory of its former crystalline structure. This is seen, e.g., by observation of spherulitic films on a hot stage under the light microscope where the spherulites may appear at the original position when heating the material beyond the melting point and cooling again [21]. By differential scanning calorimetry it can be shown that the onset of crystallization is at higher temperatures in the case of samples molten from the crystalline state as compared to "virgin" samples [22]. This effect is observed for melt temperatures between the calorimetric melting point and some upper limiting temperature, which corresponds to the equilibrium melting temperature of infinitely extended crystals. It is assumed that metastable remnants of crystals in the melt act as nuclei for crystallization when cooling. Whether these nuclei still possess crystalline order or whether there is only a certain degree of orientational order left because of the slow relaxation of the polymer chains is not clear. In the latter case one might speak of true precursors and not nuclei. With respect to this uncertainty about the microscopic structure there is a clear relationship between the cases of precursors in low molecular weight systems and in polymers. Recently there have been a large number of investigations on the influence of shearing the melt for a defined period of time before cooling it [23,3]. In all cases a dramatic decrease in crystallization time as compared to the quiescent sample was observed. This effect clearly shows that directional correlations between the polymer chains in the melt do not decay immediately but serve as precursors for nuclei and crystals. The case of flow induced crystallization – without stopping shear before crystallization sets in – lies outside the scope of this contribution and will not be discussed further.

A different type of precursor has been proposed by Strobl. He does not address the formation of the primary nuclei but the process of growth of polymeric lamellae. It is suggested that the initial step is the creation of a mesomorphic layer (the precursor) which thickens and solidifies through a cooperative structural transition. The transition produces a granular crystalline layer, which transforms in the last step into homogeneous lamellar crystallites [24]. A similar multi-step approach to crystallization was proposed by Sasaki et al. on the basis of small angle X-ray scattering and spectroscopic data on crystallizing polyethylene [25]. A theoretical approach to disordered precursors ("bundles" of polymer chains) to crystallization has been put forward by Allegra [26]. For a discussion of Strobl's block model the reader is referred to three comments on his approach [27–29].

The boundary between the notion of presursors and the discussion of transient states within the concept of Ostwald's rule of stages, which will be discussed below, is fluid. The phenomenon of precursors is considered here separately because the rule of stages presupposes the existence of well-defined phases. From the above it is clear that precursors do not fit into the concept of phase separation. Besides, the question of how to define a cluster within the framework of statistical thermodynamics is by no means trivial to solve [30].

2.4   Spinodal vs. Binodal Phase Separation

When describing particle formation or crystallization one uses in most cases classical nucleation theory thus assuming that the supersaturated system resides in a metastable state and that phase separation proceeds via nucleation and growth. But if the supersaturation (supercooling) is sufficiently high the border to spinodal decomposition is crossed where phase separation occurs instantaneously without actual nucleation. In the case of low molecular weight systems it is claimed that "spinodal decomposition has never been observed in solutions made up of a solid solute and a liquid solvent because of the large width of the metastable zone that must be crossed without nucleation being induced" [31]. It might be assumed that this statement is due to the fact that spinodal decomposition is very hard to detect in low molecular weight systems because of the short time scales involved; the structures, which initially form, transform quickly to more compact shapes because of the high surface energy involved [32,33]. Recent investigations indeed give evidence for the existence of structure formation due to spinodal decomposition [34,35]. But we are still far from understanding the demarcation between binodal and spinodal phase separation in low molecular weight systems if there is any [32].

The situation is similarly unsatisfying in the field of polymer crystallization. There is a number of publications claiming that spinodal decomposition of ordered and less ordered domains occurs before the actual crystallization sets in [36–38]. Olmsted at el. proposed a theoretical approach as support for the interpretation of the experimental data in favor of spinodal decomposition [39]. The authors argue that the coupling of density and chain conformation can induce a liquid-liquid phase separation within the liquid-solid boundary curve if the polymer melt is quenched below the spinodal curve. Presently, one of the main concerns is the question whether the experimental data "prove" the existence of a spinodal phenomenon or whether other crystallization mechanisms can explain the experimental observations as well [40].

2.5   Computer Simulations of the Early Stages of Particle Formation and Crystallization

With the aid of computer simulations it is meanwhile possible to gain insight into the process of structure formation during nucleation in particle formation and crystallization. The main limitation lies in the fact that one is still restricted to idealized systems with simplified interaction potentials such as the Lennard-Jones-α-β-potential (LJ α-β) for the description of the interaction between two monomers: $V(r) = -a/r^{\alpha} + b/r^{\beta}$. Using this potential with $\alpha=6$ and $\beta=12$ Frenkel et al. studied the homogeneous nucleation of particles [41,42]. It was shown that aggregates with dimensions below the critical size exhibit a liquid-like internal ordering of the constituents. When the critical size is exceeded the face-centered cubic structure prevails. The surface of the particles still exhibits a liquid-like structure whereas between the surface and the core a shell with body-centered

cubic structure exists. The phase behavior of protein molecules can be approximately described by a LJ 3-6 potential. In this case it was found that under
certain conditions amorphous nuclei develop. The cores of these particles crystallize upon further growth [41–43]. This particle formation mechanism is linked to
a relatively low barrier of activation energy thus possibly facilitating the growth
of protein crystals.

In a similar way crystallization of polymers from dilute solutions has been
considered by Liu and Muthukumar using the LJ 6-12 potential between the
monomers and Langevin dynamics [44]. They found that for the case of relatively
short chains straight segments form rapidly and assemble in small domains while
over longer times the chain readjusts itself into a more perfect crystal. This point
is akin to some aspects of precursor formation during polymer crystallization and
is reminiscent of the structural changes occurring during particle formation discussed in the last paragraph. When studying the deposition of disordered chains
onto crystal surfaces of polymer lamellae it is found that the attachment and
re-ordering process of the chains is far more complicated than assumed in classical growth theories, cf. [4]. Similar attempts to the simulation of crystallization
have been undertaken by Meyer and Müller-Plathe [45] [2] and Fujiwara and Sato
[46] using molecular dynamics simulations. On the basis of kinetic Monte Carlo
simulations Doye and Frenkel proposed a selection mechanism for the lamellar
thickness [47]. Their approach is more restricted than the investigations cited
before since the geometry of the system is fixed to a three dimensional lattice
structure. All these simulations show that the formation of crystalline lamellae
occurs via structures that are by far more disordered than those considered in
classical theories for polymer crystallization.

2.6 Control of Crystallization by Additives

In the case of particle formation of atomic, ionic or low molecular weight systems from a liquid phase additives (molecules or particles) can affect and control
the process with respect to particle size and structure in several ways: heterogeneous nucleation on additives, colloidal stabilization or controlled aggregation
of intermediate stages, and control of crystal growth. These points are discussed
at length in [1]. Only some aspects will be highlighted here.

Whether single molecules or dissolved polymers can act as nucleation centers
for particle formation is still a matter of a controversial debate. The situation
is clearer in the case of heterogeneous nucleation, i.e. nucleation on substrates
with dimensions above about ten nanometers [48,49]. But the transition to the
first case is fluid as, e.g., in the case of the transition of a polymer from the coil
to the collapsed state as substrate for heterogeneous nucleation.

The situation is still less satisfying when discussing the mode of action of
nucleating agents used to accelerate crystallization in polymers. Many systems
are used in commercial products such as minerals, organic salts, and organics

---

[2] See the chapter of Meyer on page 177 in this book.

– mostly in particulate form. Usually it is assumed that the enhancing effect is due to epitaxial interactions, i.e. matching of the crystal structure of one of the lattice planes of the polymer crystal with the surface structure of the additive but this assumption has been proven only in some cases [50,51]. A complete understanding is still lacking.

Growth, crystal structure and form of low molecular weight crystals can be affected by the presence of foreign ions or molecules. When these ions/molecules are built into the growing crystal lattice they are effective down to concentrations of 0.1 to 1 mmol/l. If the foreign substances interact with the growing surfaces they are effective even at concentrations of 10^{-7} mol/l [52,53]. Usually, polymer crystallization from the melt is far less sensitive to foreign substances which is quite fortunate for application because many commercial types of polymers contain a number of additives for different purposes as, e.g., color. stabilization against degradation, and improved flow properties.

2.7 Ostwald's Rule of Stages

In 1897 Ostwald suggested that the phase that nucleates upon supersaturating (undercooling) is not necessarily the one that is energetically the most favorable but the one that is closest to the disordered state ("Ostwald's rule of stages") [54]. The system evolves via phase transformations to increasingly stable phases [55]. A number of approaches have been published to explain this phenomenon [52,56,57]. Furthermore, it was shown how a metastable intermediate phase can hinder the evolution of the thermodynamically stable phase [58]. In the section on computer simulations typical examples of particle formation via intermediate states were presented. Similar phenomena are often found in precipitation reactions [16] where the precipitated material first appears in a strongly hydrated, amorphous state and then transforms either to the stable crystalline form or enters further metastable modifications [59,60]

Metastable phases occur in many semicrystalline polymers. Two examples where polymorphism occurs – with implications for application-related properties – are polypropylene with three crystalline and one mesomorphic phase [61] and polyamide 6 with two crystallographic phases of varying quality [62]. There is experience of how to induce the different crystal structures but there is only sparse knowledge about the molecular processes responsible for the formation of specific crystal structures. In some cases the metastable phases in semicrystalline polymer transform into the stable form upon specific thermo-mechanical treatments.

Ostwald's rule of stages occurs in the domain of polymer crystallization in another context – in a similar way as precursors of a defined structure are observed in particle formation: It has been argued that in the case of polyethylene (for which the orthorhombic phase is the stable one under ambient conditions) crystallization starts via the hexagonal phase [63]. The crystals, in the thermodynamically metastable state, continue to grow until a nucleus for the orthorhombic phase is formed. Once the nucleation barrier is overcome the nucleus formed spreads over the whole crystal and further crystal growth is arrested. It was stated that the proposed crystallization mechanism can be applied also to other polymers [63] [3]. Whether transient structures that have been observed during fiber spinning can be interpreted within this framework remains to be discussed [64,65].

2.8   Ostwald Ripening vs. Secondary Crystallization

Despite the many similarities between particle formation in colloidal systems and polymer crystallization there is a distinct difference. Nanoparticle systems are prone to Ostwald ripening, i.e. growth of larger particles at the expense of smaller ones due to the differences in Laplace pressure and the finite solubility of the solid phase in the serum [66]. A comparable effect is found in common semicrystalline polymers only at elevated temperatures where thick crystals may grow at the expanse of thinner ones with a lowered melting point [67].

On the other hand, thinner, i.e. less stable, crystals develop often during crystallization after the formation of primary crystal lamellae (secondary crystallization) [68,69]. This effect, which in a certain sense is the reverse of Ostwald ripening, is typical for polymeric systems where crystals once they are built can not consume material indefinitely from the surrounding amorphous phase because of the constraints imposed by chain connectivity.

2.9   Discussion

Comparing the process of particle formation and crystallization in colloid physics and the crystallization of polymers from solution or from the melt it is evident that many phenomena can be treated within a common physical frame. It is to be hoped that a deeper cross-fertilization can be achieved by comparing theoretical as well as experimental approaches. It shall not be denied that much has been reached within the past years. The use of synchrotron radiation to follow structure formation in-situ is but one example. But still there are many open questions concerning the molecular processes. It is expected that computer simulations will have a major impact on the understanding of the processes both during particle formation and polymer crystallization though it is still a long way until a quantitative understanding will be reached.

References

1. D. Horn, J. Rieger: Angew. Chem. Intl. Ed. **40**, 4330 (2001)
2. G. Strobl: *The Physics of Polymers* (Springer Verlag, Heidelberg 1996)
3. D. Eder, H. Janeschitz-Kriegl: 'Crystallization'. In: *Materials Science and Technology, Vol. 18*, ed. by R.W. Cahn, P. Haasen, E.J. Kramer (VCH-Verlag, Weinheim 1997) pp. 269-342

---

[3] For details, see the contribution of S. Rastogi on page 17 in this book.

4. K. Armistead, G. Goldbeck-Wood: Adv. Polym. Sci. **100**, 219 (1992)
5. L. Mandelkern: 'The Crystalline State'. In: *Physical Properties of Polymers*, ed. by J.E. Mark (ACS, Washington 1993)
6. P.J. Barham: 'Crystallization and Morphology of Semicrystalline Polymers'. In: *Materials Science and Technology, Vol. 12*, ed. by R.W. Cahn, P. Haasen, E.J. Kramer (VCH-Verlag, Weinheim 1993) pp. 153-212
7. P.G. Debenedetti: *Metastable Liquids* (Princeton University Press, Princeton 1996)
8. J. Schmelzer, G. Röpke, R. Mahnke: *Aggregation Phenomena in Complex Systems* (Wiley-VCH, Weinheim 1999)
9. A. Laaksonen, V. Talanquer, D.W. Oxtoby: Annu. Rev. Phys. Chem. **46**, 489 (1995)
10. H. Reiss, R.K. Bowles: J. Chem. Phys. **112**, 1390 (2000)
11. P.H. Geil: Polymer **41**, 8983 (2000)
12. Van Krevelen: *Properties of Polymers, 3rd edn* (Elsevier, Amsterdam 1990)
13. L. Li, C.-M. Chan, J.-X. Li, K.-M. Ng, K.-L. Yeung, L.-T. Weng: Macromolecules **32**, 8240 (1999)
14. G.J. Vasco, L.G.M. Beekmans, R. Pearce, D. Trifonova, J. Varga: J. Macromol. Sci. Phys. **B38**, 491 (1999)
15. A. Randolph, M.A. Larson: *Theory of Particulate Processes, 2nd edn.* (Academic Press, New York 1988)
16. J.W. Mullin: *Crystallization* (Butterworth-Heinemann, Oxford 1992)
17. J. Stávek, J. Ulrich: Cryst. Res. Technol. **29**, 465 (1994)
18. J. Livage, M. Henry, J.P. Jolivet: in *Chemical Processing of Advanced Materials*, ed. by L.L. Hench, J.K. West (J. Wiley and Sons, New York 1992)
19. R.K. Iler: *The Cemistry of Silica* (Wiley, New York 1979)
20. G.D. Botsaris, R.-Y. Qian, A. Barrett: AIChE J., **45**, 201 (1999)
21. W. Heckmann: priv. commun.
22. A. Ziabicki, G.C. Alfonso: Coll. Polym. Sci. **272**, 1027 (1994)
23. G. Kumaraswamy, A.M. Issaian, J.A. Kornfield: Macromolecules **32**, 7537 (1999)
24. G. Strobl: Eur. Phys. J. E **3**, 165 (2000)
25. S. Sasaki, K. Tashiro, M. Kobayashi, Y. Izumi, K. Kobayashi: Polymer **40**, 7125 (1999)
26. G. Allegra, S.V. Meille: Phys. Chem. Chem. Phys. **1**, 5179 (1999)
27. B. Lotz: Eur. Polym. J. E **3**, 185 (2000)
28. S.Z.D. Cheng, C.Y. Li, L. Zhu: Eur. Polym. J. E **3**, 195 (2000)
29. M. Muthukumar: Eur. Polym. J. E **3**, 199 (2000)
30. I. Kusaka, D.W. Oxtoby: J. Chem. Phys. **110**, 5249 (1999)
31. A.S. Myerson: in *Molecular Modeling - Applications in Crystallization*, ed. by A.S. Myerson (Cambridge University Press, Cambridge 1999)
32. K. Binder: in *Materials Science and Technology, Vol. 5 Phase Transformations in Materials*, ed. by P. Haasen (VCH-Verlag, Weinheim 1991)
33. V. Sofonea, K.R. Mecke: Eur. Phys. J. B **8**, 99 (1999)
34. H. Haberkorn, D. Franke, Th. Frechen, W. Goesele, J. Rieger: submitted for publication.
35. J.F.M. Lodge, D.M. Heyes: J. Chem. Soc., Faraday Trans. *93*, 437 (1997)
36. A.J. Ryan, J.P.A. Fairclough, N.J. Terrill, P.D. Olmsted, W.C.K. Poon: Faraday Discuss. **112**, 13 (1999)
37. T.A. Ezquerra, E. Lopez-Cabarcos, B.S. Hsiao, F.J. Balta-Calleja: Phys. Rev. E **54**, 989 (1996)
38. G. Matsuba, K. Kaji, K. Nishida, T. Kanaya, M. Imai: Macromolecules **32**, 8932 (1999)

39. P.D. Olmsted, W.C.K. Poon, T.C.B. McLeish, N.J. Terrill, A. Ryan: Phys. Rev. Lett. **81**, 373 (1998)
40. Z.-G. Wang, B.S. Hsiao, E.B. Sirota, P. Agarwal, S. Srinivas: Macromolecules **33**, 978 (2000)
41. P.R. ten Wolde, D. Frenkel: Science **277**, 1975 (1997)
42. P.R. ten Wolde, D. Frenkel: Phys. Chem. Chem. Phys. **1**, 2191 (1999)
43. P.R. ten Wolde, D. Frenkel: Theor. Chem. Acc. **101**, 205 (1999)
44. C. Liu, M Muthukumar: J. Chem. Phys. **109**, 2536 (1998)
45. H. Meyer, F. Müller-Plathe: J. Chem. Phys **115**, 7807 (2001)
46. S. Fujiwara, T. Sato: J. Chem. Phys. **114**, 6455 (2001)
47. J.P.K. Doye, D. Frenkel: J. Chem. Phys. **110**, 7073 (1999)
48. J.W. Mullin: *Crystallization* (Butterworth-Heinemann, Oxford 1992)
49. X.Y. Liu: *Proc. 14th Int. Symp. Ind. Cryst.*, *IChemE* (Rugby, computer optical disk), 1999.
50. D. Vesely, G. Ronca: J. Microscopy **201**, 137 (2001)
51. A. Thierry, B. Fillon, C. Straupe, B. Lotz, J.C. Wittmann: Prog. Coll. Polym. Sci. **87**, 28 (1992)
52. O. Söhnel, J. Garside: *Precipitation* (Butterworth-Heinemann, Oxford 1992)
53. D.L. Klug: in *Handbook of Industrial Crystallization*, ed. by A.S. Myerson (Butterworth-Heinemann, Boston, 1993)
54. W. Ostwald: Z. Phys. Chem. **22**, 289 (1897)
55. A. Keller, S.Z.D. Cheng: Polymer **19**, 4461 (1998)
56. J. Nývlt: Cryst. Res. Technol. **30**, 443 (1995)
57. J.W.P. Schmelzer, J. Schmelzer, I.S. Gutzow: J. Chem. Phys. **112**, 3820 (2000)
58. L. Gránásy, D.W. Oxtoby: J. Chem. Phys. **112**, 2410 (2000)
59. S. Kabasci, W. Althaus, P.-M. Weinspach: Trans. Inst. Chem. Eng. **74A**, 765 (1996)
60. J. Rieger, J. Thieme, C. Schmidt: Langmuir **16**, 8300 (2000)
61. Ph. Tordjeman, C. Robert, G. Marin, P. Gerard: Eur. Phys. J. E **4**, 459 (2001)
62. N.S. Murthy, R.G. Brady, S.T. Correale, R.A.F. Moore: Polymer **36**, 3863 (1995)
63. S. Rastogi, L. Kurelec: J. Mat. Sci. **35**, 5121(2000)
64. G.E. Welsh, D.J. Blundell, A.H. Windle: Macromolecules **31**, 7562 (1998)
65. J.M. Samon, J.M. Schultz, J. Wu, B. Hsiao, F. Yeh, R. Kolb: J. Polym. Sci. B Polym. Phys. **37**, 1277 (1999)
66. P. Taylor: Adv. Coll. Interf. Sci. **75**, 107 (1998)
67. R. Hingmann, J. Rieger, M. Kersting: Macromolecules **28**, 3801 (1995)
68. Z.-G. Wang, B.S. Hsiao, B.B. Sauer, W.G. Kampert: Polymer **40**, 4615 (1999)
69. R. Kolb, C. Wutz, N. Stribeck, G. von Krosigk, C. Riekel: Polymer **42**, 5257 (2001)

3 Role of Metastable Phases in Polymer Crystallization; Early Stages of Crystal Growth

Sanjay Rastogi

Department of Chemical Engineering/Dutch Polymer Institute
P.O.Box 513, 5600 MB Eindhoven; The Netherlands
s.rastogi@tue.nl

Abstract. In this chapter we have addressed issues concerning crystallization of flexible polymers from melt. We have described several examples where crystallization proceeds via a transient phase before a thermodynamically stable phase intervenes. With several examples it has been shown that transient state of a phase strongly depends on the crystal size. A phase that is thermodynamically stable during the initial stages of crystallization becomes metastable with crystal growth. Taking linear polyethylene as an example, it is shown that crystal size can be controlled during polymerization, thus the phase behavior. These findings have implications in polymer processing like sintering of intractable ultra high molecular weight polyethylene.

3.1 Introduction

Crystallisation of polymers has been a widely studied subject and still is an important topic in view of several different models describing the complex crystallisation behaviour of long chain molecules. In the case of synthetic polymers, the first concepts of crystallisation were based on very extensive crystallisation studies performed on solution-grown single crystals where crystallisation proceeds in very dilute systems (< 0.01 wt. %). The long flexible chains crystallised from solution form platelet (lamellar) single crystals. Since the thickness of these lamellar crystals is much smaller than the length of a fully extended chain, it was proposed by Keller [1] and Fischer [2] in 1957 that chains fold back and forth, forming folded chain lamellar crystals. Detailed studies demonstrated that the crystallisation temperature, viz. the degree of supercooling, plays a prominent role in determining the lamellar thickness of solution-crystallised single crystals. It was also found that the lateral habit of single crystals is dependent on the crystallisation temperature [3]. Moreover the melting temperature of single crystals is strongly dependent on the lamellar thickness, mainly due to the small crystal size in the chain direction, viz. the fold length, related to the relatively large surface area compared with the volume of the crystals [4,5].

Detailed studies have been performed concerning crystallisation from dilute solutions involving narrow molecular weight fractions. Based on these studies, it was proposed that polymer crystallisation is a nucleation controlled process. Classical, well-established theories on crystallisation (surface nucleation) for low molar mass substances were adopted for describing polymer crystallisation [4,5].

Together with thermodynamic parameters such as supercooling, kinetic features concerning surface nucleation were included to understand the experimental observations. Several theories were proposed, for example by Frank & Tosi [4], followed by Hoffman and co-workers [5], and later by the late David Sadler [6]. The proposed theories took into account the crystal growth process after the formation of critical nuclei. However, the origin of nucleation in polymers is still a matter of debate [7], both for quiescent and notably for oriented polymer systems.

With the help of detailed Transmission Electron Microscopy, the formation of terraces in polymer crystallisation has been observed frequently [8]. The proposed screw dislocation for crystal growth in general, proposed by Frank, was used to explain such intriguing morphological features [9]. Similar to the constrained lattice, which results in dislocations in inorganic crystals, it is the surface of lamellar crystals which causes dislocations in polymers.

The studies and theories/modelling, mentioned above were based on experimental studies of solution-crystallized polymers, notably linear polyethylenes. Crystallisation from the melt is a much more complex phenomenon. The polymer chains in the melt are highly entangled and folding of chains during the nucleation and crystallisation process is hindered. However, current models invoke reeling-in of chains from the melt on the crystal surface which is in fact a disentangling process from a virtual tube based on the concept of reptation [10].

Crystallisation from the polymer melt results in the formation of spherical crystal aggregates, the so-called spherulites, notably at isothermal crystallisation at low supercoolings. The origin of spherulitic growth including nucleation, the growth of sheaf like structures in the beginning and subsequent branching and the development of spherical crystal aggregates is still not very well understood, despite many studies [11].

To bridge the gap between single crystal growth (crystallisation from dilute solutions) and growth into spherulitic structures (crystallisation from the melt), crystallisation studies on linear polyethylenes at elevated pressures and temperatures are of utmost importance since single crystals can be grown directly in the melt and their growth can be studied in-situ and can be compared with the nucleation and growth of spherulites at ambient pressures.

In the past, many experiments have been performed [16,17,19,22] concerning in-situ observation of single crystal growth at elevated pressures and temperatures using PE as a model substance. It appeared that crystal growth initially proceeded via the hexagonal phase and, moreover, that a metastable hexagonal phase could be observed at early stages of crystal growth within the thermodynamically stable orthorhombic phase [12]. The observation that a thermodynamic stable state is reached via a metastable state of matter is not unique for polymers nor a novel issue. Already in 1897 Ostwald made this observation in the case of freezing of liquids [1].

In this article we will provide experimental evidence to bridge the gap between single crystal formation at elevated pressures and temperatures and crys-

---

[1] See also the contribution of J. Rieger on page 7 in this book.

tallisation at atmospheric pressure. The main aim of this paper will be to investigate metastability in polymer systems and its influence on crystallisation at atmospheric pressure. After addressing some salient findings on crystallisation from the mesophase we will proceed with new experimental findings on the fate of the crystal after transformation from the metastable back to its thermodynamic stable state. Subsequently, the distinction between primary and secondary thickening will be made and its morphological consequences will be discussed. Finally, we will address the phenomenon of size dependent phase stability and its significance in improving the processability of the ultra high molecular weight polyethylene (UHMW-PE) powder.

Generalisation to the proposed viewpoints in the crystallisation behaviour, especially the issue of metastability, will be made by giving examples of polymers other than polyethylene, such as trans-1,4-polybutadiene, poly-di-alkyl siloxanes, nylons, poly(ethyleneterephthalate) and paraffins.

3.2 Experimental Section

3.2.1 Materials

Two types of sharp fractionated polyethylene were obtained from the National Bureau of Standards, U.S.A. The two fractions are: sample 1, NIST-SRM1483 having M_w=32000, $M_w/M_n = 1.1$ and sample 2, NIST-SRM1484 having $M_w = 120000, M_w/M_n = 1.2$. Trans-1,4-polybutadiene was synthesised in our laboratory [13]. UHMW-PE nascent powder, specially synthesised for our studies, possesses a molecular weight of 3.5×10^6 and $M_w/M_n = 5.6$. Solution crystallised films of UHMW-PE have been prepared from 1% Xylene solutions via a route well documented in literature [14].

3.2.2 High Pressure Cell

In this work a piston-cylinder type of pressure cell similar to the one used by Hikosaka [15] has been used. A sample is sandwiched in between two diamond windows which enable in-situ observation by light microscopy, Raman spectroscopy and X-ray diffraction. The pressure on the sample is generated hydrostatically by precise movement of two pistons, provided by pressure-regulated flow of nitrogen gas.

3.2.3 Polarising Optical Microscopy

Polarising optical micrographs have been taken *in-situ* during crystallisation of polyethylene under elevated pressures and temperatures. In our previous studies it has been shown that when crystallising the polyethylene at elevated pressures and temperatures, the crystals emerge and grow as isolated "cigar"-shaped birefringent entities. Such uniform objects have been shown to be in the hexagonal phase. At a certain stage of growth, the shape of the crystal changes into a blotchy structure with reduced birefringence which indicates the transformation from the hexagonal into the orthorhombic phase [16–18].

3.2.4 Wide Angle X-ray Scattering

In situ X-ray experiments were performed using monochromatic X-rays of wavelength 0.0798 nm and a high flux available on beamline ID11-BL2 on the European Synchrotron Facilities in Grenoble. The lower wavelength was required to avoid the X-ray absorption from the diamond windows. Each diffraction pattern was recorded for 15 s on a two dimensional CCD detector. Using the FIT2D program of Dr. Hammersly (ESRF), 2D X-ray patterns were transformed into one dimensional patterns by performing an integration along the azimuthal angle.

3.3   Results and Discussion

Before proceeding with our most recent observations, it is essential to recapitulate some of the salient findings which have been published in the past and which form the basis for the present discussion.

3.3.1 Polyethylene – A Summary of Previously Reported Results on Crystallisation of Single Crystals in Polyethylene Melts

With the help of in-situ optical microscopy at the requisite pressure and temperature, it has been shown that crystallisation of a linear sharp fractionated polyethylene proceeds via the hexagonal phase, even in the thermodynamically stable orthorhombic region of the pressure-temperature phase diagram (see Fig. 3.1a). It was shown that the birefringent entities observed at 45° to the polarised light, are crystals growing from the melt under isobaric and isothermal conditions. It was also possible to make a distinction between the hexagonal and orthorhombic phase [5], optically. Compared to the crystals in the orthorhombic phase, the crystals in the hexagonal phase possessed smooth and well defined boundaries. Thus, the phase transition could be followed in-situ as a sudden change in birefringence. From a series of experiments under isobaric and isothermal conditions within the thermodynamically stable region of the orthorhombic phase, the following conclusions were made.

1. Crystallisation always occurs initially via the hexagonal phase [16,17].
2. After a certain crystallisation time, crystals initially in the hexagonal phase transform into the orthorhombic phase [16,17].
3. Once a crystal is transformed, further crystal growth is arrested, at least within the experimental time [16,17] of several hours.
4. The residence time for a crystal in the hexagonal phase is dependent on the supercooling and pressure at which an experiment is performed [19].
5. Below the equilibrium triple point (Q^∞), under isobaric and isothermal conditions, after a certain crystallisation time two different populations of crystals can be observed – one in the hexagonal phase which continues to grow and the other one which has transformed into the orthorhombic phase, and stops growing. On heating the sample, the crystals in the hexagonal phase melt at a lower temperature in comparison with the ones which are in the

orthorhombic phase. The melting behaviour is contradictory to the observations above the triple point (Q^∞) – i.e. the melting of an orthorhombic
crystal occurs via its transformation to the hexagonal phase. Thus, above
the triple point, crystals in the hexagonal phase possess a higher melting
temperature [16,17].

6. The difference in the melting temperature of the orthorhombic and the
   hexagonal phase increases with decreasing pressure, below the triple point [16].
7. With a series of such experiments, a more general pressure-temperature
   phase diagram was proposed (Fig. 3.1b) with three different supercoolings
   defined below the equilibrium triple point – first a virtual transition line
   from orthorhombic to hexagonal, second from orthorhombic to liquid and
   the third from hexagonal to liquid [17].
8. Three different regions were defined in the proposed schematic phase diagram
   (Fig. 3.1b). A Region III below the equilibrium triple point Q^∞, bounded in
   between the hexagonal-liquid (h-liquid) and orthorhombic-liquid (o-liquid)
   in the phase diagram was defined as a region for no crystal growth [16,17],
   see also below.

3.3.2 Experimental Observations on a Crystal after Its Transformation from the Metastable Hexagonal Phase into the Thermodynamically Stable Orthorhombic Phase

In a series of composite figures (Figs. 2, 3, 4 and 5) taken at a fixed pressure of
3.2 kbar and at different supercoolings ($\Delta T = 2.7$ K in Fig. 3.2; $\Delta T = 4.7$ K in

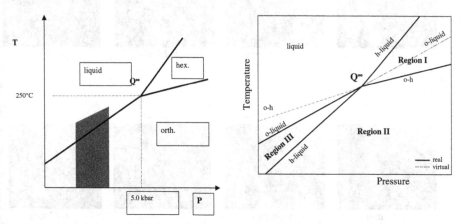

Fig. 3.1. (a) Schematic pressure-temperature phase diagram of polyethylene. Q^∞, the
equilibrium triple point (from [12]) is the intersection of orthorhombic to hexagonal,
hexagonal to liquid, and orthorhombic to liquid transition lines. The shaded region in
the figure is the $p-T$ region where experiments have been performed in this work. **(b)**
Schematic $p-T$ phase diagram according to [16]. The diagram includes metastable
(—) and virtual boundaries ($\cdots\cdots$) between the hexagonal and orthorhombic phase
as well as the liquid. Q^∞ is the equilibrium triple point. Regions I, II, III are defined
in [16].

Fig. 3.3; $\Delta T = 5.7\,\mathrm{K}$ in Fig. 3.4; $\Delta T = 6.7\,\mathrm{K}$ in Fig. 3.5), relative to the melting temperature of the hexagonal phase, the following observations can be made. Taking the criteria for optical distinction between the hexagonal (marked by ➡) and the orthorhombic phase (marked by ⇒) it can be stated that independent of the supercooling, crystallisation always starts in the hexagonal phase. The crystals in the hexagonal phase, observed as birefringent entities, possess chains running perpendicular to the lateral growth direction, continue to grow. Once the transformation from the hexagonal to the orthorhombic phase occurs, viewed as a sudden change in birefringence, further crystal growth is arrested. The issue of arrest in crystal growth will be given quantitatively in Fig. 3.8. From the series of optical micrographs it is obvious that with increasing supercooling the residence time for a crystal in the hexagonal phase decreases. The quantitative values on the residence time for the hexagonal phase has been provided elswhere [19]. Further, it has to be noted that when a crystal transforms into the orthorhombic phase, many crystals grow around the transformed crystal and after a certain time the overall growth of the multicrystal cluster becomes spherical in nature. This is clearly more evident for higher supercoolings like 5.7 and 6.7 K in Figs. 4 and 5.

The crystal that has transformed from the hexagonal to the orthorhombic phase appears to become blotchy (or block like). Many crystals seem to grow on

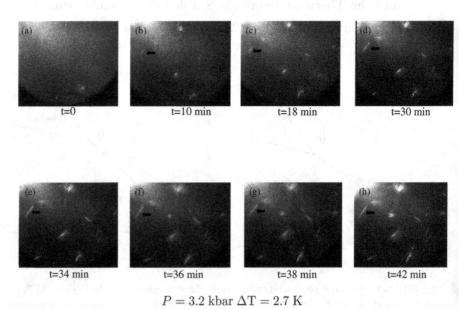

$$P = 3.2\ \mathrm{kbar}\ \Delta T = 2.7\ \mathrm{K}$$

Fig. 3.2. In-situ optical micrographs of the sharp fractionated polyethylene (NIST SRM1483, Mw = 32000) during isobaric and isothermal crystallisation at $P = 3.2$ kbar and $(\Delta T)_{\mathrm{orth}} = 2.7$ K. The bold arrows ➡ the hexagonal crystals. It can be observed that within the experimental time no transformation from the hexagonal to the orthorhombic phase took place at low supercoolings. Scale bar 50 μm.

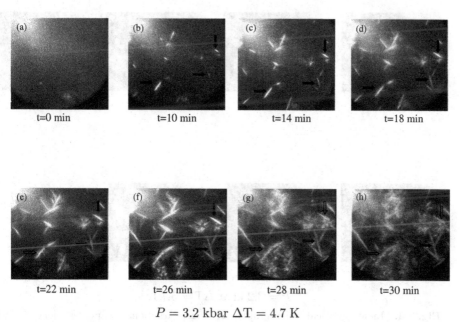

Fig. 3.3. In-situ optical micrographs of sharp fractionated polyethylene (NIST SRM1483, Mw=32000) during isobaric and isothermal crystallisation at $P = 3.2$ kbar and $(\Delta T)_{orth} = 4.7$ K. The bold arrows ➡ point the hexagonal crystals while the open ones ⇒ indicate the orthorhombic crystals. Scale bar 50 μm.

the thus transformed crystal. It appears that the just transformed crystal acts as a nucleating centre for other new crystals (Fig. 3.3e–h, 3.4e–g, 3.5b–h). It has to be noted that the crystals growing on the substrate of the transformed hexagonal-orthorhombic crystals are in the hexagonal phase[16]. *In-situ* polarising optical microscopy, involving multistage temperature cycling below the equilibrium triple point, revealed that those branched lamellae possess a hexagonal structure, at least initially [16]. During the multistage temperature cycling, under isobaric conditions, it is observed that the new crystals formed on the basal surface of the just transformed crystal from the hexagonal to the orthorhombic phase, melted at the same temperature with the other hexagonal crystals. But crystals in the orthorhombic phase melted at a higher temperature. Further, it has been noticed that unlike the orthorhombic crystals, which do not grow laterally, the new hexagonal crystals continue to grow. Moreover when the sample is left to anneal above the melting temperature of the hexagonal phase (Region III in Fig. 3.1) no crystal growth is observed. This phenomenon will be explained in detail later in this article. Since the hexagonal to orthorhombic transition is a solid-solid phase transformation and is a nucleation controlled process, the crystals growing on the orthorhombic substrate are much easier to transform into the thermodynamically stable orthorhombic phase. Fig. 3.6 shows the Transmission Electron Micrograph of isolated single crystals surrounded by lamellae spread

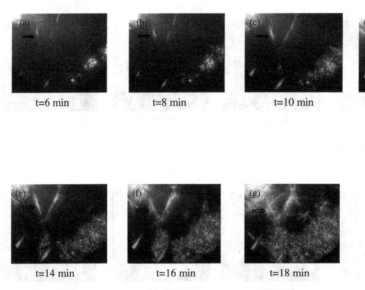

$$P = 3.2 \text{ kbar } \Delta T = 5.7 \text{ K}$$

Fig. 3.4. In-situ optical micrographs of sharp fractionated polyethylene (NIST SRM1483, Mw=32000) during isobaric and isothermal crystallisation at $P = 3.2$ kbar and $(\Delta T)_{orth} = 5.7$ K. The bold arrows ➡ point the hexagonal crystals while the open ones ⇨ indicate the orthorhombic crystals. Scale bar 50 μm.

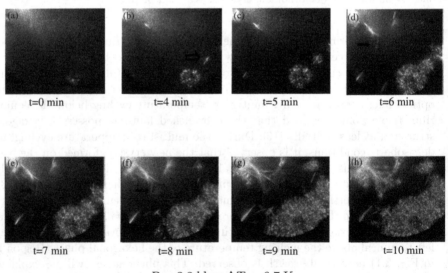

$$P = 3.2 \text{ kbar } \Delta T = 6.7 \text{ K}$$

Fig. 3.5. In-situ optical micrographs of sharp fractionated polyethylene (NIST SRM1483, Mw=32000) during isobaric and isothermal crystallisation at $P = 3.2$ kbar and $(\Delta T)_{orth} = 6.7$ K. The bold arrows ➡ point the hexagonal crystals while the open ones ⇨ indicate the orthorhombic crystals. Scale bar 50 μm.

Fig. 3.6. Transmission electron micrograph showing single crystals grown at 3.1 kbar and $(\Delta T)_{hex}$ = 4.5 K surrounded by lamellae perpendicular to the basal surface. The spread lamellae crystallised after the transformation of the single crystals from the hexagonal to the orthorhombic phase. Scale bar is 3.5 μm.

perpendicular to its basal surface, thus confirming the high nucleation activity of the once transformed single crystal.

On the other hand, the crystals which did not transform into the orthorhombic phase continue to grow and no other crystals can be observed in their vicinity (for example see Fig. 3.3g–h). It seems that these crystals are pulling chains from the melt during the growth process. At this instant it needs to be mentioned that the single crystals of polyethylene have a tapered morphology, suggesting that in the very initial stages of crystallisation chains are folded [20,16,17]. Due to the enhanced chain mobility within the hexagonal phase, which arises from a more open lattice and thus weak Van der Waals interaction between the neighbouring chains, refolding to longer fold lengths is facilitated which leads to full chain extension, a thermodynamic requirement in polymer crystallisation. This implies that the crystals in the hexagonal phase grow simultaneously along the lateral and the thickening direction. To fill in the cavities, as generated during crystal growth, mainly because of simultaneous thickening, more material is needed to be pulled from the surface of the growing crystal. This may be an explanation for the absence of new crystal formation in the vicinity of the growing crystal [19,9,21].

3.3.3 Multilayering: Primary and Secondary Thickening

A crystal in the hexagonal phase can also stop growing because of morphological reasons. For example, growth in the lateral direction may get restricted because of impingement with another crystal, growing in the melt. The crystal growth along the thickness direction may also stop because of multilayering –i.e. the growth of crystals on top of each other, due to a screw dislocation. A series of electron micrographs summarised in Fig. 3.7, suggests that with increased supercooling, at a fixed pressure, multilayering becomes more and more prominent,

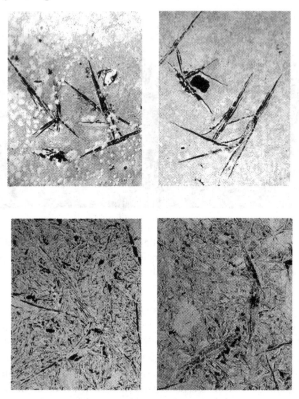

P=3.2kbar

Fig. 3.7. Transmission electron micrograph showing multilayering at different super-coolings and constant pressure of 3.2 kbar. **(a)** $(\Delta T) = 4$ K, **(b)** $(\Delta T) = 4$ K; bending of the crystals is marked by ➡ **(c)** $(\Delta T) = 8$ K, **(d)** $(\Delta T) = 10$ K. It is to be noticed that the overall shape of the multilayered crystals as viewed edge-on has resemblance with the crystals observed by optical microscopy (as birefringent entities, when viewed edge-on) when growing in the hexagonal phase. Scale bar is 1 µm.

even when a crystal is in the hexagonal phase. These observations suggest that with increasing supercooling, as the growth rate increases laterally, and along the thickening direction, screw dislocations also become prominent. It is also noticeable that once a crystal transforms from the hexagonal to the orthorhombic phase, several crystals grow on top of the transformed crystal without the uniform registration of the newly formed crystals, ultimately leading to spherical like growth of crystal aggregates (Figs. 4 and 5). A combination of the multi-layering and the spherical like growth of crystal aggregates gives an insight into the initial stages of spherulitic formation. In Fig. 3.7b, the crystal marked by the arrow shows how a lamella bends when it comes into contact with another crystal (as viewed edge-on). This indicates the flexibility of crystal growth in the hexagonal phase.

The thickening growth in an isolated single crystal corresponds to primary thickening. Once crystals are multi-layered, lying on top of each other, further thickening requires penetration of chains within the adjacent crystals. The process of further thickening in the solid state is usually referred to as secondary thickening. This is feasible only when the chains within the crystals adjacent to each other are in regular registration, thus facilitating further lamellae thickening in the preferred morphology only, i.e. the morphology having regular chain registraton with adjacent crystals. An extensive study performed by Hikosaka et al. [19,22]. showed that both the primary thickening and the lateral thickening have the same energy barrier to overcome for crystal growth.

At a fixed pressure, as the residence time for the crystals in the hexagonal phase decreases exponentially with increase in supercooling the average lamellar thickness also shows an exponential decay [19]. These results strongly suggest that the primary lamellar thickness is dependent on the crystals residence time in the hexagonal phase. Figure 3.8, shows variation in the crystal length with crystallisation time for different supercoolings at constant pressure. The slope of the curve (in Fig. 3.8) at a fixed degree of supercooling and a fixed pressure represents the crystal growth rate. It is to be noted that with an increase in supercooling, the crystal growth rate increases. Since in the hexagonal phase crystal growth occurs simultaneously along the lateral and the thickening direction, the growth rate can be measured from a series of electron micrographs showing isolated single crystals. A detailed description of the method used to quantitatively determine values for the lateral and thickening growth rates is provided in [16,19]. Table 3.1, shows numerical values for such observations. Thus giving an insight into chain mobility especially along the c-axis in the hexagonal phase. From Fig. 3.8 it is evident that once a crystal transforms into the orthorhombic phase, crystal growth stops, at least in the lateral direction. The residence time for crystals in the hexagonal phase decreases with increase in supercooling.

Table 3.1.

Pressure = 3.2 kbar

ΔT_{hex} (°C)	lateral growth rate (nm/sec)	thickening rate (nm/sec)
4.0	21.8	0.8
6.0	112.5	5.2
8.0	317.0	26.4
10.0	652.0	37.2-55.8

From the results summarised above in Sects. 3.1 to 3.3 important implications for polymer crystallisation can be drawn and these will be discussed in the following two sections.

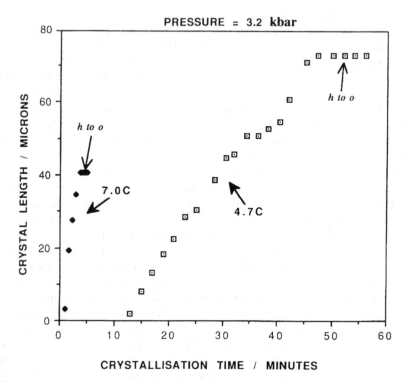

Fig. 3.8. Variation of crystal length with crystallisation time for two different super-coolings, $(\Delta T)_{hex} = 4.7$ K and $(\Delta T)_{hex} = 7.0$ K, at 3.2 kbars.

3.3.4   Size Influence in Phase Transformation: Stable, Metastable and Transient States of a Phase

In this section we will address the issue of metastability of the hexagonal phase below the equilibrium triple point. From the unique tapered morphology of the single crystals, it was concluded that at the early stages of crystallisation, due to kinetic reasons chains are in the folded state [16]. As a crystal grows, chains slide along the c-axis to thicken the crystal. When the experiments are performed in Region II (Fig. 3.1), i.e. within the thermodynamic stability region for the orthorhombic phase, a crystal initially in the hexagonal phase transforms into the thermodynamically stable orthorhombic phase. These observations strongly suggest a third parameter, i.e. the crystal size in the pressure-temperature phase diagram.

The experimental observations can be expressed mathematically.

In general, to a first approximation, the Gibbs Free energy of a crystalline phase having a surface area A and thickness l can be expressed for the orthorhombic (G_o) and

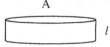

the hexagonal (G_h) phase by the following expressions:

$$G_o \cong (H_o - T \cdot S_o) + 2 \cdot \sigma_o \cdot A \tag{3.1}$$

$$G_h \cong (H_h - T \cdot S_h) + 2 \cdot \sigma_h \cdot A \tag{3.2}$$

Where, H and S are enthalpy and entropy for the specific phases, respectively and σ the end surface free energy.

The difference in Gibbs Free energy between the orthorhombic and the hexagonal phase in polyethylene can be expressed as:

$$\Delta G = G_o - G_h$$
$$= [-(H_h - H_o) + T \cdot (S_h - S_o)] - 2 \cdot (\sigma_h - \sigma_o).A$$

The orthorhombic phase will be thermodynamically stable when $\Delta G < 0$.

At the transition temperature $T_{h \to o}^{\infty}$ of a crystal having infinite thickness, $\Delta G_{h \to o}^{\infty} = 0$ and

$$H_o - T_{h \to o}^{\infty} \cdot S_o = H_h - T_{h \to o}^{\infty} \cdot S_h,$$

Therefore,

$$S_h - S_o = \frac{(H_h - H_o)}{T_{h \to o}^{\infty}}$$

Thus,

$$\Delta G = \left[-(\Delta H)_{h \to o} \cdot \frac{(\Delta T)_{h \to o}}{T_{h \to o}^{\infty}} + 2 \cdot (-\Delta\sigma) \cdot A \right] \tag{3.3}$$

Making an assumption that the volume for both the hexagonal and the orthorhombic phase is the same, to the first approximation the Gibbs free energy per unit volume can be expressed as

$$\Delta g = \frac{\Delta G}{V} = \left[-(\Delta h)_{h \to o} \frac{(\Delta T)_{h \to o}}{T_{h \to o}} + \frac{2(-\Delta\sigma)}{l} \right] \tag{3.4}$$

where Δh is the enthalpy per unit volume and $(\Delta h)_{h \to o} = h_h - h_o > 0$ because the transition is endothermic in nature as observed by high pressure DSC [23], $(\Delta T)_{h \to o} = T_{h \to o}^{\infty} - T > 0$ is a supercooling defined from the equilibrium transition temperature $T_{h \to o}^{\infty}$ for the infinite crystal size and $(-\Delta\sigma) = \sigma_o - \sigma_h > 0$ is the change in the end surface free energy for the hexagonal to the orthorhombic transition, as has been discussed earlier [20].

At a critical thickness of the crystal $l = l_{cr}, \Delta g = 0$.

For $l > l_{cr}$, Δg becomes negative thus a crystal initially in the thermodynamically stable hexagonal phase will be no longer stable and the transformation from the hexagonal to the orthorhombic phase will become possible.

Further thickening in the metastable hexagonal phase will solely be a consequence of kinetics, rather than of thermodynamics.

In our case from equation (4) we can derive the following expression:

$$T_{h \to o} = T_t^{\infty} \cdot \left[1 - \left(\frac{2(-\Delta\sigma)}{l \cdot (\Delta h)_{h \to o}} \right) \right] \tag{3.5}$$

since solid to solid transformation from the hexagonal to the orthorhombic phase is a nucleation controlled process. The rate of nuclei formation for the orthorhombic phase within the hexagonal crystal can be expressed as

$$r_{h \to o} = \frac{k \cdot T}{h} \exp \left(\frac{-\Delta G}{k \cdot (\Delta T)_{h \to o}} \right); \tag{3.6}$$

where ΔG is the nucleation barrier.

Once a nucleus is formed a metastable crystal is in a transient state. The transition time of the transient hexagonal phase into the thermodynamically stable orthorhombic phase can be expressed by the formation of the nucleus and its propagation from its origin to the overall crystal.

From the given mathematical expressions, it is evident that a crystal initially in a hexagonal phase passes through four different states before the transformation to the thermodynamically stable orthorhombic phase is reached:

(a) thermodynamically stable region for the hexagonal phase i.e. below $l < l_{critical}$

(b) metastable region for the hexagonal phase; though a crystal is thermodynamically unstable further thickening is a kinetic process i.e. $l > l_{critical}$, before a nucleus for the transformation is formed

(c) since the solid-solid phase transition is a nucleation controlled process, the residence time for a crystal to stay in the metastable hexagonal phase depends on the nucleation barrier which the crystal has to overcome to go in the thermodynamically stable orthorhombic phase.

(d) once a nucleus for the orthorhombic phase is formed, the time required for a crystal to stay in the hexagonal phase (i.e. in its transient state) depends on the propagation time of the nucleus over the whole crystal.

An intriguing possibility, having implications in condensed matter in general, arises when phase size is taken into consideration in the phase diagram. A similar hypothesis for the size dependence has been invoked in the past for the stability of different phases in pure substances. It is well established that for an infinite sphere (or crystal), an equilibrium triple point Q^{∞} can be defined as an intersection of three planes i.e. solid-vapour plane, liquid-solid plane and vapour-liquid plane in the pressure-temperature phase diagram. For a finite sphere (or crystal), the triple point lies below the equilibrium triple point in the pressure-temperature phase diagram [24].

In combination with the thermodynamic concepts laid out in the paragraph above, when an issue of metastability during crystallisation is invoked, Ostwald in 1897 stated that the thermodynamic stable state is reached through a metastable state of a matter, via a phase which grows faster [25]. For example on cooling from vapour, the first nucleus appears to be liquid even though the temperature may be well below the freezing point of the liquid. Ostwald's stage rule states that the phase transformation will always start with the phase (polymorph) which is stable down to the smallest size, irrespective of whether this is stable or metastable when fully grown. In the case where the phase transformation is nucleation controlled, a connection between the kinetic and thermodynamic considerations can be readily established.

When considering polymers a unique feature of varying size dependence with crystal growth in terms of thermodynamic stability arises, i.e. the crystal size increases with crystallisation time as shown above for polyethylene. This leads to an unique phenomenon in polymers, that a phase which is initially thermodynamically stable may no longer be stable after crystal growth, therefore passing through a metastable state before a thermodynamically stable phase intervenes. The latter will be a nucleation controlled phenomenon.

Similar hexagonal phases with enhanced chain mobility are present in trans-1,4 polybutadiene at atmospheric pressure, thus making the experimentation more convenient [26,13,27]. The monoclinic to the hexagonal phase transition is observed on heating at atmospheric pressure. The equilibrium phase transition temperature from the monoclinic to the hexagonal phase, for the infinite lamellae thickness, is found to be approximately 80°C [26]. Like polyethylene, crystallisation within the thermodynamically stable hexagonal phase leads to the formation of lamellae several hundred nanometers thick of the extended or nearly extended chain type [27]. Lamellar thickness was found to increase by several tens of nanometers in solution-crystallised mats, having an initial lamellae thickness of 10.4 nm, immediately upon heating above the monoclinic to hexagonal transition temperature [27]. For these finite crystals the monoclinic to hexagonal phase transformation occurs at 68°C. In addition, a unique phenomenon of isothermal phase reversal in the solution-crystallised mats was observed on annealing just above the transition temperature. That is, on annealing just above the monoclinic to hexagonal transition temperature at 68°C (which is below the equilibrium transition temperature of 80°C for the monoclinic to hexagonal transition) the crystals transformed initially to the hexagonal phase and subsequently back into the thermodynamically stable monoclinic phase, as shown by *in-situ* WAXS results presented in Fig. 3.9. On heating again the transformed crystals in the monoclinic phase transform back into the hexagonal phase before melting occurs. These observations when combined with SAXS strongly supported the above viewpoint that crystal size depends upon the phase transition [27].

Thanks to the synchrotron radiation facility at Grenoble, it has been possible to observe the phenomenon of isothermal phase reversal in polyethylene at the elevated pressures and temperatures even for the inaccessible experimental conditions. For our studies, solution-crystallised films of Ultra High Molecular Weight Polyethylene were prepared and used as a model system because of the well defined regularly stacked lamellae having a thickness of 12.5 nm [14]. The film was placed between two diamonds in the piston cylinder type pressure cell of Hikosaka [15]. The cell was mounted in a high-resolution powder diffraction

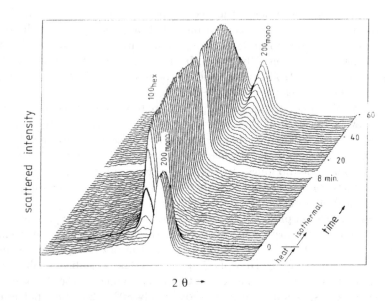

$2\theta \rightarrow$

Fig. 3.9. Wide angle X-ray diffraction experiment showing isothermal phase reversal. The monoclinic $(200)_{mon}$ and the hexagonal $(100)_{hex}$ reflections can be seen. The single crystal mats are heated at $3°C/min$ to $68.5°C$ and after that the temperature was kept constant. In the beginning the monoclinic to the hexagonal phase can be observed and subsequently the reverse from the hexagonal to the monoclinic phase occurs upon annealing.

beamline, ID11/BL2 (ESRF, Grenoble). The following routes, shown schematically in Fig. 3.10 were used for the *in-situ* pressure-temperature experiments [28].

Route 1. Isobaric heating (Fig. 3.11a)

The solution-crystallized UHMW-PE sample was heated at a fixed pressure of 1.6 kbar. The X-ray diffraction pattern at lower temperatures shows the characteristic orthorhombic (110) and (200) reflections. Upon heating, the (100) reflection of the hexagonal phase appears at approximately $195°C$, next to the (110) reflection of the orthorhombic phase. Note that the pressure of 1.6 kbar is far below the pressure corresponding to the triple point [22] and the hexagonal phase is observed within the region of the thermodynamically stable orthorhombic phase.

Route 2. Isothermal and isobaric annealing (Fig. 3.11b)

Upon annealing at $204°C$ at a pressure of 1.6 kbar (route 2 in Fig. 3.10), the (100) reflection of the hexagonal phase disappears again, after approximately 5 minutes, whereas the orthorhombic reflections gain in intensity.

Route 3. Isobaric heating (Fig. 3.11c)

During isobaric heating at a rate of $2°C/min$ (route 3 in Fig. 3.10), the

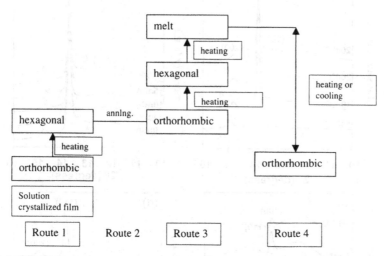

Fig. 3.10. Schematic drawing of the experimental routes adopted in the in-situ X-ray study of the solution crystallised UHMW-PE at constant pressure of 1.6 kbar, i.e. below the equilibrium triple point.

hexagonal phase re-appears and before final melting at 220°C, the characteristic (100) reflection of the hexagonal phase gains intensity at the expense of the (110) and (200) reflections of the orthorhombic phase.

Route 4. Isobaric cooling (Fig. 3.11d)

Upon cooling from the melt at the same constant pressure of 1.6 kbar, only the orthorhombic phase is visible on crystallisation. If the sample is heated once again to the melting temperature, the hexagonal phase is no longer observed (route 4 as shown in Fig. 3.10).

From this set of experiments, it can be concluded that the melting of the lamellar crystals of approximately 12 nm (initial) thickness proceeds via the hexagonal phase much below the equilibrium triple point and within the region of the thermodynamically stable orthorhombic phase. After complete melting and upon re-crystallisation from the melt, the appearance of the hexagonal phase could not be observed again. During heating and annealing of solution-crystallized samples, the thickness of the lamellar crystals increases, especially in the mobile hexagonal phase [29,37,30,17,16] with the result that crystals initially in the hexagonal phase transform first into a transient metastable hexagonal phase and subsequently back to the thermodynamically stable orthorhombic phase. The appearance and disappearance of the hexagonal phase during heating and annealing at pressures below the equilibrium triple point can be related to the initial lamellar thickness. Crystallisation from the melt usually leads to much thicker lamellar crystals, compared to solution-crystallized samples, and

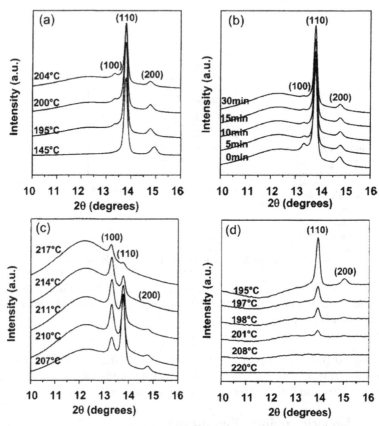

Fig. 3.11. X-ray diffractogram of solution-crystallised UHMW-PE at fixed pressure of 1.6 kbar. **(a)** Next to the (110) and (200) reflections typical for the orthorhombic unit cell of polyethylene, incoming of the hexagonal (100) reflection with increasing temperature at constan pressure can be observed. **(b)** The disappearance of the hexagonal (100) reflection during isothermal and isobaric annealing. The orthorhombic (110) and (200) reflections gain the intensity. **(c)** The melting of the crystals via the hexagonal phase can be observed, similar to the melting behaviour anticipated above the triple point. **(d)** The crystallisation from the melt directly into the orthorhombic phase can be observed; to observe the very initial stage of crystallisation, each diffraction pattern is subtracted from the diffuse melt spectrum.

consequently no crystals in the hexagonal phase are observed on cooling at the same pressure and temperature[2].

---

[2] These findings may be in disagreement with the earlier in-situ observations by polarizing microscopy discussed earlier in this article. Such disagreement in the results can be a result of a difference in pressures used for optical microscopy (always above 2.8 kbar) and the one used for in-situ X-ray measurements (pressure always below 1.6 kbar). Further, it is necessary to consider the experimental limitations of the two

These observations have been confirmed and further strengthened by performing the same set of the experiments on irradiated solution crystallised films. It was discussed above, that in polymers because of the chain mobility along the c-axis the crystal size can alter, unlike with other materials. However, the chain mobility along the c-axis can be suppressed by crosslinking the amorphous zone between the lamellae by irradiation. If the irradiation dose is sufficiently high the lamellae thickening process can be completely suppressed. With the help of Small Angle X-ray Scattering it has been shown that the crystal thickening can be fully suppressed on irradiating the solution crystallised films of UHMW-PE by 2000 kGy. Unlike in the unirradiated solution crystallised films of UHMW-PE, crosslinking of the amorphous region can inhibit lamellar thickening even in the hexagonal phase. Therefore, it would be anticipated that even below the equilibrium triple point, once the transformation from the orthorhombic to the hexagonal phase occurs, no phase reversal from the hexagonal to the orthorhombic phase during annealing under isobaric and isothermal conditions could be observed [28]. Moreover, on heating the irradiated crystals not only transform into the hexagonal phase but also melt via the hexagonal phase at pressures below the equilibrium triple point. The observed melting behaviour is very similar to the one anticipated above the triple point, thus confirming the thermodynamic stability of the hexagonal phase for the small crystals in the thermodynamic stable region of the orthorhombic phase for crystals having infinite lamellar thickness.

To summarise, with a series of experiments on polyethylene and polybutadiene, described above, it has come up strongly that on considering the effect of phase size on the phase diagram an intriguing possibility has arisen. These results suggest that the true thermodynamic stability conditions can invert with size. Specifically for a polyethylene crystal that is small enough, the hexagonal phase can be the thermodynamically stable phase, even in the pressure-temperature regime where orthorhombic is the thermodynamically stable phase for an infinite crystal-size, as shown in Fig. 3.12 by dashed lines, having its triple point Q located below Q^∞. When this is the case, true metastability need not be involved to account for the observation of a metastable phase appearing first. In fact, here the metastable phase in its very small dimensions, will be the thermodynamically stable phase with an inversion of phase stability on growth, i.e. Q moves towards Q^∞ with thickening of lamellae. These observations are in accordance with the mathematical expressions provided earlier in this section.

Further, a Gibbs Free Energy phase diagram can be provided for the experimental observations on the isothermal phase reversal in unirradiated and irradiated UHMW-PE solution crystallised films.

The melting point of lamellar crystals with average thickness l is given by the well-known melting-point depression relationship [31]:

$$T_m = T_m^\infty \cdot [1 - (2\sigma_e/(l \cdot \Delta H)]. \tag{3.7}$$

different techniques used. With optical microscopy observations are made on single crystals while X-ray studies have been recorded on bulk samples of 0.4 mm thickness.

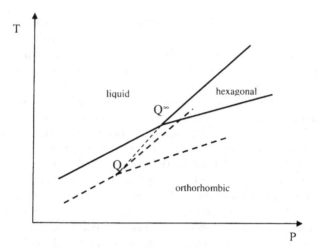

Fig. 3.12. Schematic drawing of the shift of the triple point from Q^∞ to Q due to the small initial lamellar thickness. Full lines represent the equilibrium phase diagram for the infinite size crystals with Q^∞ being its equilibrium triple point. Dashed lines represent shifted phase diagram due to the reduced initial crystal size with Q being its triple point.

As shown in Hoffman-Weeks equation [31], the observed melting point (T_m) is dependent on the thickness l, i.e. if l approaches infinity, T_m approaches T_m^∞, the equilibrium melting temperature. σ_e is the end surface free energy of the crystals.

The ratio $\sigma_e/\Delta H$ for orthorhombic crystals is greater than for hexagonal crystals [20]:

$$(\sigma_e/\Delta H)_{orth.} = 3.5(\sigma_e/\Delta H)_{hex..} \tag{3.8}$$

This implies that for a folded-chain crystal of the same average lamellar thickness l the difference ($T_m^\infty - T_m$) is larger for the orthorhombic crystals than for the hexagonal ones. Equivalently, the Gibbs free energy for a lamellar crystal with thickness l is closer to the equilibrium Gibbs free energy G^∞ in the case of a hexagonal crystal structure. Based on these experimental facts, the Gibbs free energy diagram in Fig. 3.13, could be constructed to account for the experiments described in Fig. 3.11. For the sake of simplicity, the Gibbs free energy functions in Fig. 3.13 are drawn as straight lines, which is in fact an oversimplification but not an essential requirement for the present discussion.

In Fig. 3.13, the Gibbs free energy at pressure P as a function of temperature T is shown for orthorhombic and hexagonal crystals. Below the equilibrium triple-point, the Gibbs free energy of a perfect (extended-chain) orthorhombic crystal, G_{orth}^∞, is lower than that of a perfect (extended-chain) hexagonal crystal, G_{hex}^∞. The Gibbs free energy of the folded-chain crystals is higher due to the contribution of the surface free energies. Assuming for the present discussion that the free energy curves for the folded-chain crystals are parallel with the

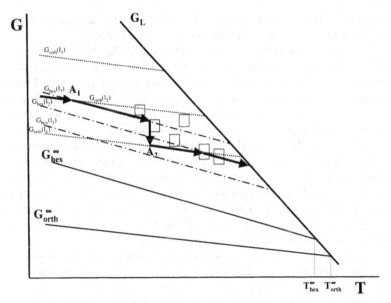

Fig. 3.13. A thermodynamic explanation for the experimental observations made during heating and annealing at the pressure of 1.6 kbar (i.e. below the equilibrium triple point). Black bold lines are equilibrium free energy lines for extended chain orthorhombic and hexagonal crystals, ($\cdots\cdots\cdots$) are the free energy lines for folded-chain orthorhombic crystals possessing different thickness $l_1 > l_2 > l_3$, ($-\cdot-\cdot-\cdot-\cdot-$) are the free energy lines for folded-chain hexagonal crystals possessing different thickness $l_1 > l_2 > l_3$.

free energy of the extended-chain crystals and taking into account, as discussed above, that for a given lamellar thickness l the free energy is closer to equilibrium for hexagonal crystals, the various curves in Fig. 3.13 become self explanatory.

If an orthorhombic folded-chain crystal with thickness l, discontinuous line in Fig. 3.13, is heated at a constant pressure P, the corresponding free energy curve crosses the free energy curves of hexagonal folded-chain crystals. At the crossing point A_1, the crystal can transform from an orthorhombic into a hexagonal crystal structure since the decrease in free energy with increasing temperature, (dG/dT), is faster in the case of hexagonal crystals. This situation is encountered during isobaric heating (see Fig. 3.11a). Upon annealing at temperature T, Fig. 3.11b, the hexagonal crystals thicken in order to decrease the Gibbs free energy towards the equilibrium value. However, during annealing and thickening at temperature T, the driving force $(G_{hex}^\infty - G_{hex})$ becomes smaller and the thickening process slows down and finally is arrested, for example at point A_2. The driving force $(G_{orth}^\infty - G_{orth})$ at temperature T and point A_2 is higher than $(G_{hex}^\infty - G_{hex})$ and, consequently, the crystal could transform back into the orthorhombic crystal structure. Upon further heating, Fig. 11c, once again, the Gibbs free energy curve corresponding to the orthorhombic crystals crosses many times the Gibbs free energy curves of hexagonal folded-chain crystals.

Consequently a transformation from orthorhombic into hexagonal crystals can appear again before the final melting into the liquid phase.

In the case of irradiated samples thickening during heating and annealing is hindered by the crosslinks present in the amorphous region of the lamellar morphology. Nevertheless, the transformation from the orthorhombic into the hexagonal phase during heating at pressures below the equilibrium triple point will occur at the crosspoint A1 (Fig. 3.13) because decrease of Free energy with increase in temperature is faster for the hexagonal crystals than for the orthorhombic ones. However, during isothermal and isobaric annealing, thickening in the mobile phase is arrested and melting proceeds via the same line corresponding to the free energy of the hexagonal folded chain crystals having thickness l_3 for the hexagonal crystal. Due to the absence of lamellar thickening the triple point of the P-T phase diagram stays "arrested" at the lower values of pressure and temperature (position Q^∞ in Fig. 3.12), implying that the shift in the triple point is related to the crystal size.

The authors hasten to add that the above explanation refers only to thermodynamic parameters. The rate of transformation from the orthorhombic into the hexagonal crystals and vice versa, however, is dependent on kinetic barriers between the two crystal structures as stated in the thermodynamical section discussed above. The transformation from the orthorhombic into the hexagonal crystal structure involves nucleation and growth [22] and consequently, the occurrence of a metastable hexagonal phase and the rate of transformation from orthorhombic into the hexagonal crystal structure and vice versa, is system dependent and will depend upon parameters like the initial morphology such as crystal thickness, molar mass and pressure. In fact, preliminary experiments on other solution-crystallized polyethylene samples, for example a fractionated sample kindly provided by the National Institute of Standard Technology, possessing a M_w of 32 kg.mole^{-1} and a molecular weight dispersion of 1.11, showed that a metastable hexagonal phase appeared at even lower pressures, for example as low as 1 kbar.

3.3.5 Implications to Crystallisation at Atmospheric Pressure

The observations summarised above, together with the Region III in the proposed schematic pressure-temperature phase diagram (Fig. 3.1b) raises further implications in our understanding of crystallisation at atmospheric pressure. Our observations have been that, even below the equilibrium triple point, crystallisation always proceeds via the phase which grows fastest even if it is not a thermodynamically stable phase, obeying Ostwald's stage rule. In the proposed schematic phase diagram in Fig. 3.1b, region III is defined as the region between the melting temperature of the hexagonal and the orthorhombic crystals within, the pressure-temperature phase diagram.

Below the equilibrium triple point in the pressure-temperature phase diagram, melting of the crystals that have transformed from the hexagonal to the orthorhombic phase occurs directly without transformation; an observation in contradiction to the anticipated melting behaviour above the triple point. The difference between the melting temperature for the orthorhombic and the hexagonal crystals increases with the decreasing pressure. Thus below the equilibrium triple point a crystallisation temperature can be defined by at least two supercoolings, one from the melting temperature of the hexagonal phase and the other from the orthorhombic phase. Since the difference between the melting temperatures for the orthorhombic and the hexagonal phase increases with decreasing pressure and the melting temperature for the orthorhombic phase is higher than the hexagonal phase, the supercooling defined from the orthorhombic phase is always higher than the one defined from the hexagonal phase. Figure 3.14 shows the opening of region III with decreasing pressure for two different sharp fractionated molecular weights of polyethylene synthesised at the laboratory scale, obtained from NIST. The data point shown in the phase diagram has been obtained by *in-situ* melting of the single crystals (extended chain as confirmed later by electron microscsopy) under isobaric conditions [17]. Since the melting points for the hexagonal and the orthorhombic crystals, within the pressure-temperature region explored so far, falls along the straight line an estimation of the difference in the melting temperatures versus pressure can be made. The estimated value for the difference in the melting temperatures of the extended chain crystals (as confirmed by Transmission Electron Microscopy) of the orthorhombic and the hexagonal phase against decreasing pressure amounts to 4.375 K/kbar. The value of 4.375 K/kbar extends from the experimentally measured triple point of approximately 3.5 kbar for NIST SRM1483 ($M_w = 32000, M_w/M_n = 1.11$). On extrapolating the value from the measured triple point to atmospheric pressure the difference between the melting temperatures for the extended chain crystals of the hexagonal and the orthorhombic phase amounts to approximately 15 K. If we consider the extension of our viewpoint on Region III (defined as no crystal growth region) to atmospheric pressure we are forced to the suggestion that at least 15 K supercooling from the equilibrium melting temperature of the orthorhombic crystals would be required for crystallisation to occur. It means that if we consider the equilibrium melting temperature of polyethylene to be 145°C, the minimum temperature required for crystallisation would be 130°C.

Crystallisation in the hexagonal phase requires much lower supercooling (like 1 K) in comparison with the orthorhombic phase. For crystallisation in the hexagonal phase the supercooling is defined from the melting temperature of the hexagonal crystals. The lower supercooling for the hexagonal crystals is justified because of the much lower end surface free energy desired for the crystallisation to occur in the hexagonal phase. On considering the extrapolation of "no growth region III" to atmospheric pressure the melting temperature of the hexagonal phase would be in the vicinity of 130°C, for infinite crystals in an unconstrained bulk.

On taking in view the decreasing residence time of crystals in the hexagonal phase with decreasing pressure, a transformed crystal at a high nucleation point for other crystals, and low energy barrier required for crystallisation to occur into the hexagonal phase, suggests a possibility of crystallisation starting via the hexagonal phase. The crystal on transformation to the thermodynam-

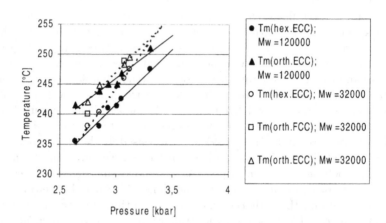

Fig. 3.14. Pressure-temperature phase diagram for two different sharp fractionated polyethylenes (NIST SRM 1483, $M_w = 32000$, $M_w / M_n = 1.11$ and NIST SRM 1484, $M_w = 120000$, $M_w / M_n = 1.2$) obtained by in-situ melting of the single crystals at the isobaric conditions.

ically stable orthorhombic phase favours the growth of other crystals, leading to formation of spherulitic morphology (as discussed earlier). The hexagonal phase leads to a unique tapered morphology at elevated pressures and evidence in single crystals for such a unique tapered morphology has been also reported atmospheric pressure [32].

Recently, in paraffins like hexadecane it has been reported that crystallisation always proceeds via a transient metastable rotator phase before transformation into the thermodynamically stable triclinic phase sets in. Sirota and co-workers [33] have reported these findings for the first time, showing the influence of a metastable phase in crystallisation of paraffins independent of even or odd number of carbon atoms. The observations with paraffins strongly support the above stated hypothesis, especially in polyethylene.

Similar observations on the influence of the mesophase in the early stage of crystallisation have been made for nylons [34], poly-di-alkyl siloxanes [35] and trans-1,4-polybutadiene [27].

Referring to a series of observations on polymers and paraffins we have tried to strengthen the viewpoint that polymer crystallisation from the melt can often proceed via a transient phase which grows faster and has a lower nucleation barrier due to low surface energy in comparison to a thermodynamically stable phase. Several other examples are also known in a range of inorganic and organic materials where a material crystallises via a metastable phase when cooled from the melt. In such a class of inorganic or organic materials a metastable phase may remain stable indefinitely and for very large undercoolings. However, if a nucleus is formed a metastable phase is transient in nature and is bound to transform into the thermodynamically stable phase. In polymers, where a folded-chain crystal is not a thermodynamic stable entity, it is forced to thicken, to minimise its surface free energy. In this process, a crystal which possess small dimensions has a stable phase, when grown, may no longer be a stable phase [36]. In this respect polymers fall into a unique class of materials.

It is important to mention at this point that our views are limited to polymer crystallisation from the melt. We do not have any evidence for crystallisation starting via a transient (or metastable) phase during growth of a crystal from solution. In that respect the existing theories of Hoffman and Weeks, Sadler-Gilmer for polymer crystallisation should hold well and good for crystallisation from solution. However, in melt crystallisation where a transient phase can play a prominent role in the early stages of crystallisation, a theory proposed by Hikosaka [37] can be easily extended for the purpose.

3.3.6 An Application of Enhanced Chain Mobility in the Transient Phase of Ultra High Molecular Weight Polyethylene

The observation of the hexagonal phase below the equilibrium phase diagram could be of great significance, mainly due to the enhanced chain mobility within this phase. Considering the feature of enhanced chain mobility within the hexagonal phase, processing of polyethylene via this phase could become feasible within an experimentally accessible pressure and temperature range, in particular for ultra-high molecular weight polyethylenes (UHMW-PEs). UHMW-PE powder is considered to be intractable via conventional processing routes for polymers due to the excessively high melt-viscosity. Polymer melts are highly viscous and the viscosity strongly increases with increasing molar mass, given by [38] [3]:

$$\log \eta_0 = C + 3.4 \log M_w. \tag{3.9}$$

Equation (9) is a universal relationship, where η_0 is the zero-shear viscosity, C is a constant depending on the polymer architecture and M_w is the weight-average molar mass. Similar to viscosity, the reptation time is also strongly dependent on the molecular mass, given by the mathematical expression [39]:

$$\tau_r \propto M^{3.4}. \tag{3.10}$$

Thus with increasing molar mass the reptation of the chain from one point to the other becomes increasingly difficult for high molar mass melts. Consequently products based on UHMW-PE, usually processed by compression-moulding, possess fusion defects within the bulk material usually referred to as grain boundaries.

On the other hand, mechanical and physical properties of polymeric materials are also highly dependent on the molar mass. In this respect, UHMW-PE is a well-known type of polyethylene possessing excellent mechanical properties such

---

[3] For a detailed discussion of the chain length dependence on polymer crystallization, see the contribution by Okui on page 343 in this book.

as wear and friction characteristics compared with any other polymer material. Due to this reason UHMW-PE is selected as the material of choice in high-performance products such as hip- and knee-joint prostheses. In both types of artificial joints, a UHMW-PE part is used as an interface between human body and metal component of the artificial joint. However, the limited lifetime of the artificial joints is related to the failure within the UHMW-PE component. Improper fusion of the UHMW-PE particles is considered as an important issue in extending the lifetime of the artificial joints.

Figures 3.14a and 3.14b show the optical micrographs (using phase contrast technique) of cut-cryo thin sections from as-recieved and used (for 7 years) UHMW-PE hip cups respectively. Grain boundaries related to the original UHMW PE powder particles can be observed in both hip-cups though it is much more pronounced in the used UHMW-PE cup, Fig. 3.15. Similar grain boundaries are seen in the UHMW-PE inlays used in knee-joint prostheses where it has been well documented that cracks propagate through the improperly fused grains causing delamination [40].

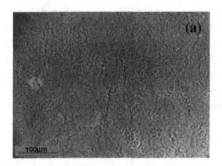 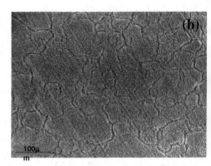

Fig. 3.15. (a and b) Optical micrographs of thin sections of new and used (7 years in the active patient) hip cups respectively. Thin sections of approximately 5 μm thickness were prepared by cryo sectioning the samples of the hip cups. Scale bar: 100 μm.

From the series of studies reported above, it has emerged that in the pressure-temperature phase diagram of polyethylene the presence of the hexagonal phase is strongly dependent on initial crystal size. Ongoing research activity in our group suggests that during polymerisation at relatively low temperatures (lower than the dissolution temperature) small metastable folded chain crystals of UHMW-PE can be obtained directly on the catalyst surface. The obtained "nascent morphology" of the folded chain crystals exhibits the hexagonal phase even at pressures of 1 kbar [28]. Taking this into account a novel route has been developed to process grain boundary free products of UHMW-PE. The optical micrographs of fully grain boundary free materials are shown in Fig. 3.16.

Since crystal formation is a process for chain disentanglement, because of the enhanced chain mobility along the c-axis in the hexagonal phase, the phase has been also used to disentangle chains in UHMW-PE. At this stage it is important to mention that the sintering of UHMW-PE powder particles is a combination of

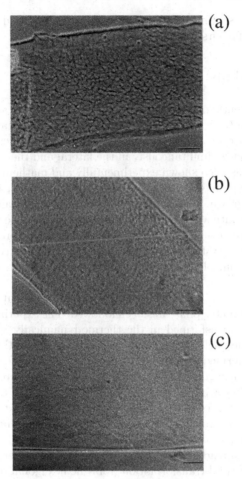

Fig. 3.16. Optical micrographs of thin sections of compression-moulded nascent UHMW-PE. (a) The compression moulded nascent UHMW-PE at 0.8 kbar, heated to 205°C before being cooled to room temperature. (b) The compression moulded nascent UHMW-PE at 1.0 kbar, heated to 205°C before being cooled to room temperature. (c) The compression moulded nascent UHMW-PE at 1.2 kbar, heated to 220°C before being cooled to room temperature. Scale bar: 50 μm.

a chain disentanglement process and crystal size finally achieved before melting – a requirement to increase the radius of gyration on melting, especially at the interface of the powder particles. This is a desired condition to overcome the problem of "grain boundary" in UHMW-PE. Ward and co-workers have also shown that the hexagonal phase can be used for disentanglement of chains in UHMW-PE [30].

If the polymerisation temperature is lower than the crystallisation temperature, a growing chain on the catalyst surface will immediately crystallise and fold, leading to disentangled chains. In this respect the disentangled nascent morphology also favours the sintering of UHMW-PE.

3.4   Conclusions

From the experimental observations reported in this article and elsewhere, it has emerged that in polyethylene even in the thermodynamic stability region for the orthorhombic phase crystallisation always starts via the hexagonal phase. The crystal growth occurs simultaneously in the lateral and the thickening direction. In this article it has been shown experimentally and mathematically that a crystal initially in the stable hexagonal phase, will be no longer thermodynamically stable when fully grown. The crystal, in the thermodynamically metastable state, continues to grow until a nucleus for the orthorhombic phase is formed. Once the nucleation barrier is overcome, the nucleus thus formed spreads over the whole crystal and further crystal growth is arrested. The residence time for the crystals in the hexagonal phase decreases with decreasing pressure and increasing supercooling.

The growth of a polyethylene crystal after its transformation from the hexagonal to the orthorhombic phase has been followed. The experimental observations are that a crystal transformed in the thermodynamically stable orthorhombic phase promotes the nucleation of many crystals. At the initial stage of crystallisation, the newly formed crystals, formed on the basal plane of an orthorhombic crystal, are in the hexagonal phase. However, for the same pressure-temperature, an isolated crystal growing in the hexagonal phase, stays in the metastable state for a longer time when compared with a crystal growing in the hexagonal phase on the surface of the transformed orthorhombic crystal. This is because in the latter a nucleus required for the transformation from the hexagonal to the orthorhombic phase already exists. These observations further strengthen the hypothesis that solid-solid transformation from the hexagonal to the orthorhombic phase is a nucleation-controlled phenomenon. Further, from a series of optical micrographs, it is evident, that no crystals are formed in the vicinity of a crystal growing within the hexagonal phase.

The issues of primary thickening and secondary thickening have been also invoked. From a series of electron micrographs presented in this article, it can be concluded that with increasing supercooling, at a fixed pressure, regular stacking of crystals becomes prominent, as viewed edge-on. These observations strongly suggest that dislocation formation on the surface of the growing crystal is much easier to form at the higher supercoolings.

The observations like multilayering in the hexagonal phase and the overgrowth of crystals on the newly transformed crystal in the orthorhombic phase, considered in the wider generality, can give further insight in the formation of axialites at low supercoolings and spherulite formation at the higher supercooling at the atmospheric pressure.

The experimental observations summarised in this article have relevance to polymer crystallisation in general, especially on crystallising from the melt. In this article, it has been proposed that the Region III (i.e. no growth region) in the pressure-temperature phase diagram for polyethylene can be extended to atmospheric pressure. Region III has been defined as the region lying between the melting temperature for the hexagonal phase and the orthorhombic phase below the triple point. A simple calculation provided in the article shows that the extension of the Region III to atmospheric pressure, accounts for 15°C difference in the melting temperature for the hexagonal and the orthorhombic crystals. On considering the no growth region at atmospheric pressure and the experimental observations that crystallisation always start in the hexagonal phase, suggests that at least 15°C supercooling will be required for the crystallisation to occur at the atmospheric pressure. It is to be noted that the supercooling here is defined from the equilibrium melting temperature for the orthorhombic crystals (145°C). The proposed viewpoint is in agreement with the well accepted experimental observations.

With the series of experiments we have stated that the crystallisation in polyethylene, from the melt, should occur via a phase which grows faster though it may not be thermodynamically stable when fully grown. At this point it needs to be mentioned that the end surface free energy and the minimum thickness required for crystallisation to occur in the hexagonal phase is lower than the orthorhombic phase. Therefore, because of both thermodynamic and kinetic reasons, the formation of the nucleus in the hexagonal phase will be relatively easier than in the orthorhombic phase.

The observations on polyethylene have been extended to other polymers like nylons, trans-1,4 polybutadiene, paraffins, poly-di-alkyl siloxanes, polyesters etc. where observations similar to those with polyethylene have been reported. The fundamental basis laid out in the article, especially the issue of phase reversal with crystal size, has been extended for applications like sintering of ultra high molecular weight polyethylene. Recently, Andrew Keller and Stephen Cheng together [41] have published a review article expressing their views on the metastable phases of polymers, having implications in the condensed matter physics.

Acknowledgement

The scientific findings presented in this article has envolved collaboration work with many others which one of the authors (S.R.) enjoyed during his stay in the University of Bristol as a research fellow with the late Professor Andrew Keller. Among the many authors to whom we are indebted are Professor M. Hikosaka (Hiroshima University) who introduced the subject to one of the authors (S.R.) during his stay in Bristol; Professor P.J. Lemstra for providing the environment for the success of the Bristol work to its present level and for his important role in the development of technological aspects of the fundamental understanding of the subject. The authors are thankful to Dr. A. Terry of beamline ID11/BL2 at the European Synchrotron Radiation Facility, Grenoble for providing help during experimentation.

References

1. A. Keller, *Philos.Mag.* **2**, (1957) 1171
2. E.W. Fischer, *Nature* **12a**, (1957) 753
3. S.J. Organ and A. Keller, *J. Polym. Sci., Part B, Polymer Physics* **24**, (1986) 2319
4. F.C.Frank and M. Tosi, *Proc. R. Soc.* **A 263** (1961), 323
5. J.D. Hoffman, C.M. Guttman and E. A. Di Marzio, *Discuss. Farady Soc.* **68** (1979), 177
6. D.M. Sadler and G. H. Gilmer, *Polymer* **25,** (1984), 446
7. M. Imai, K. Kaji, T. Kanaya, Y. Sakai, *Phys. Rev.* **B52** (1995) 12696
8. A. Keller, *Kolloidzschr. Z. Polymere* **231**, (1969), 386
9. R.G. Chambers, J.E. Enderby, A. Keller, A. R. Lang and J.W. Steeds, "Sir Charles Frank, OBE, FRS: An eightieth birthday tribute", (Adam Higler, Bristol, 1991)
10. J. Klein and R.C. Ball, *Discuss. Farady Soc.* **68** (1979), 198
11. M.I. Abo el Maaty, I.L. Hosier and D.C. Bassett; Macromolecules **31** (1998) 153
12. M.Hikosaka, K. Tsukijiama, S. Rastogi, and A. Keller, *Polymer* **33**, (1992) 12
13. Y. Engelen-Tervoort, Thesis, Technical University of Eindhoven, the Netherlands, 1991
14. S. Rastogi, A. B. Spoelstra, J.G.P. Goossens, and P.J. Lemstra, *Macromolecules* **30,** (1997) 7880
15. M. Hikosaka and T.Seto, *Jpn. J. Appl. Phys.* **21**, (**1982**) L332
16. M. Hikosaka, S. Rastogi, A. Keller and H. Kawabata, *J. Macromol. Sci.Phys.*, **B31**, (1992) 87
17. S. Rastogi, M. Hikosaka, H. Kawabata, and A. Keller, *Macromolecules* **24**, (1991) 6384
18. D. C.Basset, "Principles of Polymer Morphology", (Cambridge, Cambridge University Press, 1981)
19. M. Hikosaka, H. Okada, A. Toda, S. Rastogi, and A. Keller, *J. Chem. Soc. Faraday Trans.*, 91, (1995) 2573
20. A. Keller, M. Hikosaka, S. Rastogi, A. Toda, P.J. Barham and G. Goldbeck-Wood, *J. Mater. Sci.*, **29**, (1994) 2579; *Phils. Trans. R. Soc. London* A348, (1994) 3
21. A. Keller, M. Warner, A.H. Windle, "Self order and form in polymeric material", Chapman and Hall 1995
22. Hikosaka, M.; Amano, K.; Rastogi, S.; Keller, A. Macromolecules **1997**, 30, 2067
23. Basset, D.C.; Turner, B. *Nature* **1972** 240, 146
24. R. Defay, I. Prigogine, A. Bellemans and D.H. Evetrett, "Surface tension and absorption", (Longmans, London, 1966)
25. W. Ostwald, *Z. Physik. Chem.* **22** (1897) 286
26. J. Finter and G Wegner, Macromol. Chem. **182** (1981)1895
27. S. Rastogi and G. Ungar, *Macromolecules* **25** (1992) 1445
28. S. Rastogi, L. Kurelec and P.J. Lemstra, *Macromolecules* **31** (1998) 5023
29. B. Wunderlich and J. Grebowicz, *Adv. Polym. Sci.* **60/61**, (1984) 1
30. A.S Maxwell, A.P. Unwin and I.M. Ward, *Polymer* **37** (1996) 3293
31. J.D. Hoffmann and I.J. Weeks, *J. Chem. Phys.* **42** (1965) 4301
32. D.C. Bassett in "Self order and form in polymeric material"edited by A. Keller, M. Warner, A.H. Windle, Chapman and Hall 1995, 27
33. E.B. Sirota and A.B. Herhold Science **283** (1999) 529
34. C. Ramesh, Macromolecules **32**(1999) 3271; ibid **32** (1999) 5704
35. S. Rastogi and M. Möller, manuscript in preparation
36. E.B. Sirota, Langmuir **14** (1998), 3133

37. Hikosaka, M. *Polymer* **28** (1987) 1257; ibid *Polymer* **31** (1990) 458
38. U.W. Gedde, Polymer Physics, Chapman & Hall, London (1995)
39. P.G. De Gennes., J. Chem. Phys., **55** (1971), 572; M. Doi, S.E. Edwards, Faraday Trans. Soc., **74** (1987), 1789
40. N.J.A. Tulp, Thesis, Katholieke Universiteit Nijmegen, the Netherlands, 1993
41. A. Keller and S.Z.D. Cheng, Polymer **39** (1998) 4461

4 A Comparative Study of the Mechanisms of Initial Crystallization and Recrystallization after Melting in Syndiotactic Polypropene and Isotactic Polystyrene

Mahmoud Al-Hussein and Gert Strobl

Fakultät für Physik, Albert-Ludwigs-Universität, Hermann-Herder-Str. 3
79104 Freiburg, Germany, alhussein@amolf.nl

Abstract. The mechanisms of recrystallization after melting during a heating scan of both syndiotactic polypropylene (sPP) and isotactic polystyrene (iPS) were studied. This was done by monitoring the structure evolution during the recrystallization process and its changes during a subsequent heating scan via time-and temperature-dependent SAXS measurements, respectively. The results of this study showed that sPP samples exhibited a recrystallization mechanism similar to the multi-stage route found upon initial crystallization of semicrystalline polymers from an entangled melt. Meanwhile, a different recrystallization mechanism was shown by iPS samples. In this case, the recrystallization process proceeded as a direct growth into the melt in a one-step process. This is the first time we have observed such a mechanism which resembles the picture presented by the classical models for crystallization from an entangled polymer melt.

4.1 Introduction

Crystallization in polymers is a process that involves some structural organization of a polymer melt (melt-crystallization) or an amorphous glassy polymer (cold-crystallization) leading to a composite semicrystalline structure comprised of crystalline and amorphous regions. During this process a hierarchy of ordered structure develops which in turn controls the physical properties of the polymer. Such a structure can be varied depending on the crystallization conditions. Therefore, understanding the molecular mechanisms which are provoked by this process is essential to establish long-sought-after relationships between processing conditions, structure and property for semicrystalline polymers. In an attempt to meet this objective several models have been proposed [1–3]. Among these models the one which was developed by Hoffman and Lauritzen has been used the most by the workers in this field in a way that made it to be seemed as a conventional wisdom in this respect. However, in the recent years this model has increasingly come under criticism as previously inaccessible information become available due to the employment of new experimental techniques. One of the early observations which indicated the inadequacy of this model was presented by Kanig. The TEM images taken for the early stage morphology of a crystallizing PE did not look anything like continuous lamellae as would be expected according to this model [4]. The Hoffman Lauritzen model also stopped short of explaining the results of a series of time- and temperature-dependent SAXS measurements that we have performed on several semicrystalline polymers [10-13]. These measurements showed that simple well-defined relationships between the crystallization temperature, the melting temperature and the crystal thickness were always found for each polymer. When they are represented in the form of a state-diagram, by plotting both the melting temperature and the crystallization temperatures versus the reciprocal of the crystal thickness on the same graph, two different straight lines are found: the 'melting line' and the 'crystallization line', which intersect each other at a finite crystal thickness. The 'melting line' describes the crystal thickness dependence of the temperature of the phase transition from the crystalline to the melt state according to Gibbs-Thomson relation. Similarily the crystallization line represents a phase transition line of another transition. We interpret it as a transition between some transient mesomorphic phase and an initial granular imperfect crystalline phase. This view is supported by direct space images obtained using both TEM and AFM techniques which show a blocky–like substructure of the crystalline lamellae. Other experimental findings which cannot be explained by the Hoffman Lauritzen model include the following. IR-spectroscopic observations of chain conformations typical for the hexagonal phase at the crystallization onset of PE [5]. Direct AFM-real time observations of growing PE-lamellae showed variations in the individual growth rate between different lamellae and also for a given crystal as a function of time, they showed also structured growth faces which are far away from being flat [6,7] [1]. Finally, oriented crystal growth was achieved by crystallizing in magnetic fields [8] [2]. This cites only a few of several new experimental findings (for more see [9]). Considering all these observations, the need for a new model which would be able to reconcile all these findings seems obvious.

There are cases where crystallization might take place in conditions different from that of a quiescent entangled melt, such as that which takes place after melting during a heating scan. In this case, the crystallites present initially melt, concurrently the molten material undergoes a continuous process of recrystallization into thicker crystallites, which then melt at higher temperatures during a heating scan [14-17]. Initial crystallization from an entangled polymer melt can be thought of as a kind of separation process between crystallizable and noncrystallizable chain segments, which eventually form the crystalline and the amorphous regions, respectively. Whether this is due to only pushing of the entanglements and noncrystallizable entities aside by the growing crystallites [18] or to a disentanglement process, i.e. pulling out chains from their mutual entanglements, the final product of the crystallization process is a 'phase-separated' structure of crystalline and amorphous regions. When such a structure melts, the phase-separated state is preserved for a certain period of time which re-

---

[1] for more details, see Chap. 6.
[2] for more details, see Chap. 5.

sults in a locally disentangled melt before it recovers its initial entangled state. Hence, if such a state lasts until recrystallization starts to take place, it would set in a locally disentangled melt, which is different from the initial homogeneous entangled polymer melt. This might have considerable consequences on the crystallization mechanisms and the structure evolution. In order to evaluate this effect, we have performed a study on the crystallization after melting, i.e. recrystallization for two different polymers, namely syndiotactic polypropylene (sPP) and isotactic polystyrene (iPS). In this communication the results of this study will be reported first. Then a comparison will be drawn between structures produced by recrystallization and those by initial crystallization in order to quantify the effect of the initial melt state on the crystallization mechanisms.

4.2   Experimental

4.2.1   Sample Characteristics and Preparation

The study was performed on two polymers: syndiotactic polypropylene (sPP) and isotactic polystyrene (iPS) of molecular weights 176,800 and 400,000, respectively. Isothermally melt crystallized samples were prepared from sPP, whereas cold crystallized (annealed from the glassy state) samples were prepared from iPS. First, compact, void-free bars were prepared from the as-received material using a small-scale compression mould. This was done by melting the as-received pellets of sPP at a temperature of 190 °C and the powder of iPS at 250 °C for 30 minutes under a slight pressure of 0.04 MPa, under vacuum. The bars were then left to cool slowly to room temperature. Then the bars were wrapped with either an acetate foil, in case of sPP, or an aluminium foil, in case of iPS, and loaded into a cylindrical sample holder which was then placed inside a temperature-controlled cell into the SAXS camera. The foil served as a seal to stop any flow inside the SAXS camera when the polymer melted. The temperature of the cell was controlled by an Eurotherm temperature controller, which made it possible to perform both isothermal crystallization and heating to the melt experiments inside the cell.

For melt crystallization of sPP, each sample was melted at a temperature of 190 °C under no pressure for 20 minutes. Then the temperature was dropped, as quickly as possible, to a desired crystallization temperature, T_c. For iPS, cold crystallized samples were prepared by melting the sample first at 250 °C for 20 minutes under no pressure and then quenching it into an ice/water mixture. Then the sample was transferred as quickly as possible to the temperature-controlled cell of the SAXS camera which was pre-set at a desired T_c. When the samples completed their crystallization, they were subsequently heated to the melt in a stepwise manner.

4.2.2   SAXS Measurements

Instrumentation. Both time- and temperature-resolved SAXS curves were recorded to follow the structure evolution during an isothermal crystallization and its changes during a subsequent heating to the melt. The measurements were performed using an evacuated compact Kratky camera attached to a conventional Cu X-ray tube. The scattering curves were registered using a position sensitive detector within a few minutes counting time.[3]

Data Analysis. The analysis of the SAXS data was based on the two-phase model which is explained in more detail elsewhere [19]. In brief, the SAXS curves were desmeared first using an algorithm developed in our group [20]. Then they were Lorentz corrected by multiplying the intensity data by q^2. The correlation function, $k(z)$, and its second derivative, $k''(z)$, were calculated from the corrected SAXS intensity curves, after subtraction of the background and the liquid-like contributions, using the equations:

$$k(z) = \frac{1}{r_e^2 (2\pi)^3} \int_0^\infty \cos(qz) 4\pi q^2 \Sigma(q) dq \tag{4.1}$$

$$k''(z) = \frac{2}{r_e^2 (2\pi)^2} \int_0^\infty [\lim_{q \to \infty} q^4 \Sigma(q) - q^4 \Sigma(q)] \cos(qz) dq \tag{4.2}$$

where q is the scattering vector, $q = 4\pi \sin \theta_B / \lambda$, θ_B denotes the Bragg angle, $\Sigma(q)$ is the differential cross section per unit volume and r_e is the classical electron radius. In order to reduce the effect of both the noise and the truncation which arises from the finite measuring range, the expression between the brackets in (2) was multiplied by a Gaussian function of appropriate width . The interface distribution function (IDF) is directly related to the second derivative of the correlation function $k''(z)$. For such a two-phase layered system, the second derivative of the correlation function can be expressed as a series of the distributions of the distances between the interfaces[21]:

$$k''(z) = \frac{O_{ac}}{2} (\rho_e^c - \rho_e^a)^2 [h_a(z) + h_c(z) - 2h_{ac}(z) + h_{aca}(z) + h_{cac}(z)....] \tag{4.3}$$

where O_{ac} is the area of the interface separating the crystalline and the amorphous regions per unit volume, ρ_e^c and ρ_e^a are the electron densities of the crystalline and the amorphous phases, respectively, h_a and h_c are the distributions of the thickness of the amorphous and the crystalline layers, respectively and h_{ac} is their sum which is identical to the long period. Furthermore, the asymptotic behaviour of the scattering curve for such a two-phase system can be described by:

$$\lim_{q \to \infty} \Sigma(q) = r_e^2 \frac{P}{(q/2\pi)^4} \tag{4.4}$$

where P is the Porod coefficient and given by:

$$P = \frac{1}{8\pi^3} O_{ac} (\rho_e^c - \rho_e^a)^2 \tag{4.5}$$

---

[3] For an illustration of structure analysis in crystallized polymers, see Fig. 15.1 in Chap. 15 on page 275.

As both sPP and iPS have crystallinities lower than 50%, which has been checked by comparing the DSC heat of fusion with the linear crystallinities, and do not show a solid state lamellar thickening, the peak position of the $k''(z)$ curve has been assigned to the crystal thickness. The long period within the lamellar stack, L, was calculated from the Bragg peak of the Lorentz corrected curves.

4.3   Results

Figs. 4.1 and 4.2 show examples of the desmeared Lorentz corrected SAXS intensity curves at different times during an isothermal crystallization, and at different temperatures during a subsequent heating to the melt, for both sPP and iPS. The quality of the SAXS curves during the isothermal crystallization of the iPS samples collected over a time of 100 seconds was not as good as that of the sPP curves due to its lower scattering efficiency. However, when the counting time was increased up to 15 minutes for the subsequent heating measurements,

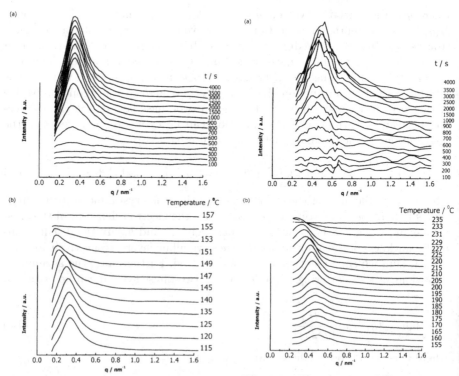

Fig. 4.1. The desmeared Lorentz corrected SAXS curves for a sPP sample during (a) an isothermal crystallization at 115 °C (b) a subsequent heating

Fig. 4.2. The desmeared Lorentz corrected SAXS curves for a iPS sample during (a) an isothermal crystallization at 155 °C (b) a subsequent heating

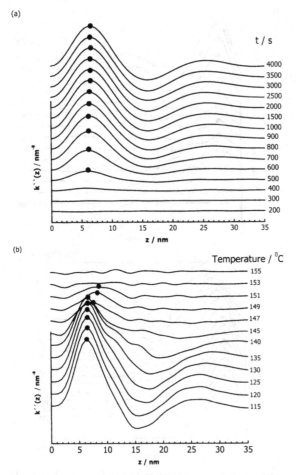

Fig. 4.3. The second derivative of the correlation function for a sPP sample during (a) an isothermal crystallization at 115 °C(b) a subsequent heating

the quality was improved considerably, as it can be seen in Fig. 4.2b. During the isothermal crystallizations of both polymers, the intensity increased with time, meanwhile the peak position was almost constant during the whole crystallization time. During the heating scan, the peak position kept constant up to a certain temperature and then it was shifted to lower values of q for both polymers.

Figs. 4.3 and 4.4 show examples of the $k''(z)$ curves during an isothermal crystallization and a subsequent heating for both polymers. The marked peak represents the crystal thickness, d_c, for both polymers as explained earlier. The long periods, calculated from the SAXS curves of Figs. 4.1 and 4.2, and the crystalline thicknesses extracted from the curves of Figs. 4.3 and 4.4, are shown in Figs. 4.5 and 4.6, respectively. One can see that, apart from a short ini-

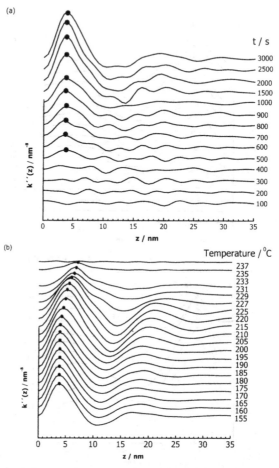

Fig. 4.4. The second derivative of the correlation function for an iPS sample during (a) an isothermal crystallization at 155 °C (b) a subsequent heating

tial period, L was constant during the whole isothermal crystallization for both polymers. During heating, however, it started to increase continuously after a certain temperature. The crystal thickness showed a similar behaviour also. It kept constant throughout the isothermal crystallization, showing no signs of either lamellar thickening or the development of a multi-lamellar population, for the two polymers. On heating, d_c kept constant first up to a certain temperature and then it started to increase. For iPS, the increase in L and d_c started at about the same temperature, for sPP, however, there was a clear difference, the crystal thickening started at a higher temperature. It is also worth noting that the crystal thickening during the heating scans was accomplished by two different fashions in the two polymers. sPP showed a discrete (jump-like) increase, while a smooth continuous increase was shown by iPS.

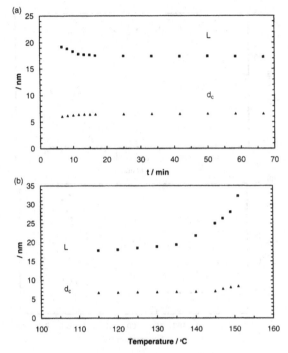

Fig. 4.5. The development of the long period, L, and the crystal thickness, d_c, for a sPP sample during (a) an isothermal crystallization at 115 °C(b) a subsequent heating

The results of such measurements for samples crystallized over a wide range of T_c are summarized in the form of state diagrams as shown in Fig. 4.7. It can be seen from these diagrams that the variation of d_c with temperature is T_c-dependent. For sPP, all samples crystallized at temperatures lower than 130 °C showed the same behaviour of a jump-like increase after a certain temperature. Samples crystallized at temperatures of 130 °C or higher, however, showed no change in d_c up to the melt. In the case of iPS, samples crystallized at temperatures lower than 220 °C showed a continuous increase in d_c after a certain temperature, while those crystallized at 220 °C or higher showed no changes prior melting.

The behaviour which was shown for moderate and low crystallization temperatures can be understood in terms of melting-recrystallization processes which take place during a heating scan. As the temperature was increased during a heating scan, the crystal thickness kept constant up to the melting temperature of the initial lamellae, T_f, concurrently recrystallization took place and resulted in thicker lamellae as demonstrated by the increase in d_c at increasing temperatures, whether it was jump-like or continuous. Recrystallization is a heating-rate dependent process and the location of the melting endotherm of the recrystallized lamellae is also set by the heating rate. At the effective heating rate used in the SAXS experiments, samples crystallized at high T_c's, higher than 130 °C

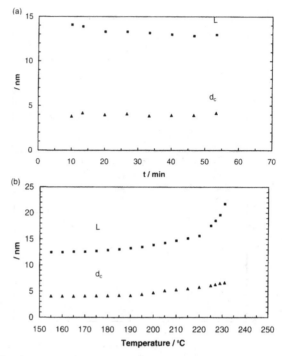

Fig. 4.6. The development of the long period, L, and the crystal thickness, d_c, for a iPS sample during (a) an isothermal crystallization at 155 °C (b) a subsequent heating

for sPP and higher than 220 °C for iPS, did not have time to recrystallize after melting.

In order to evaluate the mechanisms of recrystallization in both polymers, we carried out a number of 'temperature-jump' experiments. In these experiments, samples from both polymers were first isothermally crystallized at different crystallization temperatures. After the crystallization was completed, the temperature was raised quickly to a new temperature (temperature-jump) in the melting-recrystallization range of the lamellae formed upon initial crystallization. Then the structure formation during the recrystallization process, at the new temperature, was followed by time-dependent SAXS measurements. For sPP, two different temperature-jump experiments were carried out, namely 115 to 147 °C and 115 to 149 °C. For iPS, the experiments 155 to 205 °C and 155 to 215 °C were performed. Here the first temperature represents the isothermal crystallization temperature and the second temperature represents the jump temperature at which recrystallization took place. Fig. 4.8 shows typical examples of the kinetics of the recrystallization for both polymers. It can be seen that the whole recrystallization process can be followed in case of sPP, whereas it was too rapid in iPS case as it was completed within the minimum time available to record a SAXS curve, which was 100 seconds. One can also follow the structure development, as demonstrated by d_c, during the recrystallization process. This

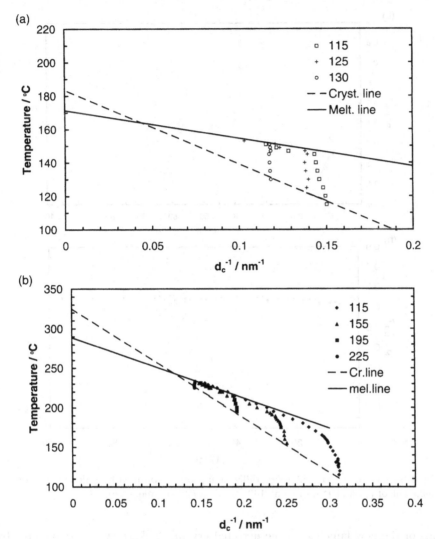

Fig. 4.7. Variation of d_c^{-1} during heating scans subsequent to isothermal crystallizations at the indicated T_c's for (a) sPP (b) iPS

was more obvious for the sPP case as shown in Fig. 4.9. The crystal thickness showed a gradual increase in the early stages of the recrystallization, which then kept more or less constant afterwards. This might be due to the fact that at the onset of the recrystallization process, there would be only a small number of the new recrystallized thick lamellae together with some persisting (non molten) thinner initial lamellae. Therefore, at this stage the detected crystal thickness would be the superposition of both the initial and the recrystallized lamellae thicknesses. At increasing time, more new thick lamellae would be produced by the recrystallization process. Consequently, the contribution of the crystal thick-

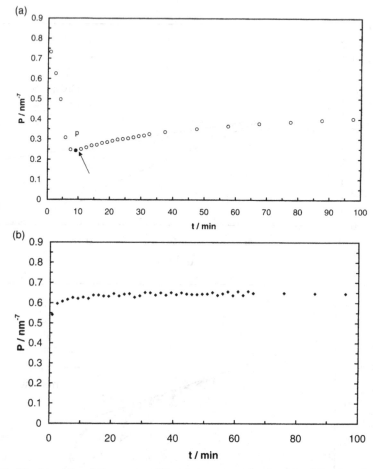

Fig. 4.8. The kinetics of the recrystallization process obtained by following the Porod coefficient of (a) a sPP sample at 149 °C (b) a iPS sample at 205 °C

ness of the new lamellae to the detected crystal thickness would dominate. To monitor the evolution of the thickness of the recrystallized lamellae only during the recrystallization process, the interface distribution curve, $k''(z)$, associated with the persisting lamellae, of point p in Fig. 4.8a, was subtracted from the subsequent interface curves during the recrystallization. As shown by (2), $k''(z)$ varies linearly with both h_c and h_a and normally the cross terms are insignificant and do not affect the first peak, therefore, the subtraction can be done without any loss of generality. Point p was chosen as it represents the point from which onward the recrstallized lamellae started to grow. The results are shown in Fig. 4.10. One can see that now the crystal thickness is kept constant throughout the recrystallization process, demonstrating that the new recrystallized lamellae show no lamellar thickening and supporting the interpretation given for the crystal thickness variation in Fig. 4.9a.

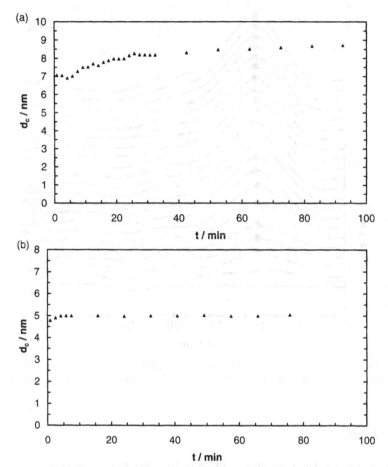

Fig. 4.9. The development of the crystal thickness, d_c, during (a) a recrystallization of sPP at 149 °C (b) a recrystallization of iPS at 205 °C without the subtraction of the persisting lamellae effect

In order to get some clues about the recrystallization mechanisms in both polymers, state diagrams similar to those of the initial crystallization were constructed. This was done by plotting the recrystallization temperature versus the reciprocal of the crystal thickness of the recrystallized lamellae on the same graph which showed also the 'melting line', T_f versus d_c^{-1}, and the 'crystallization line', T_c versus d_c^{-1}, of the initial crystallization of both polymers as shown in Fig. 4.11. It can be seen that for iPS the recrystallization points fell very close to the 'melting line' while for sPP, they were closer to the 'crystallization line'.

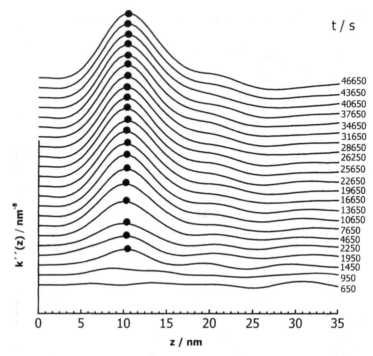

Fig. 4.10. The second derivative of the correlation function for a sPP sample during an isothermal recrystallization at 149 °C after the subtraction of the persisting lamellae effect

4.4   Discussion

Before we discuss the mechanisms of recrystallization, we first start with a short review of the classical crystallization models. The two most common theoretical approaches for polymer crystallization are the surface nucleation model, formulated by Hoffman and Lauritzen [1], and the entropy-barrier model, formulated by Gilmer and Sadler [2]. Both of them are based on the consideration of kinetic processes on the lateral growth faces. They assume that the growth faces are comprised of a range of crystallites of different thicknesses, each of which has a growth rate which depends on its thickness. The dominant thickness would be that which maximize the growth rate. This thickness is slightly greater than the minimum thickness for which the crystal is thermodynamically more stable than the melt. On the other hand, the growth rate itself is considered to result from the competition between the thermodynamic driving force for crystallization and a free energy barrier. The origin of this barrier is different according to the two models. In the surface nucleation model, it is due to the increase in the surface free energy as a molecule is deposited along the growth face. The entropy-barrier model, however, assumes that in addition to any surface free energy terms, attachment of a long-chain molecule in orderly fashion is slowed

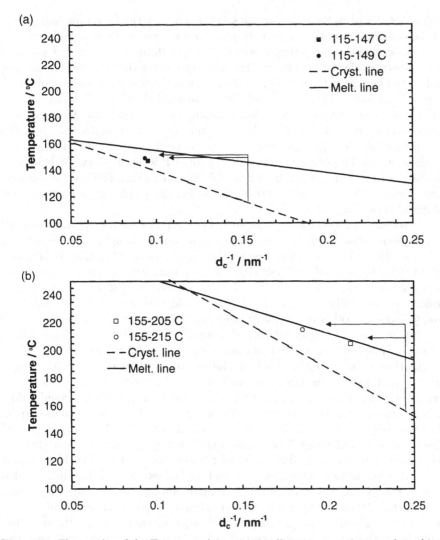

Fig. 4.11. The results of the T-jump melting-recrystallization experiments plotted into the state diagram for (a) sPP (b) iPS

down as the molecule would explore many partially attached configuration first, whereby only a few of them allow a further growth.

Despite this difference between the two models, both of them agree qualitatively on the main scheme of forming a lamellar crystallite from an entangled polymer melt. Once a nucleus is produced within an entangled melt, it would grow into a lamellar crystallite by attaching stems along the growth face continuously. Therefore, each crystallite grows directly into the melt in a single step. This picture, as depicted by both models, does not reconcile with our main findings as explained in the introduction section, whereby the formation and growth of the lamellar crystallites is a multi-stage process rather than a single one.

Considering the state-diagrams of the recrystallization of both iPS and sPP, one finds two different behaviours. For iPS, one can see that the recrystallization points lied very near to the 'melting line'. This is the first time we have observed such a behaviour. This is very much similar to what the classical models of crystallization predict and therefore the multi-stage route is no longer followed when recrystallization occurs in iPS. Thus, we conclude that recrystallization in iPS proceeds as a direct growth into the melt, without passing through any intermediate states. The situation was different for sPP where it showed a behaviour similar to that found normally for initial crystallization. The recrystallization points were away from the 'melting line' and closer to the 'crystallization line' of initial crystallization.

As a concluding remark, we comment on the rationale behind these two different mechanisms. A major candidate which can be thought of as responsible for this difference, is the initial melt state prior to crystallization. It is anticipated that the state of the melt right after melting will be different from that of a quiescent entangled melt. When a crystallized sample melts, the regions which were crystallites before the melting would have less concentration of entanglements and other non crystallizable entities than those regions which were amorphous. This results in a locally disentangled melt which lasts for a certain period of time before it recovers its initial entangled state. This period is given by the time needed for the chains to diffuse (reptate) thermally between each other and eventually the entangled melt state will be restored [22]. This time is expected to depend on the chain structure, as linear flexible chains would diffuse easier than semi-rigid chains with bulky side groups. Therefore, on would expect that the partially disentangled melt state would last longer in iPS than in sPP. From the recrystallization kinetics presented in Fig. 4.9, it is evident that the recystallization process in iPS was indeed faster than that in sPP. Therefore, it seems reasonable to assume that recrystallization sets in a locally disentangled melt in the case of iPS, while it sets in a more homogeneous entangled polymer melt in the sPP case. This might indicate a direct effect of the initial melt state prior to crystallization on both the crystallization mechanisms and the structure evolution during an isothermal crystallization. One may conclude that the multistage route is the general route for polymers which crystallize from an entangled melt or an amorphous state,while in the special case of a partially disentangled melt, the crystallization proceeds as a direct growth process.

References

1. J. D. Hoffman, G. T. Davis, J. I. Lauritzen: In Treatise on Solid State Chemistry (Plenum, New York 1976) 497.
2. D. M. Sadler, G. H. Gilmer: Polymer **25**, 1446 (1984)
3. K. Armitstead, G. Goldbeck-Wood: Adv. Polym. Sci. **100**, 219 (1992)
4. G. Kanig: Colloid Polym.Sci.**269**, 1118 (1991)
5. K. Tashiro, S. Sasaki, N. Gose, M. Kobayashi: Polymer **30**, 485 (1998)

6. J. K. Hobbs, T. J. McMaster, M. J. Miles, P. J. Barham: Polymer **39** 2437 (1998)
7. J. K. Hobbs, A. D. Humphries, M. J. Miles: Macromolecules **34** 5508 (2001)
8. H. Ezure, T. Kimura, S. Ogawa, E. Ito: Macromolecules **30** 3600 (1997)
9. G. Strobl: Eur. Phys. J. E **3** 165 (2001)
10. G. Hauser, J. Schmidtke, G. Strobl: Macromolecules **31**, 6250 (1998)
11. B. Heck, T. Hugel, M. Iijima, E. Sadiku, G. Strobl: New J. Phys. **1**, 17 (1999)
12. M. Iijima, G. Strobl: Macromolecules **33**, 5204 (2000)
13. Q. Fu, B. Heck, G. Strobl, Y. Thomann: Macromolecules **34**, 2502 (2001)
14. P. J. Holdsworth, A. Turner-Jones: Polymer **12**, 195 (1972)
15. P. J. Lemstra, T. Kooistra, G. J. Challa: J. Polym. sci. A-2 **10**, 823 (1972)
16. J. Boon, G. Challa,D. W. Krevelen: J. Polym. Sci. Polym. Phys. Ed. **6**, 1791 (1968)
17. D. J. Blundell, B. N. Osborn: Polymer **24**, 953 (1983)
18. R. K. Bayer: Colloid Polym. Sci. **272**, 910 (1994)
19. G. Strobl, *The Physics of Polymers* (Springer, Berlin 1996)
20. G. Strobl: Acta Crystallogr. **26**, 267 (1970)
21. W. Ruland: Colloid Polym. Sci. **255**, 417 (1977)
22. M. Doi, S. F. Edward: *The Theory of Polymer Dynamics* (Oxford Scienc Publication, New York 1986) 191.

5 Polymer Crystallization Viewed by Magnetic Alignment

Tsunehisa Kimura

Department of Applied Chemistry, Tokyo Metropolitan University, 1-1
Minami-ohsawa, Hachoiji, Tokyo 192–0397, Japan
kimura-tsunehisa@c.metro-u.ac.jp

Abstract. Observation of the magnetic alignment in polymeric materials provides a
unique means to detect and analyze anisotropic structures, or mesophase, existing in
polymer melts and/or forming during phase transitions because the anisotropic structures usually exhibit diamagnetic anisotropy which is a driving factor of the magnetic
alignment. In this chapter, we first describe general concepts of diamagnetic alignment,
followed by an example of the detection and analysis of a mesophase occurring during
crystallization from melt. Also, mesophase existing in a melt is demonstrated by means
of magnetic alignment. Discussion is given about a possible alternative mechanism of
magnetic alignment on the basis of the orientation dependent shift of melting point
under magnetic fields.

5.1 Introduction

One of the points about recent discussion on polymer crystallization is whether
mesophases are involved in the crystallization process. Experimental and theoretical works have been reported so far regarding the existence and formation
mechanism of mesophases [1–8].

Since polymeric systems are complicated, their phase could not be identified
by only one global minimum in the phase space, but they might be described
by one of local minima in the vicinity of the global minimum. Upon phase transition, starting from an initial state, the system passes through a number of
local minima with relatively high energy before reaching one of local minima
corresponding to the final phase. In this view, it is intuitively acceptable that a
polymer melt passes through mesophases before it reaches a crystalline phase.

Complexity of polymeric systems in association with long relaxation times
causes memory effects. Polymer chains assume a random conformation in a dilute
solution, where all memories, possessed before dissolution, are lost. However, in
most other situations including crystalline phases, liquid crystalline phases, glass
phases, and even molten phases, polymer chains could retain their memories
that are inherited from the preceding processing and thermal histories, or even
a history traced back to polymerization. Memory of ordered structures in solid
phases could be an origin of mesophases in molten states or in concentrated
solutions for example. They may disappear upon sufficient heating or dilution.
It is not known at present whether these mesophases are a thermodynamically
stable one or just a transient one because the relaxation time is very long. We

believe that melting is an important issue when crystallization from melt is studied because the crystallization process bears whole previous history.

Initial stages of crystallization have been studied by means of various techniques, including diffraction methods [9–11], spectroscopic methods [12–15], and others [16–21]. In addition to these techniques, we demonstrate in this chapter that observation of magnetic alignment of polymeric materials provides a novel means to detect and analyze mesophases. It is well known that liquid crystals undergo magnetic alignment [22,23]. The alignment is described as a rotation of anisotropic phase having diamagnetic anisotropy. Magnetic alignment of liquid crystalline polymers can be understood in a same manner [24,25]. Mesophases possess diamagnetic anisotropy due to their structural anisotropy. Then, if they are formed during crystallization, they could align under magnetic fields and hence be detected.

Use of magnetic field also provides a means to analyze mesophases from a different aspect than alignment. Under a high magnetic field, the onset of crystallization shifts to higher temperatures depending on mutual orientation of ordered structures with respect to the applied magnetic field [26]. The value of the shift reflects the entropy change of the transition, enabling to elucidate the ordered structures involved in the transition.

5.2   Principle of Diamagnetic Alignment

In this section are summarized basic concepts about the magnetic energy, the magnetic torque, and the motion of a particle under a magnetic field, which are necessary to understand the subjects in the subsequent sections.

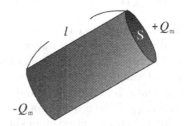

Fig. 5.1. Magnetic charges on two ends of a rod, separated by distance l.

Let us consider a cylinder on two ends of which there are magnetic charges, $+Q_m$ and $-Q_m$ [Wb] as shown in Fig. 5.1. The charge density $\mathbf{P}_m$ on the top surface is defined as

$$\mathbf{P}_m = Q_m/S [\text{Wb/m}^2] \tag{5.1}$$

where S is the area of the top surface and the direction of $\mathbf{P}_m$ is parallel to the cylinder axis. The magnetic dipole moment $\mathbf{p}_m$ is then defined as

$$\mathbf{p}_m = Q_m l = \mathbf{P}_m S l = \mathbf{P}_m V [\text{Wb} \cdot \text{m}] \tag{5.2}$$

where l is the length and V is the volume of the cylinder. The quantities defined as $\mathbf{P}_m/\mu_0$ and $\mathbf{p}_m/\mu_0$, where μ_0 is the magnetic permeability of vacuum ($4\pi \times 10^{-7} [\text{Wb}/(\text{A m})]$) are referred to as magnetization $\mathbf{M}$ and magnetic moment $\mathbf{m}$, respectively, each related by

$$\mathbf{m} = V\mathbf{M}. \tag{5.3}$$

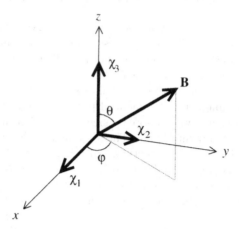

Fig. 5.2. Cartesian coordinates used to calculating magnetization M of a particle having magnetic anisotropy.

In the case of diamagnetic materials, the magnetization **M** is induced by an applied external magnetic field **H**. There is no magnetization in the absence of the external field. The strength of the induced magnetization is proportional to the field strength:

$$\mathbf{M} = \chi \mathbf{H} \tag{5.4}$$

where the constant χ is (volumetric) diamagnetic susceptibility which is negative and dimensionless. If a material is magnetically anisotropic, χ is not a scalar but a tensor. All three principal components χ_1, χ_2, and χ_3 of the tensor have different values in general. However, for the materials with axial symmetry such as nematic liquid crystals and polymer fibers, only two of them are relevant. Namely, $\chi_\parallel$ in the direction of the axis and $\chi_\perp$ in the direction normal to the axis. Taking the coordinate system shown in Fig. 5.2, we obtain the magnetization **M** induced by the magnetic field, $\mathbf{H} = (H_x, H_y, H_z)$ as follows [22]:

$$
\begin{aligned}
\mathbf{M} &= \chi \mathbf{H} \\
&= \chi_\perp H_x \mathbf{i} + \chi_\perp H_y \mathbf{j} + \chi_\parallel (\mathbf{H} \cdot \mathbf{n}) \mathbf{n} \\
&= \chi_\perp (H_x \mathbf{i} + H_y \mathbf{j}) + \chi_\parallel (\mathbf{H} \cdot \mathbf{n}) \mathbf{n} \\
&= \chi_\perp (\mathbf{H} - (\mathbf{H} \cdot \mathbf{n}) \mathbf{n}) + \chi_\parallel (\mathbf{H} \cdot \mathbf{n}) \mathbf{n} \\
&= \chi_\perp \mathbf{H} + (\chi_\parallel - \chi_\perp)(\mathbf{H} \cdot \mathbf{n}) \mathbf{n} \\
&= \chi_\perp \mathbf{H} + \chi_a (\mathbf{H} \cdot \mathbf{n}) \mathbf{n}
\end{aligned}
\tag{5.5}
$$

where **n** is a unit vector parallel to the axis direction and $\chi_a = \chi_\parallel - \chi_\perp$ is the anisotropic diamagnetic susceptibility.

The applied magnetic field $\mathbf{H}$ exerts a magnetic torque $\mathbf{N}$ on the induced magnetic moment. This torque is expressed as

$$
\begin{aligned}
\mathbf{N} &= \mathbf{m} \times \mu_0\mathbf{H} = V\mathbf{M} \times \mu_0\mathbf{H} = V\chi\mathbf{H} \times \mu_0\mathbf{H} \\
&= V\mu_0\chi_a(\mathbf{H} \cdot \mathbf{n})(\mathbf{n} \times \mathbf{H}) \\
&= V\mu_0\chi_a H^2 \cos\theta \sin\theta\, \boldsymbol{\omega} \\
&= V\mu_0^{-1}\chi_a B^2 \cos\theta \sin\theta\, \boldsymbol{\omega},
\end{aligned}
\tag{5.6}
$$

where $\boldsymbol{\omega}$ is a unit vector normal to $\mathbf{n}$ and $\mathbf{H}$, and θ is the angle between $\mathbf{n}$ and $\mathbf{H}$. Approximation $B = \mu_0 H$ is applied to derive the last line, where $B[\text{Wb/m}^2]$ is the magnetic flux density. The same relation is applied hereafter. The dimension of $\mathbf{N}$ is [N m], if V is in [m^3] and B is in [T = Wb/m^2].

The increase of energy of a particle of volume V when it is put in a magnetic field H is given

$$
\begin{aligned}
G &= -\int_0^H \mathbf{m} \cdot \mu_0 d\mathbf{H} = -\int_0^H V\mathbf{M} \cdot \mu_0 d\mathbf{H} = -\int_0^H V\chi\,\mathbf{H} \cdot \mu_0 d\mathbf{H} \\
&= -V\mu_0 \int_0^H (\chi_\perp\mathbf{H} + \chi_a(\mathbf{H} \cdot \mathbf{n})\mathbf{n}) \cdot d\mathbf{H} \\
&= -\tfrac{1}{2}V\mu_0\chi_\perp H^2 - \tfrac{1}{2}V\mu_0\chi_a(\mathbf{n} \cdot \mathbf{H})^2 \\
&= -\tfrac{1}{2}V\mu_0^{-1}\chi_\perp B^2 - \tfrac{1}{2}V\mu_0^{-1}\chi_a(\mathbf{n} \cdot \mathbf{B})^2.
\end{aligned}
\tag{5.7}
$$

The first term in the last line shows that the magnetic energy increases in the case of diamagnetic materials ($\chi_\perp < 0$) and the second term indicates that the energy depends on the direction of the anisotropic axis with respect to the magnetic field (Fig. 5.3). If χ_a is positive, for example, the parallel alignment is energetically favorable to the perpendicular alignment. The derivative of G with respect to the angle θ leads to the magnetic torque given in Eq (6). For an isotropic material with $< \chi_a >_{av} = (\chi_{\parallel} + 2\chi_\perp)/3$ the above equation is reduced to $G = -(1/2\mu_0)V < \chi >_{av} B^2$. If the field strength is not homogeneous, a magnetic force $\mathbf{F}$ exerts on the material,

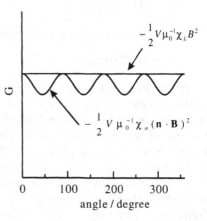

Fig. 5.3. Magnetic free energy as a function of the angle between the external magnetic field and the axis parallel to the $c//$ direction

which is given by the derivative of G with respect to the coordinate (x, y, z), ie., $F = -\nabla G$. If $\mathbf{B} = (0, 0, B_z(z))$, then $F_z = \mu_0^{-1}V < \chi >_{av} B_z(dB_z/dz)$.

The difference in magnetic energy between the two alignment states, parallel and perpendicular to the field direction, is the driving force of the alignment. In a condensed phase, this difference in magnetic energy competes with the thermal energy kT that disturbs alignment. In order for the magnetic energy to overwhelm the thermal energy, the volume V of a particle should be sufficiently large. For an estimation of the critical volume, the following criteria could be used for convenience:

$$V > 2kT\mu_0/|\chi_a|B^2 \tag{5.8}$$

In Fig. 5.4, the critical volume expressed in linear dimension, for typical values of χ_a is plotted against the magnetic flux density B.

Magnetic rotation of an anisotorpic particle in a viscous medium is described by the balance of the magnetic torque [24,27] and the hydrodynamic torque as long as the particle size satisfies Eq (8):

$$L\frac{d\theta}{dt} = -\frac{1}{2\mu_0}V\chi_a B^2 \sin 2\theta \tag{5.9}$$

Here, L depends on the volume V, the viscosity η of the medium, and the shape of the particle. This equation is solved to give a solution:

$$\tan\theta = \tan\theta_0 \exp(-t/\tau) \tag{5.10}$$

where θ_0 is the initial angle between $\mathbf{n}$ and $\mathbf{B}$, and the alignment rate τ^{-1} expressed as

$$\tau^{-1} = (V/L)\chi_a B^2/\mu_0 \tag{5.11}$$

In the case that a particle is a sphere, $L = 8\pi\eta a^3$, then Eq (11) becomes

$$\tau^{-1} = \chi_a B^2/(6\eta\mu_0) \tag{5.12}$$

It should be noted that the alignment rate does not depend on the volume of the particle. In Fig. 5.5 is plotted τ for various values of the viscosity. In the case that the particle is a prolate ellipsoid, the alignment rate is a complicated function of the aspect ratio of the ellipsoid. Larger aspect ratios cause increase in time required for the alignment [27].

In this section, the condition necessary for the diamagnetic alignment has been described. Polymeric fibers and organic crystals are good examples that satisfy this condition. Diamagnetic alignment is well known in liquid crystals and liquid crystalline polymers, but it also occurs to polymeric systems in general if there are mobile anisotropic domains larger than a critical size. This means that anisotorpic domains occurring during a phase transition are detected by means of magnetic alignment.

5.3   Mesophase Detected by Magnetic Alignment

It is well known that liquid crystalline polymers [22–24,28–33] align under a magnetic field due to their anisotropic diamagnetic properties. However, few people imagined magnetic alignment of crystalline polymers because in their melt, there are no anisotropic structures necessary for the alignment, and in their solid phase, the viscosity is too high for crystallites therein to align.

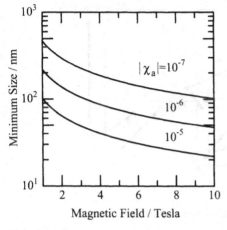

Fig. 5.4. Critical size of a particle necessary for its magnetic energy to exceed the thermal energy plotted as a function of the field strength.

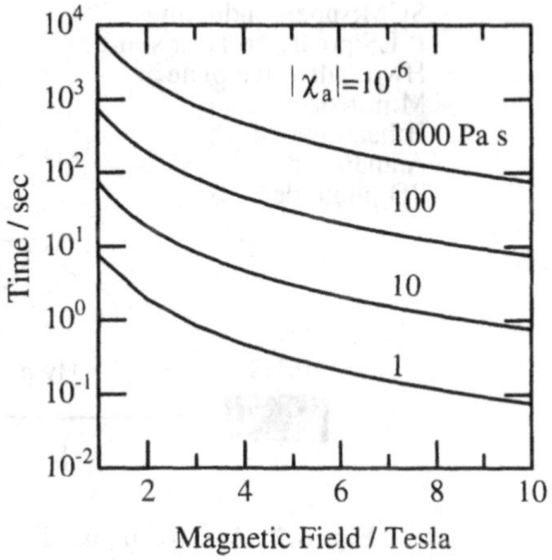

Fig. 5.5. Time necessary for the rotation as a function the field strength.

We have so far reported that four crystalline polymers, poly(ethylene-2,6-naphthalate) (PEN) [34,35], isotactic poly(styrene) (iPS) [36], low molecular weight isotactic poly(propylene) (iPP) [37], and poly(ethylene terephthalate) (PET) [38], undergo magnetic alignment during melt crystallization. Magnetic alignment of paraffin has also been reported by us [39] and by Sirota [40]. We have found that the alignment starts before the formation of crystals detected by wide-angle X-ray diffraction. Recently, detailed studies [41] on iPP and PET have been carried out and revealed that the origin of the magnetic alignment is even traced back to anisotropic structures existing in the molten state above the melting point. The detail will be described in the later sections. Also reported are alignment of high molecular weight iPP [42] induced by the magnetic alignment of a nucleating organic crystal followed by epitaxial growth of polymer crystals on it, and alignment of poly(carbonate) [43] encouraged by reaction with organic salts. Cellulose triacetate cast from a solution has also been reported to undergo magnetic alignment [44].

In Fig. 5.6, a thermal history applied to obtain magnetic alignment is shown [45]. A polymer film is heated from room temperature to the maximum melting temperature (T_{max}) and melted for a while, then brought to the temperature (T_c) at which isothermal crystallization is carried out. Isothermal crystallization is not necessarily re-

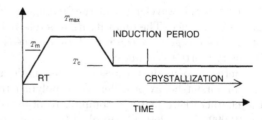

Fig. 5.6. A general thermal scheme applied to attain magnetic alignment. T_m: melting point; T_{max}: melting temperature; T_c: crystallization temperature.

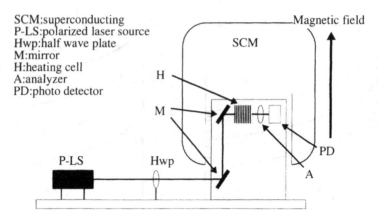

Fig. 5.7. Apparatus used to measuring in-situ birefringence

quired: cooling from T_{max} to the room temperature is in some cases enough to attain alignment. If T_{max} is too high, alignment does not occur in the subsequent crystallization process. This fact is related to the melt structure discussed in the subsequent section.

The development of the alignment is detected by an *in-situ* magnetic birefringence measurement as well as spectroscopic, optical, and X-ray diffraction methods carried out on the samples quenched at various periods of isothermal crystallization time. Figure 5.7 displays schematically a home-built apparatus for the in-situ magnetic birefringence measurement. The impinging light from a polarized He-Ne laser source is reflected by two mirrors to be led to the center of a vertical magnetic field (6 T) generated by an Oxford superconducting magnet, where a film sample sandwiched with two glass plates is set in a heating unit. The light passing through the sample is analyzed by an analyzer, finally reaches a photo-detector. The impinging polarizing light and the analyzer makes an angle of 90° (crossed polars) to each other, makes an angle of 45° with respect to the direction of the magnetic field. In this optical setup, the transmitting light intensity is expressed as

$$I^2 \approx \sin^2(\pi d \Delta n / \lambda) \tag{5.13}$$

where λ is the wavelength of the impinging light, d the sample thickness, and Δn birefringence. The quantity $d\Delta n$ is referred to as retardation. If the increase of Δn is so large that the retardation exceeds $\lambda/2$, then the transmitting light intensity starts to oscillate.

Here we focus on the results of magnetic alignment of PET [38]. Although the alignment behavior differs from polymer to polymer, PET exhibits most of the characteristic features common to other polymers. Pellets of PET (Mw = ca. 10,000, T_m = 260 °C) supplied by Asahi Chemical were dried and pressed at 290 °C for 5 min and quenched into ice water to obtain a pressed film. The films with thickness of 50 μm and 100 μm were prepared. The former was used for the *in-situ* birefringence measurement, and the latter was used for preparing samples for which the isothermal crystallization were interrupted at various periods of crystallization time.

Figure 5.8 shows the temporal change of transmitting light intensity observed for the samples melted at $T_{max} = 270\ °C$ followed by crystallization at 250 °C in the magnet and outside the magnet. Time zero corresponds to the time at which the temperature reaches the crystallization temperature. A sinusoidal change in transmitting light intensity, reflecting the increase in birefringence, is observed for the measurement under the magnetic field (6 T).

To confirm that this sinusoidal behavior is really attributed to the magnetic alignment, samples quenched at various periods of crystallization time were subjected to the optical azimuthal measurement. The azimuthal angle dependence of the transmitting light intensity is described by Eq (13). The azimuthal patterns of the samples quenched at the crystallization times of 10 and 20 min shown in Fig. 5.9 satisfy Eq (13), demonstrating that the increase in transmitting light intensity around crystallization time of 10 min is due to the alignment. At prolonged crystallization times, the angular dependence becomes unclear because the multiple scattering due to the crystal formation occurs or the retardation happens to be close to the wave length of the impinging light.

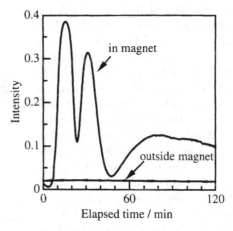

Fig. 5.8. Temporal change of the transmitting light intensity measured for PET sample, caused by increase in birefringence.

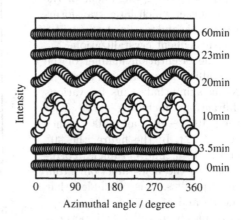

Fig. 5.9. Optical azimuthal scans for the PET samples heat-treated in the magnetic field.

Figure 5.10 displays the X-ray diffraction patterns obtained for the quenched samples. No crystallites are formed until the crystallization time of ca. 20 min. This result, in conjunction with the result in Fig. 5.9 leads to the conclusion that it is not a crystalline phase detectable by X-ray diffraction but it is a mesophase that is responsible for the magnetic alignment.

Alignment of crystallites formed at later stages of crystallization is investigated by means of X-ray azimuthal scans along the (100) plane. Figure 5.11 shows that the samples quenched at 23 min and 60 min exhibit an orientation. The samples quenched at earlier stages do not exhibit a crystal orientation because the crystallites detectable by X-ray diffraction are not formed yet. Since the crystal formation occurs in the later stage of crystallization where the viscosity starts to increase, it is difficult to conclude that the crystal alignment is due to the rotation of crystallites formed. Instead, the observed crystal alignment could be attributed to the preceding alignment of the mesophase, which later transforms to crystals and/or acts as nuclei for the subsequent crystallization.

The orientation manner of the crystallites is investigated using the X-ray azimuthal scans for various diffractions including those from the (010), (−110), and (100) planes (Fig. 5.12). In the figure, the 90° and 270° azimuthal angles correspond to the field direction. From these profiles, it is concluded that the b^*- and a^*-axes are aligned approximately parallel and perpendicular, respectively, to the magnetic field. As a result, the c^*-axis is concluded to be aligned roughly perpendicular to the field (Fig. 5.13). These assignments of orientation, of course, are just approximations because the crystal type of PET is triclinic. The azimuthal profiles shown in the figure differ from those of liquid crystalline polymers where the azimuthal peaks of the magnetically aligned samples are in general much sharper. This is partially due to the cylindrical symmetry of the a^*- and b^*-axes around the field direction. These axes are distributed randomly on the plane normal to the field direction.

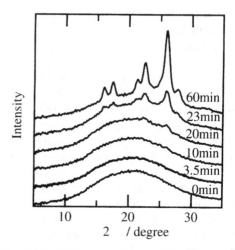

Fig. 5.10. Wide-angle X-ray diffraction patterns for the PET samples heat-treated in the magnetic field.

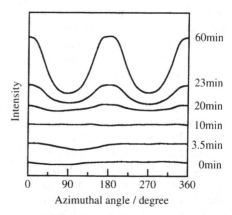

Fig. 5.11. X-ray azimuthal scans for the (100) plane of the PET heat-treated in the magnetic field.

The resultant crystal alignment of PET provides an insight into the structure of mesophase that is responsible for the magnetic alignment. The main source of the diamagnetic anisotropy of PET could originate from that of the benzene ring in the main chain. A benzene ring is energetically stable when the ring plane lies parallel to the magnetic field. In the crystalline phase, benzene rings are packed with their plane lying approximately within the bc-plane. Therefore, the diamagnetic susceptibility in the direction of the a-axis is the negative largest, causing the alignment of the a-axis in the direction perpendicular to the applied magnetic field. Of course, the a-axis

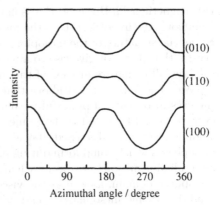

Fig. 5.12. X-ray azimuthal scans for the (010), (-110), and (100) planes of the PET sample heat-treated in the magnetic field for xx min.

does not necessary coincide exactly with one of the three principal axes of diamagnetic susceptibility tensor because the crystallographic axes and susceptibility axes do not coincide in general. Suppose that the mesophase is of nematic nature, that is, a chain (or a number of chains forming a mesophase domain) has a cylindrical symmetry around the chain axis. Unlike in the crystalline phase, benzene rings in the nematic phase are twisted to each other around the chain axis (Fig. 5.14). As a result, the susceptibility perpendicular to the cylinder axis becomes the negative largest as average. As a result, the c-axis aligns parallel to

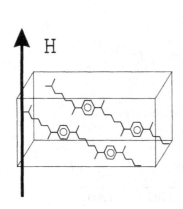

Fig. 5.13.

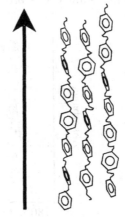

Fig. 5.14. Schematic representation of the PET chain conformation assumed in the nematic phase.

the magnetic field. However, this is not the case for what we observe in Fig. 5.13. Therefore, the hypothesis that the mesophase is of nematic is ruled out. To conclude, the structure of the mesophase might be described as a sort of disordered crystal that does not give rise to a wide angle X-ray diffraction, but still retains the crystalline nature in view of the magnetic anisotropy.

We have demonstrated above that the analysis of the alignment manner of the crystal provides a powerful means to elucidate mesophase structures. However, it should be noted that the mesophase structure determined by magnetic alignment, such as discussed above, is just one among many possible types of mesophases which could form during phase transition. As described before, magnetic alignment starts to occur when the size and magnetic anisotropy of the mesophase domain, as well as the viscosity of the surrounding, reach a suitable condition. Some mesophases could not be detected under a given condition of the magnetic alignment. Or mesophases detected might differ depending on the thermal history experienced before the formation of the mesophase. In fact, in the case of PEN, the orientation distribution of crystal differs depending on the thermal history [46].

Several spectroscopic studies have been reported on the conformation changes occurring prior to the crystallization [13,15]. The mesophase formation during the induction period is supported by an *in-situ* Fourier transform infrared (FT-IR) spectroscopic study on PEN [14]. IR spectra assignable to the pure crystal and pure amorphous phases were identified using difference spectra between highly crystalline and highly amorphous samples. It allowed us to decompose a spectrum into the spectra corresponding to the crystalline, amorphous, and residual phases. We assume that the last phase corresponds to a mesophase. Figure 5.15 shows the spectra of these three phases. Figure 5.16 shows the result of the in-situ measurement of crystallization carried out without magnetic field. A press film sample was brought to a melt at 300 °C, followed by isothermal

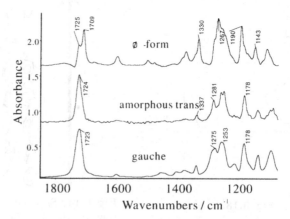

Fig. 5.15. FT-IR spectra of PEN, each corresponding to (1) crystal phase, (2) amorphous trans conformation, and (3) gauche conformation.

crystallization at 255, 250, and 245 °C. Time zero corresponds to the time at which the temperature reached the crystallization temperature. We find in the figure that the mesophase formation precedes the crystal formation.

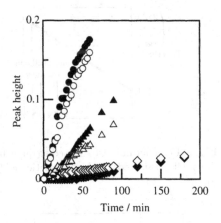

Fig. 5.16. FT-IR absorbance corresponding to crystal phase and amorphous trans conformation, where contribution by gauche conformation is subtracted.

Finally, some comments are addressed regarding what was not mentioned above about the magnetic alignment of other polymers as well as PET. First, PET, once heated above T_{max} = 280 °C for 5 min, does not undergo magnetic alignment in the subsequent cooling process down to the room temperature at 5 °C/min. [41] This is related to the melt structure described in the next section. Second, magnetic alignment of iPP depends on molecular weight. A low molecular weight iPP aligns but a high molecular one does not. The same statement applies to poly(ethylene). Poly(ethylene) does not align, but its low molecular equivalence, paraffin, aligns very easily. However, iPS (Mw of 400,000 seems high enough) undergoes magnetic alignment. The reason is not fully clarified at present, but it seems that the tendency of spherulite formation plays an important role because spherulites prevent or destroy the alignment. If there are enough number of nuclei to prevent a formation of large spherulites, alignment could occur for polymers having strong tendency to form spherulites.

5.4   Molten States

In the previous section, we mentioned that if the maximum melting temperature T_{max} is too high, no magnetic alignment occurs in the subsequent crystallization process. This suggests that the ordered structures in the solid phase could persist even in the melt up to a temperature above the melting point.

Detailed analyses have been carried out on PET and low molecular weight iPP [41]. Here we focus on the results obtained for iPP. A similar analysis applies to PET as well. Pellets of low molecular weight iPP (SA002V, melt flow index = 240 g/10 min, supplied by Japan Polychem) were vacuum dried and heat-pressed at 200 °C for 10 min, followed by quenching in ice water to obtain a film of 200 μm thick. The crystallization of iPP is so quick that the obtained film was partially crystallized. The film sample thus prepared was heat-treated following the scheme shown in Fig. 5.17 under magnetic field (6 T). The sample was heated at 5 °C/min from room temperature to the various melting temperatures T_{max}s ranging from 180 to 280 °C (T_m = 168 °C). As soon as the temperature reached

T_{max}, the sample was removed from the magnet and then quenched in ice water. Obtained samples were subjected to the optical and X-ray analyses.

Figure 5.18 shows the result of optical azimuthal scans. The maximum melting temperature T_{max} is indicated in the figure. Sinusoidal profiles indicate the orientation (13). Magnetic alignment is observed for the samples prepared with $T_{max} < 240$ °C. Because the quenching is carried out outside the magnetic field, the alignment observed should be attributed to that existing in the melt at respective T_{max}s. Figure 5.19 shows optical azimuthal scans for the samples heated in the magnet at max = 180 and 240 °C for various melting periods, followed by quenching outside the magnet. For the heat-treatment at 180 °C (Fig. 5.19a), the alignment remains even after heating for 12 h. The strength of the transmitting light intensity is a function not only of the degree of orientation but also of any factors including film thickness, value of birefringence, and scattering of light, but the increase of the intensity observed in Fig. 5.19a, especially the intensity change between 0 and 10 min, could indicate the development of the orientation. On the other hand, in the case of the heat-treatment at 240 °C (Fig. 5.19b), the alignment initially observed disappears after further heating for 30 min. The stability of the anisotropic phase exhibited at 180 °C is lost at 240 °C. At present, we cannot tell whether the anisotropic phase at 180 °C is thermodynamically stable or it will just relax after very long time.

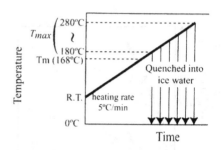

Fig. 5.17. Thermal history used to quench the melt subjected to the magnetic field (6T).

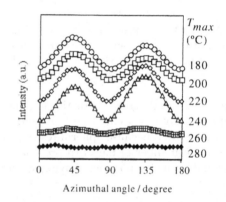

Fig. 5.18. Optical azimuthal scans of the iPP samples quenched from the melts at various T_{max} values.

Figure 5.20 shows the X-ray azimuthal scans along various (hk0) planes of the iPP crystal brought to a melt at 200 °C in the magnetic field followed by quenching. The profile for the (040) plane indicates that the b^*-axis aligns approximately in the field direction, and hence the a^*- and c^*-axes aligned approximately normal to the field. Since the detail of the three principal axes of diamagnetic susceptibility tensor of the iPP crystal is not known nor identified is the main source of the magnetic anisotropy in the iPP chain, the discussion made for PET regarding the mesophase structure does not apply.

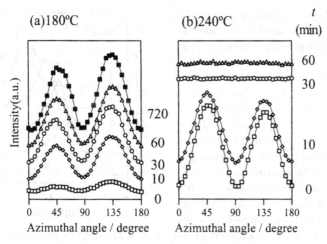

Fig. 5.19. Temporal change of the alignment in iPP melts at different temperatures.

However, if iPP chains in the mesophase have symmetry around the chain axis, that is, if the mesophase is of nematic, the azimuthal profiles for the (hk0) planes should all give the same pattern. This is not the case for the profiles shown in Fig. 5.20. It is therefore concluded that the iPP mesophase is of disordered crystal. Incidentally, it is indicated by the analysis of the azimuthal profiles for the (hk0) planes for a magnetically aligned paraffin sample (carbon numbers centered at C = 36) [39] that the mesophase responsible for the magnetic alignment of this sample is nematic, that is, the mesophase has an axial symmetry along the chain axis. Rheological study on the nematic nature of paraffin is also reported [47,48].

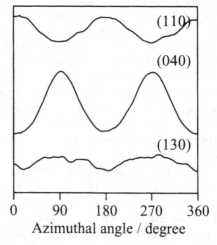

Fig. 5.20. X-ray azimuthal scan of magnetically aligned iPP for various planes indicated in the figure.

The existence of the mesophase in the molten state is suggested indirectly by crystallization behavior monitored by DSC carried out outside the magnetic field. Figure 5.21 shows the DSC thermograms measured upon cooling at 5 °C/min from various melting temperatures (T_{max}) indicated in the figure. The peak top of the exothermic peak due to crystallization shifts drastically to lower temperatures when the initial melting temperature exceeds 240 °C. This temperature coincides with the temperature up to which the mesophase is observed by magnetic method (Fig. 5.18).

From the experimental results described here, combined with those described in the previous section, a picture of crystallization in view of the magnetic alignment could be summarized as follows: Upon heating of a sample from room temperature, even a melt-quenched amorphous sample becomes to form crystalline phase due to cold crystallization before the temperature reaches the melting point. At the melting point, the crystallites formed start to melt, but some order remains up to temperatures somewhat above the melting point. This ordered phase could be referred to as mesophase. During this period, the crystallites or mesophase domains dispersed in an isotropic melt undergo magnetic alignment. The mesophase is stable in the vicinity of the melting point, but relaxes to an isotropic phase upon further increase in temperature. As a result, the degree of orientation decreases. If the heating is interrupted before the complete transformation to the isotropic phase, followed by cooling, then the crystallization starts to proceed in the aligned mesophase. If the cooling starts from higher temperatures where there are no remaining mesophase domains, a larger quench depth is necessary for the homogeneous nucleation to start. The nuclei formed in this process could be scarce and large spherulites are formed. As a result, an alignment in the initial stage, if it may occur, is obscured.

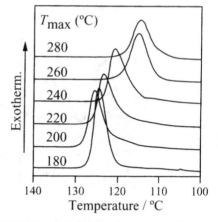

Fig. 5.21. DSC thermograms of iPP during cooling after heated up to various T_{max}s indicated in the figure.

The remaining ordered structures in a melt inherited from a solid phase is not a unique source of the mesophase. The alignment observed in the melt is very weak compared to that observed during induction period. The transmitted light intensity under crossed polars in the magnet observed for the melt is less intense than that during induction period described in the previous section. Because no crystals are formed during the induction period, at least in view of wide angle X-ray diffraction, a higher transmitted light intensity in the induction period should be attributed to the growth of the mesophase already existing or the formation of new mesophases. In fact, the in-situ FT-IR study on PEN [14] indicates the formation of mesophase during the induction period.

5.5   Magneto-Clapeyron

Magnetic fields affect the phase transition temperatures. Although this effect is small, a work with high-resolvable calorimetry in high magnetic fields [49] has been reported recently in an attempt to elucidate structure formations under magnetic field.

We consider anisotropic-isotropic transitions, including a transition from a liquid crystalline phase to an isotropic phase, a melting of crystal, etc. The shift ΔT of the transition temperature T_m under a magnetic field B is expressed by the orientation-dependent magneto-Clapeyron equation: [26]

$$\Delta T = \frac{(\cos^2 \theta - 1/3)\chi_a B^2}{2\mu_0 \overline{\Delta H}} T_m \tag{5.14}$$

Here we assume that an anisotropic phase under consideration has an axial symmetry. Molar diamagnetic susceptibilities parallel and normal to the anisotropic axis are defined as $\chi_\parallel$ and $\chi_\perp$, respectively, and $\chi_a = \chi_\parallel - \chi_\perp$. Also we assume that the average of the diamagnetic susceptibility in the anistorpic phase is equal to that of the isotropic liquid phase. $\overline{\Delta H}$ is the molar latent heat of the transition. The value of ΔT depends on the angle θ of the axis with respect to the magnetic field.

For a material with $\chi_a > 0$, a parallel alignment ($\theta = 0$) is more stable than a perpendicular alignment, and hence a material of the parallel alignment melts at a higher temperature and that of the perpendicular alignment melts at a lower temperature than the melting temperature without the magnetic field. If a powder of the material is heated, its exothermic peak remains at T_m but the peak width might be broaden. If an isotropic phase is cooled slowly, the formation of an anisotropic phase with parallel alignment occurs first, at a higher temperature, resulting in a shift of the onset of the exothermal peak. The amount of the shift is about ΔT with $\theta = 0$. In actual cases, the formation of the anisotropic phase occurs under supercooling. Even in this case, the onset of the exothermal peak would shift to a higher temperature in comparison to the case without magnetic field. The driving force of crystallization is larger for the parallel alignment than that for the perpendicular alignment.

Estimation of ΔT was made for some materials using literature values of χ_a and $\overline{\Delta H}$. The ΔT value of *trans* azobenzene is ca. 0.1 mK under 10 T, while that of p-azoxyanisole, a liquid crystal, is ca. 5 mK. The difference is mainly due to the difference between their latent heats. As seen from Eq (14), the smaller the $\overline{\Delta H}$, the larger the ΔT. Recently, DSC measurement on a normal paraffin (n-$C_{32}H_{66}$) was carried out in the magnetic field of 5 T. A melt of the sample was cooled at 0.2 mK/s and the exothermic behavior was recorded. The shift of the onset was about 40 mK (at 5 T) which is about three orders larger than the expected value. This observation strongly suggests that a smaller $\overline{\Delta H}$, that is "weak first order" [2] transition most probably due to mesophases, is relevant. Upon cooling, a mesophase which is more stable in the magnetic field (in the present case, one with the c-axis aligned normal to the field direction) first starts to form from an isotropic phase, causing the shift of the onset temperature.

5.6   Concluding Remarks

It was demonstrated that high magnetic fields provide a useful means to elucidate what occurs during the polymer crystallization. The analyses made in this chapter are based on the principle that a structure with diamagnetic anisotropy undergoes magnetic alignment and that the transition temperatures are affected by the external magnetic field. The magnetic alignment observed during the induction period strongly suggests the existence of mesophases. The magnetic alignment observed in the molten state indicates that these mesophases are traced back to the melt or even to the solid structure prior to the melting. The unexpectedly high value of the temperature shift of the onset of crystallization observed for the paraffin crystallization in the magnetic field could suggest that a same sort of mesophase formation could take place also in polymeric systems.

Acknowledgement

The author thanks Drs. T. Kawai and M. Yamato for their critical reading of the manuscript.

References

[1] G. R. Strobl: *The physics of Polymers* (Springer, Berlin, 1996)
[2] G. R. Strobl: Eur. Phys. J. E **3**, 165 (2000)
[3] P. D. Olmsted, W. C. K. Poon, T. C. B. McLeish, N. J. Terrill, A. J. Ryan: Phys. Rev. Lett. **81**, 373 (1998)
[4] M. Imai, K. Mori, T. Mizukami, K. Kaji, T. Kanaya: Polymer **33**, 4451 (1992)
[5] M. Imai, K. Kaji, T. Kanaya: Phys. Rev. Lett. **71**, 4162 (1992)
[6] M. Muthukumar, P. Welch: Polymer **41**, 8833 (2000)
[7] Y. A. Akpalu, E. J. Amis: J. Chem. Phys. **111**, 8686 (1999)
[8] Y. A. Akpalu, E. J. Amis: J. Chem. Phys. **113**, 392 (2000)
[9] T. A. Ezquerra, E. Lopez-Cabarcos, B.S. Hsiao, F. J. Balta-Calleja: Phys. Rev. E **54**, 989 (1996)
[10] B. S. Hsiao, Z.-G. Wang, F. Yeh, Y. Geo, K. C. Sheth: Polymer **40**, 3515 (1999)
[11] M. Imai, K. Kaji, T. Kanaya, Y. Sakai: Phys. Rev. B **52**, 12696 (1994)
[12] J.-M. Huang, P. P. Chu, F.-C. Chang: Polymer **41**, 1741 (2000)
[13] K. Tashiro, S. Sasaki, N. Gose, M. Kobayashi: Polym. J. **30**, 485 (1998)
[14] F. Kimura, T. Kimura, A. Sugisaki, M. Komatsu, H. Sata, E. Ito: J. Polym. Sci., Polym. Phys. **35**, 2741 (1997)
[15] T. Kimura, H. Ezure, S. Tanaka, E. Ito: J. Polym. Sci.,Polym. Phys. Ed. **36**, 1227 (1998)
[16] F. J. Balta-Calleja, M. C. Garcia-Guterrez, D. R. Rueda, S. Piccarolo: Polymer **41**, 4123 (2000)
[17] J. Kajaks, A. Flores, M. C. Garcia-Guterrez, D. R. Rueda, F. J. Balta-Calleja: Polymer **41**, 7769 (2000)
[18] T. Okada, H. Saito, T. Inoue: Macromolecules **25**, 1908 (1992)
[19] K. Fukao, Y. Miyamoto: Phys. Rev. Lett. **79**, 4613 (1997)
[20] G. Kanig: Kolloid Z.u.Z. Polymere **261**, 373 (1983)
[21] G. Kanig, Colloid Polym. Sci. **269**, 1118 (1991)
[22] P. G. de Gennes: *The Physics of Liquid Crystals*, 2nd edn. (Clarendon Press, Oxford 1993)
[23] S. Chandrasekhar: *Liquid Crystals*, 2nd edn. (Cambridge University Press, Cambridge 1992)

[24] J. S. Moore, S. I. Stupp: Macromolecules 20, 282(1987)
[25] T. Kimura, H. Sata, E. Ito: Polym. J. 30, 455 (1998)
[26] T. Kimura:Jpn. J. Appl. Phys. 40, 6818 (2001)
[27] T. Kimura, M. Yamato, W. Koshimizu, M. Koike, T. Kawai: Langmuir 16, 858 (2000)
[28] W. R. Krigbaum: in: Polymer Liquid Crystals ed. by A. Ciferre, W. R. Krigbaum, R. B. Meyer (Academic Press, London, 1982) Chapter10
[29] G. Maret, K. Dransfeld: 'Biomolecules and Polymers in High Steady Magnetic Fields'. In: Topics in Applied Physics. Strong and Ultrastrong Magnetic Fields and Their Application ed. by F. Herlach (Springer-Verlag, Berlin, 1985) Vol. 57, pp. 143-204 Chapter 4
[30] G. Maret, A. Bl;mstein: Mol. Cryst. Liq. Cryst. 88, 295 (1982)
[31] T. Kimura, T. Maeda, H. Sata, M. Yamato, E. Ito: Polym. J. 27, 247 (1995)
[32] F. Hardouin, M. F. Achard, H. Gasparoux, L. Liebert, L. Strzelecki: J. Polym. Sci., Polym. Phys. Ed. 20,975(1982)
[33] A. Anwer, H. Windle: Polymer 32, 103 (1991)
[34] H. Sata, T. Kimura, S. Ogawa, M. Yamato, E. Ito: Polymer 37, 1879 (1996)
[35] H. Sata, T. Kimura, S. Ogawa, E. Ito: Polymer 39, 6325 (1998)
[36] H. Ezure, T. Kimura, S. Ogawa, E. Ito: Macromolecules 30, 3600 (1997)
[37] T. Kawai, T. Kimura: Polymer 41, 155 (2000)
[38] T. Kimura, T. Kawai, Y. Sakamoto: Polymer 41, 809 (2000)
[39] T. Kimura, Masafumi Yamato, Wataru Koshimizu, Takahiko Kawai: Chem. Lett. 1999, 1057 (1999)
[40] H. H. Shao, H. Gang, E. B. Sirota: Phys. Rev. E 57, R6265 (1998)
[41] T. Kawai, Y. Sakamoto T. Kimura: Mater. Trans. JIM 48, 955 (2000)
[42] T. Kawai, R. Iijima, Y. Yamamoto, T. Kimura: Polym. Prepr. Jpn 48, 3402 (1999)
[43] H. Aoki, M. Yamato, T. Kimura: Chem. Lett. 2001, 1140 (2001))
[44] T. Kimura, M. Yamato, S. Endo, F. Kimura, H. Sata, H. Kawasaki, Y. Shinagawa: J. Polym. Sci. B. 39, 1942 (2001)
[45] T. Kimura, H. Ezure, H. Sata, F. Kimura, S. Tanaka, E. Ito: Mol. Cryst. Liq. Cryst. 318, 141 (1998)
[46] H. Sata: Study on Magnetic Orientation of Aromatic Polymers. PhD Thesis, Tokyo Metropolitan University, Tokyo (1998)
[47] J. K. Kruger, L Peetz, W. Wildner. M. Pietralla: Polymer, 21, 620 (1980)
[48] J. K. Kruger: Solid State Commun. 30 43 (1979)
[49] H. Inaba, K. Tozaki, H. Hayashi, C. Quan, N. Nemoto, T. Kimura: submitted

6 Following Crystallization in Polymers Using AFM

Jamie K. Hobbs

University of Bristol, H.H. Wills Physics Laboratory, Tyndall Avenue, Bristol BS8 1TL. UK. Jamie.hobbs@bristol.ac.uk

Abstract. Atomic Force Microscopy (AFM) allows polymer crystallization and melting to be followed in-situ, in real-time, at the nanometre level. An overview is given of the additional information that can be obtained using this technique compared to other forms of microscopy. Some recent data are presented that have particular relevance to suggestions that polymer crystallization occurs through a series of stages of intermediate order. Although real-space imaging does not supply convincing supporting evidence of theories suggesting the existence of mesophases prior to crystallization, significant discrepancies with the predictions of classical theories do exist. The most striking of these is the local variation in growth rate from individual lamella to lamella, and, in time, for each lamella. This appears to be a manifestation of local fluctuations in the shape of a growing polyethylene lamella. The possibilities of following melting subsequent to crystallization, and the additional information that can be gained about the structure formed, is also explored.

6.1 Background

The crystallization of synthetic polymers from the melt has been the subject of intensive study and debate for more than 50 years. Interest has centred on the influence of morphology on properties, on attempts to enhance and control processing and processability, as well as on efforts to gain a fundamental understanding of the complex problem of the transformation of a disordered and entangled melt into a highly ordered crystalline structure. Throughout these studies the various microscopical techniques have provided invaluable information, supplying much of the data that has shaped theoretical advances. However, in many cases, understanding has been hindered by an inability to image crystallization at sufficiently high resolution to observe the fundamental length scales of the system, such as the size (radius of gyration) of the molecules and the thickness of the crystals, while crystallization is actually happening. This is particularly problematic in the case of crystallization from the melt as, until the transformation is complete, there is always a large reservoir of material available that will solidify on cooling (prior to ex-situ examination) and hence obscure the morphology that has developed. This situation has changed with the advent of scanning probe microscopes, most notably atomic force microscopy (AFM), which are starting to enable the real space examination of polymer crystallization in-situ, in real time, over length scales from nanometres to microns.

Our knowledge of polymer crystallization, and the theories that have been developed to explain it, is based on a number of key observations: polymer "chain-folding" [1] and the existence of a characteristic crystal thickness dependent on the supercooling at which growth occurred [2]; the shape of the crystals themselves as they grow e.g. [3]; and the relationship between crystallization temperature and growth rate (the crystallization kinetics) [4,5]. In the majority of cases polyethylene has been used as a model for crystallizable flexible chain polymers. Several models for crystal growth have been developed, based on secondary nucleation [6,7], or more recently, on entropic barriers [8], although new developments have largely been led by experimental discoveries (for a review see [9]).

There has recently been a resurgence of interest in polymer crystallization, both experimentally and theoretically, due to the new information available from time resolved synchrotron x-ray experiments and the rapidly increasing power of computer simulation. Based on x-ray data implying the formation of order in the melt prior to crystallization [10], a new model for polymer nucleation has been suggested [11], in which primary crystallization is preceded by spinodal decomposition into phases rich or depleted in conformations close to those in the crystal. This model has also been applied to oriented melts, promising a unified scheme for polymer nucleation. However, interpretation of the experimental data on which the model is based is currently an area of active debate [12]. Similarly, by detailed interpretation of SAXS data, it has been suggested by G. Strobl [13] that crystallization does not occur spontaneously at a clear crystal-melt interface, but rather through a semi-ordered mesophase, followed by the formation of "blocks" which then migrate into the lamella [1]. In both of the above cases the materials studied have not been the traditional linear sharp fractions of polyethylene but rather materials with compositions closer to those used industrially. Indeed, it has been suggested that the behaviour of polyethylene is the exception. Both of these new hypotheses suggest that crystallization occurs through progressive degrees of order; a subject that has also been much discussed by the late A. Keller [14] in the context of size dependent phase transitions.

Before detailing the contribution to this area that can be made by AFM, a brief overview of use of optical and electron microscopy when applied to these questions will be given with the intention of both pointing the interested reader towards the relevant literature, and of putting the use of AFM in context, rather than of providing an exhaustive review.

Optical microscopy, although fundamentally limited in resolution by diffraction, has provided a wealth of information on polymer crystallization. Early work on spherulitic crystallization was able to link the large-scale optical morphology, such as the Maltese cross and regular banding, to the underlying variations in local crystal orientation [15–18]. Later the use of optical microscopy to measure rates of spherulitic growth provided much of the data that underpins the Lauritzen Hoffman theory of polymer crystallization [4,6,19]. The spherulitic radial growth rate was found to be constant at a given temperature for a given sharp molecular weight fraction of a polymer, and a direct link was made between this

---
[1] See Chap. 4.

rate and the net rate of attachment of chains to the growing crystal front. The use of optical microscopy to measure rates of growth is now a standard tool (e.g. [20]), with the resultant data providing the input for microscopic models of polymer solidification under processing conditions (e.g. [21]).

Electron microscopy, with its improved resolution relative to optical microscopy due to the angstrom wavelength of the electrons, has the capability of probing many of the important length scales of polymer crystallization. However, it was clear very early that high energy electrons cause considerable damage to most organic molecules and even at very low dosage can cause cross-linking, changing the crystalline structure of polymers as well as changing the nature of the molten polymer. TEM requires very thin films to allow transmission of the electrons, so the polymer must either be crystallized in such a film, microtomed after crystallization, or else a replica taken of the crystal surface. The latter two possibilities necessitate ex-situ study, although they have provided a broad base of information of polymer microstructure.

The examination of microtomed films usually requires an initial staining stage so as to induce a sufficiently large electron density difference between the crystalline and amorphous phases [22]. Standard replication techniques require the sample to be etched so as to preferentially remove the amorphous material, inducing a significant topography representative of the crystal morphology [23]. In studies using replication and staining the extent to which the observed morphology, regardless of quenching method, represents the morphology during crystallization must remain an open question. These methods therefore provide most reliable information when it is the final structure that is of interest. The use of very thin films and defocus imaging of the actual polymer film was pioneered by Petermann, and was particularly successful when applied to isotactic polystyrene [24,25], which has a relatively high resistance to beam damage. In-situ low dosage studies are technically very difficult and have only been successfully applied to a few polymers. Lying somewhere between these extremes of ex-situ and in-situ is the work by Phillips where the structure of *cis*-polyisoprene is frozen in apparently instantaneously by the application of a gaseous staining agent [20]. Staining during growth is a potentially powerful technique, although the reactions that occur are complex and can lead to artefacts. Also, once stained, the material is frozen, so it is not possible to follow time evolution.

Scanning probe microscopy has two principal advantages compared with other types of microscopy. Firstly, it by-passes the diffraction limit common to all microscopes that use lenses, enabling very high resolution images to be obtained (atomic resolution in some special cases), without the necessity to use high energy particles [26]. However, perhaps most importantly for soft matter, it is possible to image non-conducting, relatively soft materials (under the right conditions very low concentration gels can be imaged [27]), without causing damage to the sample. This means that processes can be followed in-situ, and, as images can be collected in tens of seconds, in real-time. It is this innovation, and its application to polymer crystallization, that is the subject of this paper.

Early studies using AFM to follow polymer crystallization [28,29] used a polymer that happened to crystallize at room temperature at a relatively slow rate, polyhydroxybutyrate-co-valerate. Soon afterwards, the use of hot-stages was introduced, and the crystallization [30,31] and now melting [32] of several more polymers could be followed. Some principal findings of these studies were an apparently discontinuous growth rate, when measured on the scale of a single lamella, and large-scale motion of material in front of growing spherulites. The number of groups working in this area has increased significantly, especially with the release of commercial heating stages, and there are now a growing number of publications e.g. [33–40] following growth in a variety of different polymers at temperatures up to 250°C.

In the following sections the approach used for high temperature imaging in our laboratory will be discussed, some of the problems and solutions that we have used when attempting to follow polymer crystallization and melting, and finally some data that have been obtained during fairly extensive studies of the crystallization and melting of polyethylene. The data presented here will be confined to polyethylene as this is the material that we have studied most extensively, in an attempt to tie in the new data obtainable with AFM with the enormous wealth of complementary information available on this most studied of polymers.

6.2   The Use of AFM to Follow Crystallization

When using an AFM to follow the crystallization or melting of a polymer, the first, and in many ways the most difficult problem that needs to be overcome is how to heat the sample in a controlled manner without damaging the microscope, and with sufficient stability to enable imaging with nanometre resolution.

Damaging the microscope is a significant risk as all commercial AFMs use piezoelectric devices to move either the scanning tip over the sample or the sample under the tip, and many piezoelectric materials cannot be used at elevated temperature. So the scanner itself has to be protected from the heat, which is usually done by inserting insulation between the scanner and the heater/sample. In the work presented here, a "double-glazed" cantilever holder is used with a DI Veeco Dimension D3100 microscope, inserting an insulating air-gap between the tip and the scanner [40]. As in this microscope the tip is scanned over the sample, it is possible to place a standard heating stage under the tip. The set-up is illustrated schematically in Fig. 6.1. Other techniques have been effectively utilised elsewhere (e.g. [32,36,37]).

As it is usually necessary to use tapping mode when imaging soft materials, a method of inducing an oscillation of the cantilever is necessary. This is usually done with another piezo, although this can cause problems when at elevated temperatures, due to a strong temperature dependence of piezoelectric response, the possibility of damaging the piezo, and the necessity to have conducting links relatively close to the cantilever. We have chosen to use a magnetic drive, either by coating the cantilever directly with cobalt, or by using a magnetic actuator

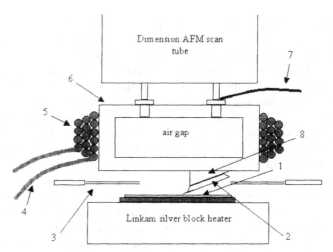

Fig. 6.1. A schematic diagram showing the modified cantilever holder used for hot-stage experiments. 1. The sample mounted on a glass slide. 2. The cantilever substrate. 3. The aluminium and Kapton^{TM} heat shield. 4. Copper wire from coil to current driver. 5. Copper coil. 6. Glass cantilever holder. 7. Thermocouple. 8. Magnetic actuator if the cantilever is not coated with cobalt.

placed under the cantilever substrate, and then feeding the drive signal that usually drives the piezo, through a voltage to current to converter, into a solenoid wrapped around the double-glazed cantilever holder. This allows the cantilever to be driven remotely, without any conducting links passing to the cantilever side of the double-glazing, and hence improves the temperature stability of the heater set-up. This method, coupled with a simple heat shield to prevent convective heating around the edge of the cantilever holder, allows temperatures in excess of 170°C to be safely accessed, with minimal degradation in image stability.

Figure 6.2 shows a slow scanned, high-resolution image of polyethylene lamellae after growth, taken at a temperature of ~130°C, showing very good image stability. At these high temperatures the microscope is more sensitive to air movements than usual, and the wavy nature of the lamellae at the bottom of the image is an artefact caused by some drift during imaging caused by air movements.

When the sample and heating-stage are heated, they expand, and this causes a motion of the sample relative to the cantilever, which, even if undergoing the same heating ramp, will respond mechanically in a different way from each other. Therefore, imaging during a temperature ramp is more difficult inherently than imaging isothermally, and requires regular shifts in the position of the scanner in order to keep the same area under the scanning tip. Also, temperature stability is very important, as variations in temperature will cause shifts of the sample relative to the cantilever, making the image very difficult to interpret. Fortunately, commercial Linkam heaters have remarkably good stability, and have been used throughout our studies.

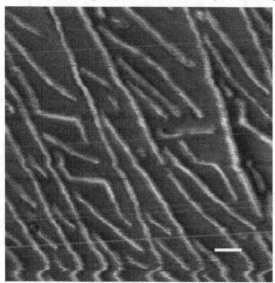

Fig. 6.2. A phase image taken at 512×512 pixels showing polyethylene lamellae (bright) in a matrix of molten polymer (dark) at 135°C at a scan line speed of 1Hz. Even though the image took 8.5 minutes to collect, there is relatively little noise or drift. The scale bar represents 100 nm. Black to white represents a change in phase of 30°.

AFM is a technique that obtains information about a surface through the mechanical response of a cantilever to the interaction of a sharp probe with the surface under study. If a surface with inhomogeneous mechanical properties is imaged, the interaction of the probe with different parts of the surface will vary, for instance by sinking further into softer materials for the same imaging force. Also, the probe itself has a certain size, and interaction area with the surface, which is generally unknown, and will be convoluted with the true topography of the surface in the final image. These are factors that have to be borne in mind when analysing AFM data. Finally, the fact that only the surface, or very near surface region [33], can be imaged must be taken into account when analysing the data obtained. The extent to which the behaviour at the surface differs from that in the bulk will depend on the polymer under investigation, the temperature of the study, the particular process that is of interest, the environment and many other factors.

In order to obtain useful information about a process it is usually necessary to collect several images during the time that the process occurs. The fundamental barrier to high-speed scanning in conventional tapping mode AFM lies in the long response time of both the conventional micro fabricated cantilever and in the piezoelectric scan tube, owing to their inertia. Current AFMs have a realistic maximum scan speed in the region of 20–30 seconds when operated in tapping mode, in air, on a scan size of ∼1 μm. This significantly limits the current applicability of AFM to many processes in polymers, although advances are being made (e.g. [41]). Even at these scan speeds, if there are sharp features or changes in properties, for instance when the cantilever moves from the melt to a crystalline lamella, it takes several pixels before complete control is regained, and a large error signal is obtained (the amplitude signal). This amplitude image therefore highlights such features and can be a useful presentation of the data. In some of the images shown in this paper there are regions to the right of sharp features (when the tip is being scanned from left to right) where the cantilever is not tracking the surface properly for several pixels. These regions show up as either a light or dark "shadow". These occur due to the fast scan rates used to enable as much data as possible to be obtained during the process under investigation, and are felt to be a necessary evil of pushing AFM to the limits of its usable range of scan speeds. Fast scan rates, and the reduction in cantilever control that they entail, also means that topographic data will often be more heavily convoluted with changes in mechanical properties than would usually be the case, and phase data, which usually highlights differences in material properties, will contain more topographic information.

6.2.1   Sample Preparation

Samples of two different polyethylenes, an NBS sharp fraction Mw 119.6 k, and a whole polymer, rigidex 50, were prepared by dissolving in xylene and casting from solution onto hot glass slides at ∼150°C. After holding for two minutes the samples were quenched to room temperature. The samples were then heated to 160°C to melt them, held for two minutes, and cooled to the temperature of interest. Imaging parameters were adjusted so as to just maintain contact with the surface while applying the minimum possible force. Imaging with high forces significantly changed the behaviour of the polymer, particularly by inducing nucleation, as has been observed previously [33] for contact mode imaging.

The accurate measurement of the temperature of the sample surface is influenced by the distance from the heat shield, the alignment of the probe relative to its holder etc., all of which vary from experiment to experiment, making a reliable calibration problematic. The temperature difference between the sample surface and the heater was estimated to be 3.5–7°C for the experiments reported here. The temperatures quoted in the text are the temperature given on the Linkam and are therefore ∼4°C higher than the estimated sample temperature. This issue is discussed more extensively in [40].

Although the cantilever is not actively heated, the growth behaviour of the polymer in neighbouring regions was checked by increasing the scan size, to ensure that the behaviour of the area scanned was not different from that of the surrounding area. As the growth rate of polyethylene is very strongly dependent on temperature, this is a good test of the local temperature control.

6.3   Results and Discussion

In the following sections several examples of observations made using AFM during the crystallization and melting of polyethylene will be presented and discussed, with an emphasis on the impact of the new observations on our understanding of polymer crystallization. Particular attention will be paid to the relation between these new observations and the recently emerging theoretical and conceptual developments in polymer crystallization.

6.3.1   The Crystal Melt Interface

The ultimate goal of an in-situ study of polymer crystallization is to image the individual polymer chains as they re-organise from melt to crystal. With current technology this is not possible. However, much new information can be gained by imaging the growing crystals at high resolution. This section will detail our recent work on the high resolution imaging of the crystal growth front and our interpretation of the insight that gives into the behaviour on a molecular scale.

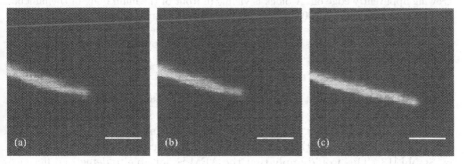

Fig. 6.3. A series of AFM phase images showing the growth tip of a lamella viewed edge on. (b) was the taken during the same scan as (a) but the data was collected when the probe was moving from right to left rather than left to right, so as to ensure that image of the interface between crystal and melt was not influenced by the scanning parameters. (c) is the same lamellae imaged 99 seconds later. Black to white represents a change in phase angle of 35°. Scale bars represent 100 nm.

Figure 6.3 shows several images of the growing tip of a polyethylene lamella viewed edge on. The images are phase images, which are sensitive to differences in adhesive and mechanical properties, the contrast occurring due to the soft and sticky nature of the melt compared to the hard, non-adhesive crystal [42–44]. In all cases there is a sharp interface between the melt and the crystal, occurring certainly within 20 nm. As explained above, and in other chapters in this book, there are several new theories of polymer crystallization all of which propose that crystalline order is reached through progressive stages of increasing order. In all cases, we might expect to be able to see evidence of intermediate phases with the AFM.

Figure 6.4a shows a single image of a lamella growing edge on in which there is a gap in the crystal – there appears to be a block of crystal growing ahead of the main growth front. This occurs occasionally, though we have not observed it when the lamellae are growing strictly edge on. Figure 6.4b shows a lower magnification image of the same crystal later in its growth, showing a large portion

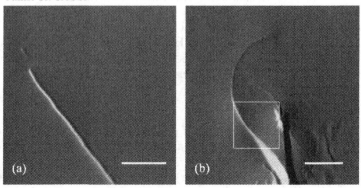

Fig. 6.4. A pair of amplitude images of polyethylene lamellae. (a) shows a lamella growing apparently edge on. A block is visible in front of the main body of the lamella. The scale bar represents 300 nm. (b) shows a lower magnification image of the same lamella some time later after it has curved round to the right – the box shows the area of image (a). The scale bar represents 1μm. Black to white indicates a change in amplitude of 0.5V.

of the lamella is lying flat on in the surface of the sample, implying that this crystal is perhaps growing at some acute angle to the surface. Two explanations of this morphology present themselves. It might be that the crystalline block has actually grown in front of the main lamella, in a manner similar to that suggested by Strobl [13]. This would imply that the AFM is not able to detect a difference between the initial mesophase and the melt, but does detect the transformation of that mesophase into block like structures of the final crystalline phase. An alternative explanation is that the observation of a block is an artefact caused by only imaging the surface of the melt. If the growth front is irregular on the scale of tens of nanometres, the intersection of that growth front with the surface might leave pockets of melt along the length of the lamella until these grow out. The degree of roughness necessary to form the appearance of blocks at the surface becomes less if the lamella is not growing strictly perpendicular to the surface, as could be indicated by Fig. 6.4b.

Figure 6.5 shows several images of lamellae growing flat on in the plane of the surface, in which a growth front roughness, with a wavelength of ∼100 nm and an amplitude of at least 10 nms can be seen. In Figs. 6.5b and c this is clearest in the lamella to the top left of the image, which is arrowed. Such rough growth fronts are typical of all the samples of polyethylene studied, even though the growth front is a smooth curve (in agreement with TEM studies) on longer length scales. When viewed flat on, no observation of blocks of crystalline material appearing in front of the main lamella have been observed. The appearance of blocks illustrated in Fig. 6.4 is most likely due to the roughness of the growing crystals and their intersection with the surface.

It is clear from Figs. 6.3–6.5 that there are two very different phases present during growth. After growth, there is no change in properties of the crystalline phase observed in our AFM studies to date. The melt appears to be homoge-

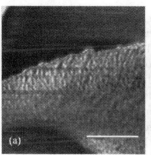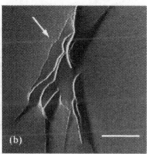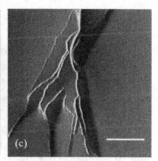

Fig. 6.5. A series of images in which lamellae are lying substantially in the plane of the surface. (a) is a phase image; black to white represents a change in phase of 20°; scale bar represents 500 nm. (b) and (c) are amplitude images; black to white represents a change in amplitude of 0.3V; scale bars represent 1μm.

neous, and appears to transform directly into a more or less homogeneous crystal. Clearly these statements have to be made within the context of the resolution of the AFM to differences in properties. This is currently an inexact science, and is further complicated by the fast scan speeds necessary to follow the process of crystallization, and the high temperatures that might increase the noise on the signals. However, it can be said that we see no convincing evidence of either mesophases occurring as an intermediate phase between the crystal and the melt, or of local variations in the density of the melt, in the actual images obtained with the AFM. If such phases do exist, they look either like the melt or like the crystal to the AFM probe.

In [40] we have shown that when opposing lamellae meet, having grown from two neighbouring shish-kebab crystals, they sometimes change direction so as to avoid hitting each other. We proposed several explanations of this behaviour, including the possible presence of a partially ordered state in front of the growing crystal. There is, therefore, some circumstantial evidence that could be interpreted as supporting the presence of mesophases during crystallization. If they do exist, we have not been able to image them directly using AFM.

The observation of a rough crystal growth front, on a scale of tens of nanometres, is in contradiction to the predictions of Lauritzen Hoffman (L-H) theory. Sadler Gilmer rough surface growth theory predicts a surface that is rough on a molecular scale, meaning with a high density of steps, but does not necessarily imply a growth front of the type observed. L-H theory predicts a smooth surface at these temperatures, and although it has been adapted to include the possibility of a continuously curving growth front, it cannot explain this observation. The observed growth front could be rough either due to fluctuations, or due to an instability most probably caused by diffusion. More data is required to determine between these possible causes. It should also be noted that the shape of the growth front varies with time. Points along a radius that are in advance of the average growth front at one point in time, might be behind the average at another point in time. Figures 6.5b and 6.5c show two images of the same growing crystal aggregate, illustrating this change in shape with time, particularly visible in the crystal to the top left of the aggregate, arrowed in Fig. 6.5b. If a radial growth rate from the middle of the growing crystal is measured, it is not a constant either from one radius to another, or with time. This is again in contradiction to many theories of polymer crystallization. It is in agreement with previous observations [40] of lamellae growing edge on from an oriented nucleus, in which a variation in growth rate from lamella to lamella, and with time for each lamella, was observed. These variations in growth rate, and the growth front roughness, are therefore most likely manifestations of the same phenomena. It appears that the value for growth rate that is measured optically, and is a constant at a particular temperature, is the average over many growth rates that actually occur over shorter length scales than that resolvable optically.

Secondary nucleation theory, in the form of L-H theory, cannot explain these observations. Growth at these temperatures should be in regime I (or possibly just in regime II), and the fluctuations caused by the nucleation events should average out over several stems or tens of stems, that is a few nanometres at most. The rough growth front might be evidence of the role played by diffusion in the growth process, possibly the diffusion of matter towards the growth front, or of entanglements and chain defects away from the growth front. This is an area that requires more study.

6.3.2   On the Surface Texture of Polyethylene Lamellae

Figure 6.6 shows a series of images of the same polyethylene lamella taken at different temperatures after initial crystallization at 133°C. The images are phase images, with contrast due to both mechanical properties and surface texture. There clearly is some microstructure within the lamellae, with variations in phase angle of between one and four degrees over distances of 10–100 nm. The structure under discussion here is the fine, somewhat globular texture within the large, flat expanses of crystal. The typical length scale of this structure is between 10 and 30 nm. Although there is some noise on these images, many of the same features can be seen throughout the temperature cycle, and were present in the sample from the point of crystallization, all the way through cooling and finally re-heating towards the melting temperature. For instance, the circular, or horseshoe shaped feature arrowed in 6a. It seems unlikely that these features are due to defects in the crystal structure, as they would be expected to anneal out, especially on heating. They do not behave in the way that is predicted of the blocky textures proposed by Strobl [13]. Another possibility is that molten polymer is trapped on the top surface of the lamella, and relaxes into this rough texture. This seems unlikely as the texture survives cooling, when the increased driving force for crystallization would be expected to force more of the molten polymer into the crystal. Such gradual growth of the crystal is in fact seen if Figs. 6.6a and 6.6b are compared. Chains that are partially crystallized would be pinned in place more tightly and might be responsible for the rough texture. We are currently unable to explain the surface texture of the lamellae, but we

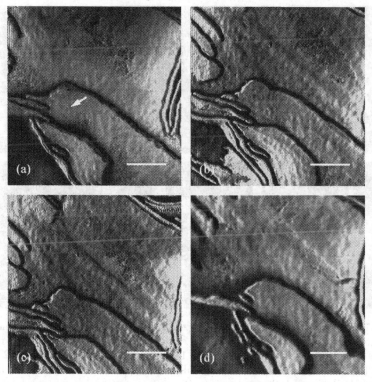

Fig. 6.6. A series of images of a lamellar aggregate taken during a cooling and then heating cycle, showing the fine surface structure that is frequently found on polyethylene lamellae. (a) was taken at 133°C, the crystallization temperature, just after primary crystallization. Although the texture is visible everywhere, the arrow indicates an example that can be clearly seen in all the images. (b) was taken after cooling at 5°C/min to 123°C. (c) was taken after cooling at 5°C/min to 118°C. (d) was taken after rapid heating to 132°C and then heating at 1°C/min to 138.9°C, just prior to complete melting – part of the lamella has already started to melt. In all cases scale bars represent 300 nm. In (a) black to white represents a change in phase of 20°. In all the other images it represents a change in phase of 10°.

can say that it is not an artefact of AFM imaging, and that it does not appear to be evidence of different degrees of stability within the crystal.

6.3.3 The Melting of Isolated Lamellae

Figure 6.7 shows a series of images of a lamellar aggregate as the sample is heated towards the melting point, in which the lamellae lie largely in the plane of the surface of the sample. The lamella does not break up into separate pieces as it melts but rather melts backwards from the initial crystal melt interface, that is from the growth front. Melting is not the reverse of crystallization – this crystal grew with an approximately continuously curved growth front, but

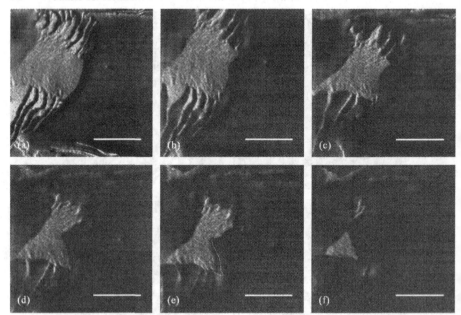

Fig. 6.7. A series of phase images showing the gradual melting of a polyethylene lamellar aggregate. Black to white represents a change in phase from 0 to 45°. (a) taken at 130°C. (b) taken at 134.6°C. (c) taken at 135.5°C. (d) and (e) taken at 136.2°C. The line on (e) shows the position of the crystal-melt interface in (d). (f) taken at 137.1°C. The scale bar represents 1μm.

melts preferentially in the middle. In all cases where polyethylene lamellae are surrounded by material that has not crystallized, they melt back from the growth front (that is, when crystallization is not allowed to reach completion).

This behaviour gives us several insights into the re-organisations that occur following crystallization, which is one of the themes of this book. In the previous section, a surface texture was described. A similar texture is observable on these crystals, and again it can be seen that the texture does not influence the melting behaviour. The texture is not an indication of different degrees of stability within the crystal itself, and neither does the texture change significantly with temperature.

The melting back from the crystal front shows that the edge of the crystal is less stable than the interior. This is perhaps as expected   melting is expected to start from a surface, and in the case of polymers the edge surface, in which the chains lie, is likely to be more mobile than the top surface, the re-organisation of which requires considerable co-operative motion. During the experiment shown in Fig. 6.7, the heating of the sample was stopped at several temperatures during the heating cycle, and it was observed that melting slowed considerably. Figures 6.7d and e were taken at the same temperature with a time interval of 2 min 34s. In Fig. 6.7e the melting front has retreated slightly compared to its position in Fig. 6.7e - the line indicates the position of the crystal-melt interface in Fig. 6.7d. The gradual melting back from the front does not appear to be due to superheating of part of the crystal. Different parts of the crystal either have different stabilities that depend on age due to a gradual change in crystal perfection, or else the rest of the crystal that has not melted is continuously re-organising and thickening during heating.

Crystal thickening during heating is known to occur in polyethylene [45], although studies of crystal thickening in isolated lamellae grown from solution have shown a distinct preference for thickening to start at the edge of the crystal, where mobility is highest [46]. If this solution behaviour were mimicked in the case of isolated lamellae in the melt, the edges would melt last, unless the thickening rate is substantially quicker than the rate of melting. The fact that the surface texture remains unchanged with age and with temperature implies either that the crystal is not thickening, or else that the surface texture is not connected with the crystal itself, but is some layer lying on top of the crystal.

The use of AFM to follow crystal re-organisations and melting on heating provides the opportunity to probe the stability of the as formed crystal, and how this varies with crystallization temperature and the local crystal and melt environment. This is an area with great potential, and will be discussed in more detail elsewhere [47].

6.4   Conclusions

The use of AFM as a tool to probe the crystallization of polyethylene has been discussed, and several examples presented where new insights into the crystallization process have been gained. The particular strength of AFM is its ability to image crystallization as it occurs, without changing the process, with nanometre resolution. It has been shown that this not only allows the evolution of a process to be followed, but also provides unequivocal information about the morphology at any point in time.

Within the resolution limit of the images obtained, crystallization occurs as a transformation directly from a one component melt to a one component crystal, without intermediate phases that look significantly different to the AFM from these principal phases. Currently it is unclear to what extent this conflicts with the predictions of recent theories of polymer crystallization based on transformation through progressive degrees of order. Future developments both in quantifying and reducing noise on the AFM data, and in quantifying the predictions of the theories, should allow AFM to play a central role in this area.

The observations of crystallization at the scale of tens of nanometres are in contradiction to the predictions of secondary nucleation theory, as growth is found to be sporadic in time and in space, and the growth fronts of individual lamellae are found to be significantly more irregular than predicted by theory. In particular, a stronger role for a diffusive process is indicated even at these very low supercoolings.

It is clear from the small snapshot of work presented here that in-situ AFM studies can make a significant and possibly decisive contribution to our understanding of polymer crystallization. As AFM technology progresses towards faster scan-rates, easier use at elevated temperatures and lower imaging forces, the technique will inevitably become more widely used, opening up nanometre length scales to routine real-space, real-time investigation.

References

1. A. Keller, Phil. Mag. **2**, 1171 (1957)
2. P.J. Barham, R.A. Chivers, A. Keller, J. Martinez-salazar, S.J. Organ, J. Mater. Sci. **20: (5)**, 1625 (1985)
3. S.J. Organ, A. Keller, J Mater. Sci. **20(5)**, 1571 (1985)
4. J.D. Hoffman, L.J. Frolen, G.S. Ross, J.I. Lauritzen, J. Res. Nat. Bur. Stand. A: Phys and Chem. **79A**, 671 (1975)
5. D.J. Blundell, A. Keller, Polymer Letts **6**, 433 (1968)
6. J.D. Hoffman, R.L. Miller, Polymer **38(13)**, 3151 (1997)
7. F.C. Frank, M. Tosi, Proc R Soc **A263**, 323 (1961)
8. D.M. Sadler, Nature **326**, 174 (1987)
9. K.A. Armistead, G. Goldbeck-Wood, Advances in Polymer Science **100**, 219 (1992)
10. N.J. Terrill, P.A. Fairclough, E. Towns-Andrews, B.U. Komanschek, R.J. Young, A.J. Ryan, Polymer **39(11)** 2381 (1998)
11. P.D. Olmsted, W.C.K. Poon, T.C.B. McLeish, N.J. Terrill, A.J. Ryan, Phys. Rev. Lett. **81**, 373 (1998)
12. Faraday Discussions No. 112, "Physical Chemistry in the Mesoscopic Regime" 64-75 (1999)
13. G. Strobl, Eur. Phys. J. E **3**, 165 (2000)
14. A. Keller, M. Hikosaka, S. Rastogi, A. Toda, P.J. Barham, G Goldbeck-wood, J. Mater. Sci., **29(10)**, 2579 (1994)
15. J.J. Point, "Bulletin de l'Academie royale de Belgique" 982 (1955)
16. A. Keller, J. Pol Sci. **39**, 151 (1959)
17. H.D. Keith and F.J. Padden, J. Pol Sci **39**, 101 (1959)
18. F.P. Price, J. Pol Sci **39**, 139 (1959)
19. J.D. Hoffman, J.I. Lauritzen, G.T. Davis, in Treatise on Solid State Chemistry (1976) Ed N.B. Hannay, (Plenum Press, NY **3**)
20. P.J. Phillips, Rep. Prog. Phys. **53**, 549 (1990)
21. H. Zuidema, G.W.M. Peters, H.E.H. Meijer, Macromol. Theory Simul., **10(5)**, 447 (2001)
22. G. Kanig Prog Colloid Polym Sci **57**, 176 (1975)
23. D.C. Bassett, "Principles of polymer morphology" (Cambridge: Cambridge University Press 1981)
24. J. Petermann, M. Miles, H. Gleiter, J. Pol Sci. Pol. Phys. **17**, 55 (1979)
25. T.X. Liu, J. Petermann, C.B. He, Z.H. Liu, T.S. Chung Macromolecules **34(13)**, 4305 (2001)
26. G. Binnig, H. Rohrer, Ch. Gerber, E. Weibel. Appl Phys Lett; **40**, 178 (1982)
27. J.K. Hobbs, A.K. Winkel, T.J. McMaster, A.D.L. Humphris, A.A. Baker, S. Blakely, M. Aissaoui, M.J. Miles, Macromolecular Symposia **167**, 1 (2001)
28. T.J. McMaster, J.K. Hobbs, P.J. Barham, M.J. Miles, Probe Microscopy, **1**, 43 (1997)
29. J.K. Hobbs, T.J. McMaster, M.J. Miles, P.J. Barham, Polymer **39(12)**, 2437 (1998)
30. R. Pearce, G.J. Vancso, Polymer **39**, 1237 (1998)

31. J.M. Schultz & M.J. Miles, J Polymer Sci. B **36**,2311 (1998)
32. R. Pearce, G.J. Vancso, J. Polymer Sci. B **36**, 2643 (1998)
33. G.J. Vancso, L.G.M. Beekmans, R. Pearce, D. Trifonova, J. Varga, J. Macromol. Sci. - Phys **B38(5&6)**, 491 (1999)
34. L. Li, C-M. Chan, K.L. Yeung, J-X Li, K-M. Ng, Y. Lei, Macromolecules **34**, 316 (2001)
35. Y.K. Godovsky, S.N. Magonov, Langmuir **16**, 3549 (2000)
36. Y.K. Godovsky, S.N. Magonov, Polymer Science Series A**43(6)**, 647 (2001)
37. C. Basire, D.A. Ivanov Phys Rev Lett **85(26)**, 5587 (2000)
38. A.K. Winkel, J.K. Hobbs, M.J. Miles, Polymer **41(25)**, 8791 (2000)
39. J.K. Hobbs, M.J. Miles, Macromolecules **34**, 353 (2001)
40. J.K. Hobbs, A.D.L. Humphris, M.J. Miles, Macromolecules **34**, 5508 (2001)
41. M.B. Viani, T.E. Schaffer, G.T. Paloczi, L.I. Pietrasanta, B.L. Smith, J.B. Thompson, M. Richter, M. Rief, H.E. Gaub, K.W. Plaxco, A.N. Cleland, H.G. Hansma, P.K. Hansma Review of Scientific Instruments **70(11)**, 4300 (1999)
42. J.P. Cleveland, B. Anczykowski, A.E. Schmid, V.B. Elings, Appl. Phys. Letts. **72(20)**, 2613 (1998)
43. J. Tamayo, R. Garcia, Appl. Phys. Letts. **73(20)**, 2926 (1998)
44. G. Bar, Y. Thomann, R. Brandsch, H.J. Cantow, M.H. Whangbo Langmuir **13**, 3807 (1997)
45. A. Keller, A. O'Connor, Disc. Faraday Soc. **25**, 114 (1958)
46. J.K. Hobbs, S.J. Organ, M.J. Miles, in preparation
47. J.K. Hobbs and M.J. Miles, in preparation

7 Atomic Force Microscopy Studies of Semicrystalline Polymers at Variable Temperature

Dimitri A. Ivanov[1,*] and Sergei N. Magonov[2]

[1] Laboratoire de Physique des Polymères, CP 223, Université Libre de Bruxelles,
Boulevard du Triomphe, B-1050 Brussels, Belgium,
Tel.: +32 2 650 57 59, fax +32 2 650 56 75,
divanov@ulb.ac.be

[2] Digital Instruments/Veeco Metrology Group,
112 Robin Hill Rd., Santa Barbara, CA 93117, U.S.A.

Abstract. The capabilities of Atomic Force Microscopy, AFM, in studies of polymer phase transitions and, more generally, the semicrystalline polymer morphology are described. It is shown how variable temperature AFM can provide unique information on the organization of semicrystalline polymers at the nanometer scale and its evolution in the course of crystallization. The practical examples selected for this work illustrate applications of AFM to structural studies of homopolymers and polymer blends crystallized in the bulk, in thin films and in solutions. The investigated systems include solution grown single crystals of polyethylene, cold-crystallized poly(ether ether ketone), as well as melt-crystallized poly(ethylene terephthalate), poly(trimethylene terephthalate), syndiotactic polystyrene, poly(ε-caprolactone), isotactic and syndiotactic polypropylene. The issues related to AFM image analysis and its quantitative comparison with the results of complementary techniques such as small-angle X-ray scattering (SAXS) are addressed. More specifically, it is shown how AFM can provide statistically meaningful parameters for the semicrystalline structure and an accurate choice of a structural model for the interpretation of SAXS data.

7.1 Introduction

During the last 10–15 years, the development of scanning probe microscopy and its applications to studies of different materials has been in focus of many researchers. The invention of scanning tunneling microscopy, STM, in 1982 was the first milestone [1], which opened direct access to atomic-scale surface structures. In STM, a measurement of the current flowing between a sharp metallic tip and a conducting or semiconducting surface, is used for atomic-scale imaging of surface structures. The ability to get atomic resolution on samples in ambient conditions and even under liquid substantially enhances the capabilities of STM. Yet, a severe limitation related to the requirement of sample conductivity motivated researchers to seek for a more universal tip-sample interaction, which can be applied for high-resolution surface imaging. This demand has been satisfied

---

* Corresponding author.

with the appearance of atomic force microscopy, AFM [2]. Attractive and repulsive forces acting between a sharp probe and the sample surface are employed for surface imaging with this technique. Several innovations such as optical level detection [3], use of microfabricated probes and new imaging modes [4] made AFM a well-established surface characterization tool applicable to a very broad spectrum of materials ranging from solid inorganic crystals to partially or fully disordered organic liquid-like systems. Nowadays, AFM is used in many laboratories specialized in various fields of materials science. Such rapid expansion of this technique can be explained by further important instrumentation developments significantly extending the capabilities of AFM and related methods beyond simple topography measurements, which now also include measurements of local mechanical, adhesive, thermal, as well as magnetic and electric properties. One of such very recent instrumentation developments in this field consists in the heating stage accessory designed for in situ work between room temperature and approx. 250 °C [5]. Although the advantages of in situ high-temperature AFM are evident, especially for applications to organic materials, such studies have been limited so far by the absence of appropriate thermal accessories [6]. It is, therefore, expected that numerous AFM results on the thermal behavior of organic surfaces will soon be generated using this enhanced technical feature.

Of special interest for the present work are studies of soft materials, such as synthetic and natural polymers [7]. The nondestructive character of AFM presents an important advantage in studies of such systems, as they can be destroyed during preparation or observation with classical techniques such as Transmission Electron Microscopy, TEM. It should be noted that studies of surfaces with AFM do not require any special sample preparation such as metal or carbon deposition or micro-sectioning or staining. Therefore, AFM can provide a great wealth of information on the structure of such systems, without causing any damage to the sample. In particular, the evolution of the sample structure at the nanometer scale can be studied in situ using appropriate conditions, e.g., under controlled atmosphere or in liquids. The softness of materials to be studied could be even advantageous for this technique since, in some instances, it could help extracting structural information from AFM observations at different levels of the tip-sample interaction. For example, low-force imaging is optimal for the correct determination of surface profiles of such samples, whereas imaging at elevated forces can be useful for surface compositional analysis and for recognition of individual components in heterogeneous polymer materials [8] or biological systems [9].

The objective of the present work is to give a brief description of the capabilities of AFM in studies of semicrystalline polymer morphology and phase transitions. The structure of this chapter is the following: At first, some technical issues related to the tip-sample interaction in tapping mode, image resolution and requirements as to sample preparation will be shortly addressed. In the main part of the review, several examples of morphological analysis of polymer surfaces including homopolymers and polymer blends crystallized in the bulk, in thin films and solutions will be presented to illustrate the unique information AFM can deliver. In particular, the structural data provided by AFM will be quantitatively compared with the results of conventional techniques such as small-angle X-ray scattering, SAXS. It will be shown, for example, how the direct space information obtained from AFM images can be used to correctly model the SAXS curves.

7.2   Studies of Polymers in Tapping Mode

Description of any scanning probe microscopy technique logically includes a physical model for the tip-sample interaction and different technical issues related to the detection of this interaction and scanning operation of the probe over the sample surface. In the contact mode initially introduced for AFM a sharp tip fixed at the extremity of the cantilever comes into direct contact with the sample, which causes the cantilever deflection. A laser beam reflected from the backside of the cantilever is used for measurements of its deflection. A microscope mode of AFM is realized when the tip being scanned over the sample surface is controlled with a feedback mechanism, which adjusts the vertical sample position to keep the tip-sample interaction force constant [10]. This operation results in a topography image, which presents a three-dimensional surface profile. Surface structures with characteristic shapes and dimensions are recognized in such images given in color or grayscale codes, where brighter zones typically represent elevated surface features. Changes of the cantilever's parameters such as deflection, frequency or phase are usually recorded during scanning and they are also collected in AFM images. Such images might reflect not only topographic features but also local sample properties (friction, adhesion, stiffness, etc.). Therefore, in its applications to polymers, AFM combines the possibilities of probing local material properties and surface imaging. It is worth noting, however, that visualization of surface structures is the major AFM application, in which AFM complements optical and electron microscopy. As will be illustrated below, AFM is also invaluable in combination with diffraction techniques for justification of the structural models used to analyze the reciprocal space data.

7.2.1   Tip-Sample Forces

Already in early AFM studies of polymer samples, it had been noticed that contact mode operation might cause modifications or even damage of the surface of soft materials. In this mode the probe scans the surface while remaining in permanent contact with it. In this case, the cantilever feels not only normal forces, which are employed for feedback control, but also the lateral tip-sample forces. The latter could be used for studies of friction, yet the same forces generate shearing deformation that leads to modification and damage of soft materials. This circumstance has limited AFM studies of polymer materials until the appearance of a new oscillatory mode, known as tapping mode [4], which was introduced to avoid the described effect of lateral forces. In tapping mode, the probe oscillates close to its resonant frequency and the variation of the cantilever amplitude due to tip-sample force interactions is employed for surface profiling. In this mode, the lateral tip-sample forces are largely eliminated, however, consideration of the tip-force effects is still an important issue governing the image resolution and AFM capabilities for exploring polymer samples. It is worth noting that in contrast to the contact AFM mode, in which the cantilever deflection is directly related to the tip-sample force, the oscillatory nature of tapping mode makes the description of the tip-force interactions more involved. The changes of the cantilever amplitude do not describe the probe interaction with the sample to a full extent. Changes of resonant frequency and phase of the oscillating probe are most sensitive to the tip-sample forces and can discriminate between the overall attractive and repulsive force regimes. Therefore, AFM phase images, which express local differences between phases of a free-oscillating probe and the one interacting with the sample, became useful for many AFM applications, e.g. for compositional mapping of heterogeneous polymer systems [11]. A general correlation between the phase contrast and the energy dissipated by the probe interacting with the sample per oscillation cycle has been discussed [12]. Unfortunately, this does not help to explicitly correlate the phase changes in terms of differences in specific material properties such as adhesion, stiffness or viscoelasticity. A detailed description of the tip-sample interaction in the tapping mode experiment on a polymer sample is extremely complex. It should take into account a large number of phenomena and factors including the cantilever vibration amplitude, cantilever stiffness, sample stiffness and adhesion, tip shape, penetration of the tip into the sample, viscoelasticity (which is, most likely, nonlinear under tapping mode conditions). Although numerous experimental and theoretical efforts are directed towards understanding these issues, imaging of model samples with components having different physical properties (stiffness, density, viscosity, adhesion, etc.) might be a good starting point helping to rationalize the tip-sample interactions and their correlation with the oscillation parameters (amplitude, frequency, phase, etc.). In one of such attempts [8], the surface of a blend of two polyethylene materials with different densities (0.92 and 0.86 g/cm^3) was examined at different free oscillation amplitudes and set-points. It was found that the difference in the component's density resulted in an apparent height variation caused by depression of the tip into the low–density component. This effect is most pronounced when the initial cantilever amplitude is sufficiently large ($80 - 100$ nm) and the set-point amplitude is around $40 - 50\%$ of the initial amplitude. Such imaging at elevated forces, which is also known as hard tapping, is usefull when the compositional mapping of multi–component systems is desirable [13]. In the described case, the apparent contrast in the height image reveals the presence of components with different density and stiffness (the mechanical moduli differ approx. by a factor of 1000). In studies of other heterogeneous materials, in which the difference in the properties of components is not so drastic, the contrast in phase images is used for the identification of components.

Contrary to hard tapping, imaging at a low-level of tip-sample forces (light tapping) is realized when the set-point amplitude is close to the amplitude of the free-oscillating probe (A_0). The tip-sample forces are even smaller when A_0 is small. Unfortunately, the latter is not always achievable because of capillary or adhesive force. Practical use of light tapping is obvious: such imaging conditions are applied when high-resolution profiling of the top surface is needed for example for imaging of single macromolecules deposited on the substrate or for correct estimate of surface roughness of technologically important surfaces. When a top surface layer is in the rubbery state, imaging in light tapping reveals the top layer morphology. However, with the increase of the force the tip will penetrate the top layer and reveal the morphology of the sub-surface material. This effect was observed in studies of block copolymer films [14] and polyethylene films coated with low-molecular weight material [8]. In imaging of single dendrimer macromolecules, which are seen as 5-nm spheres, a transition from light to hard tapping revealed a harder core of these spheres related to a higher density in the macromolecule center [15].

7.2.2 Image Resolution

Image resolution is an important characteristic of AFM. In contact mode, the images of flat surfaces of organic and inorganic crystals can correctly reproduce the periodic atomic structures of the surface. However, the image resolution of real polymer samples is a more complex issue, which is determined not only by the tip-sample contact area, but also by the sample nature and roughness. For example, the width of isolated macromolecules deposited on a flat surface is over-estimated when measured from AFM images taken even in light tapping. This is due to the effect of dilation of a real profile of the extended macromolecules by the tip geometry, which gives rise to an artificial width increase of about $\sqrt{2Rh}$, where R is the tip radius and h the height of the step. For the same reason, the estimates of the macromolecule width become more precise when performed from images of compact macromolecular stacks. AFM measurements of the contour length of such objects are more accurate, and they could be even applied to study molecular weight distribution of polymers [16,17]. Object orientation is also of importance for determination of its dimensions from AFM images. For example, to get a correct estimate of the interlamellar distance in a lamellar stack (also known as long period, L_B, determined by SAXS), one should select the stacks with the smallest intercrystalline spacing. These objects, given the sufficient image statistics, should be close enough to edge-on orientation. It is worth noting that by varying the tip-sample force during imaging of lamellar structures one can clarify the thickness of the most compact, crystalline regions (denoted here as L_c) and also the size of a less compact amorphous exterior of crystals ($L_a = L_B - L_c$) [8]. Since, as was mentioned before, the morphological details pertinent to the bulk polymer structure are often hidden underneath a featureless top layer of amorphous material, the crystalline lamellae are visualized with higher contrast in the images obtained in hard tapping.

7.2.3 Sample Preparation

Sample preparation is one of the crucial steps of microscopic studies with AFM. It is clear that, for example, the sample surface roughness can present a practical limitation for AFM. Since most piezo actuators used in atomic force microscopes have only a few microns vertical range, AFM studies are restricted to surfaces with corrugations not exceeding this value. Also imaging of fine features with steep slopes is quite difficult because of imperfections of the feedback and cross-talk between motion of vertical and lateral actuators. Therefore, flat surfaces are most preferable for examination of fine morphological features of polymer samples. Many commercial films having surface roughness of a few hundreds of nanometers can be examined without any surface preparation. Films prepared by casting polymer solutions on flat substrates or by hot pressing between smooth plates such as mica sheets are much smoother, and they are well suited for high-resolution AFM studies. In preparation by hot pressing, the polymer film is separated from the molding plates at room temperature and the opened surface is examined with a microscope. Films of semicrystalline polymers with T_g higher than room temperature prepared by this method might be quite flat but constrained upon cooling in-between the molding plates. In such a case, thermal annealing of free-standing films close to T_g will be essential for obtaining a sample with a more equilibrated surface structure. However such annealing is usually accompanied by surface roughening.

AFM is primarily a surface characterization technique, and access to the bulk polymer morphology requires surface etching, ultramicrotomy or a combination of both techniques. Chemical or plasma etching removes a top surface layer whereby amorphous material is usually etched faster than crystalline structures. Etching recipes are documented for many semicrystalline polymers [18] and can be successfully applied for AFM sample preparation. Ultramicrotomy conventionally used for preparation of thin (30–100 nm) sections of polymers for TEM studies is also applicable for AFM studies. It should be noted that ultramicrotoming of polymers with T_g above room temperature is usually done at room temperature, whereas polyolefins and other polymers with T_g below room temperature require the use of a cryo-stage at temperatures 10–20 degrees lower than T_g. Roughness of polymer surfaces prepared by sectioning with a diamond knife, which is by far more appropriate for this procedure than a glass knife, is typically of several hundreds of nanometers. However, the quality of cuts may be subject to variations due to the material properties, and therefore the lamellar structure of semicrystalline polymers is not always resolved in AFM images of microtomed surfaces.

7.3 Practical Examples of Imaging of Semi-crystalline Polymers at Variable Temperature

The sensitivity of AFM to local material properties opens access to studies of polymer crystallization. In this section, several examples illustrating capabilities of AFM in such studies will be given. It should also be noted that the morphological observations of semicrystalline polymers reported in this work cannot be fully interpreted in the frame of the classical one-dimensional picture of the semicrystalline structure composed at its elementary level of interleaved sheets of amorphous and crystalline regions. In this respect, AFM data providing evidence for the existence of other types of organization in semicrystalline polymers, e.g., the much-debated granular morphology, will deserve special attention. The section will start with the most studied group of flexible chain polymers showing some examples where AFM can bring additional information with respect to conventional techniques of morphological analysis. In the second part, the morphology of the somewhat less studied family of semirigid chain semicrystalline polymers will be addressed.

7.3.1 Flexible Chain Polymers

7.3.1.1 Solution Grown Single Crystals of Polyethylene, PE

Generally, the crystallization of polymers proceeds by multiple folding of chain macromolecules to form two-dimensional crystalline lamellae. This process is well defined in dilute polymer solutions where only few chain entanglements are present. Individual lozenge-shaped lamellae of PE grown in solutions are probably the best-explored polymer crystals [18,19]. These crystals, which form initially as hollow pyramids, collapse into two-dimensional platelets while being collected on the substrate. The length of a straight segment of polymer chains defines the lamellar thickness, which is usually in the 10 12 nm range. Prior to AFM, TEM in combination with X-ray diffraction has been applied for the examination of single PE crystals. and the basic structural features of these objects have been explored in depth. A combination of TEM imaging and electron diffraction made it possible to observe the sectorization and determine the polymer chain orientation inside individual crystal sectors. However, many questions related to the organization of single crystals remain open. One of such questions concerns the structure of the lamellar surface, which is presumably formed by chain folding according to the adjacent reentry model. There are also alternative models of this structure with different degrees of order.

With the advent of AFM, researchers were motivated to use its advantages over TEM such as precision of height measurements and high-resolution imaging of top surface features in studies of polymer single crystals. Indeed, single crystals of PE have been examined in AFM images with great structural detail such as sectors, screw dislocations and overgrowth [20]. More recently, 8 10 nm size grains were observed in low-force images of the crystal surface acquired in tapping mode, whereas traces of lattices relevant to crystallographic register of polymer chains were detected in images obtained in contact mode with relatively high forces [21]. AFM imaging of single PE crystals has also been performed at elevated temperatures at which lamellar thickening usually takes place [22]. It should be noted that this process is caused by unfolding of individual chains from a kinetically favorable folded state to an energetically favorable extended-chain

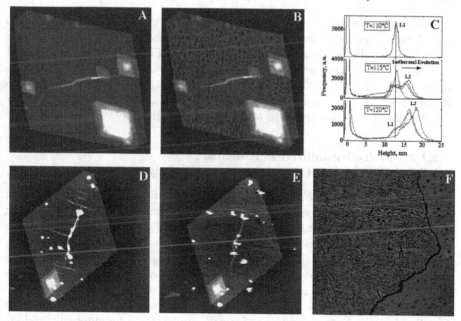

Fig. 7.1. Height images $(14 \times 14\,\mu m^2)$ of dry single crystals of polyethylene measured at room temperature (A) and 115 °C after 1.5 hr annealing (B). (C) Height histograms corresponding to AFM images of dry single crystals of PE, which show the evolution of lamellar thickness after annealing at different temperatures. (D)–(E) Height images $(10 \times 10\,\mu m^2)$ of a not completely dry PE single crystal. Images obtained at room temperature before and after annealing at 105 °C, respectively. (F) Part of the PE single crystal given in Fig. 7.1E showing edge-on lamellae $(4 \times 4\,\mu m^2$ scan).

conformation [23]. AFM observations of crystal annealing reveal formation of holes surrounded by thickened regions, which is consistent with the earlier TEM data [19]. In high temperature AFM images shown in Fig. 7.1A·B it can be seen that holes appear simultaneously with thickening of the adjacent locations. Thickening proceeds gradually after a step-wise change at 115 °C, as reflected in the height histograms in Fig. 7.1C. Thus, the sensitivity of AFM to height measurements makes it possible to monitor chain unfolding in situ with exceptional precision. A detailed comparison of the thickening mechanisms taking place in single crystals of PE and regular linear alkanes of different length [24] can be useful in understanding the role played in these processes by the crystal/amorphous interphase and the polymer nature of the reorganizing species [1].

As shown in the early TEM studies [19], annealing of single PE crystals containing traces of solvent (xylene) proceeds differently, and the thickening can be accompanied by reorientation of chains, which become aligned along the crystal surface.

---

[1] see also Chap. 6 for more details.

These observations are in good agreement with the AFM data presented in Fig. 7.1D–F. Indeed, upon annealing of a not completely dry PE crystal, the formation of holes (some of them are barely seen in Fig. 7.1E) takes place simultaneously with polymer chain reorientation as evidenced in Fig. 7.1F by stacks of the edge-on lamellae [25]. In this case, AFM offers the possibility to study the dynamics of chain reorientation in single PE crystals, which can provide deeper insight into the elementary steps of the chain unfolding [26].

7.3.1.2   Melt-Crystallized Polyolefins:
HDPE, LDPE, Isotactic and Syndiotactic Polypropylene

Polymer crystallization from the melt is a much more complex phenomenon compared to solution crystallization, and its mechanisms are far from being completely understood. It is documented that crystal growth from the melt leads to formation of characteristic morphological patterns, among which spherulitic morphology is probably the best known. AFM images of various polyethylene and polypropylene samples can illustrate different aspects of crystalline morphology and its dependence on regularity of individual polymer chain structure. It is expected, for example, that extensive folding of macromolecular chains into lamellae and compact arrangement of lamellae inside the spherulites will be characteristic of high-density polyethylene, HDPE. This is, indeed, what is confirmed by AFM images obtained on microtomed and etched surfaces of this polymer shown in Fig. 7.2B–H. The semicrystalline morphology of the sample reveals banded spherulites, as observed with polarized optical microscopy (Fig. 7.2A). The optical contrast reflects variations in birefringence [27], which is the result of different local orientations of polymer chains. AFM images of the same surface (Fig. 7.2B) also show banded patterns, with the height contrast corresponding to the surface corrugation of the etched sample [28]. At higher magnifications (Fig. 7.2C–D), one can notice that elevated regions are mainly composed of stacks of relatively flat-lying lamellae. By contrast, the structures visible in the valleys between the bright bands can be identified with edges of the lamellae oriented perpendicular to the sample surface.

Of special concern are granular structures, which are better resolved in the phase image in Fig. 7.2D. The granules can be seen on lamellar edges and also on the surface of flat-lying lamellae. During the past 20 years, there have been a number of electron microscopy observations of similar structures in semicrystalline polymers [29,30]. However, the question of whether these structures are real or originate from sample preparation, which typically includes etching or staining, has not yet been clarified. Additional evidence for the existence of such structures has been obtained in AFM studies of polymer surfaces, which were not subject to etching or staining [28], [31]. For example, a large-scale image of the surface of the melt-crystallized HDPE reveals spherulites with a relatively smooth surface (Fig. 7.2E). The spherulitic bands are formed by crystals oriented in tangential and radial directions (Fig. 7.2F–G), where, at higher magnification, individual fibrils clearly exhibit a grainy surface (Fig. 7.2H).

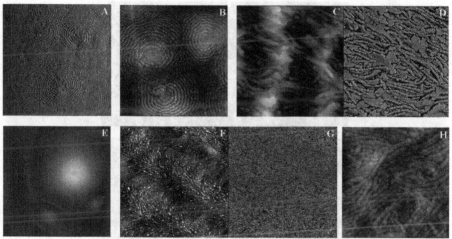

Fig. 7.2. (A) Optical micrograph of a microtomed sample of HDPE after permanganate etching, micrograph size is $35 \times 35 \, \mu m^2$. (B) AFM height image of the same surface ($35 \times 35 \, \mu m^2$ scan). (C)–(D) Height and phase images ($1.5 \times 1.5 \, \mu m^2$) of a part of the banded HDPE spherulite shown in (B). (E) Height image of a banded HDPE spherulite recorded on the surface prepared by hot molding between mica sheets ($35 \times 35 \, \mu m^2$ scan). (F–G) Height and phase images ($6 \times 6 \, \mu m^2$) of a part of the spherulite shown in (E). (H) High-resolution height image of the same spherulite as in (e) ($700 \times 700 \text{nm}^2$ scan).

In contrast to HDPE, highly branched chains of low-density polyethylene, LDPE, are less prone to multiple folding with adjacent chain reentry. At the scale of several microns, melt-crystallized samples of LDPE reveal ill-defined spherulites, also known as sheaf-like structures (Fig. 7.3A–B). However, at higher-magnification, similarly to HDPE, the surface of these sheafs appears to be decorated with numerous grains, especially when imaging is performed in light tapping (Fig. 7.3C). With the increase of the tip-sample force, the image taken at the same location shows only slightly curved fibrils (Fig. 7.3D), which is similar to typical morphological patterns observed with TEM (Fig. 7.3E). The described changes are fully reversible, and grainy morphology is restored when the tip-sample force is lowered again. Therefore we speculate that the fibrils consist of a more rigid core and a rubbery-like exterior [28], where crystalline stems form the core, while less ordered or amorphous material with T_g far below room temperature constitutes the exterior. Although an ideal lamella can be considered as a two-dimensional platelet, one can easily imagine its degeneration into a fibril (i.e., "one-dimensional" object) or a grain (Fig. 7.3F) in the presence of structural defects or geometrical constraints (cf. Fig. 7.1F). In this case, the observed grains would reflect the micro-blocks of the underlying crystalline structure. It is however, not clear whether the observed granular morphology is characteristic of the polymer organization in the bulk or only at the surface. In many instances, the grainy appearance of the polyolefin polymer surfaces can be

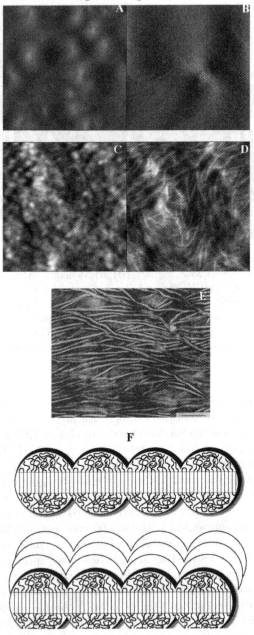

Fig. 7.3. (A)–(B) Large-scale height images of the surface of LDPE hot-molded between mica sheets, 60×60 and $12 \times 12\,\mu m^2$ scans, respectively. (C)–(D) Height images ($680 \times 680\,nm^2$) of the same part of the sheaf-like spherulite shown in (B) recorded in light (C) and hard tapping (D). (E) Typical TEM image of the lamellar structure of PE ($640 \times 640\,nm^2$ size micrograph). (F) Schematic drawing showing the proposed structure of fibrils and lamellae observed in AFM images of melt-crystallized polymers.

smeared out by increasing the tip-sample force, which probably indicates that this structure is more related to the surface organization. Fibrils and grains can be found not only on surfaces of melt-crystallized bulky polymer samples, but also in thin and ultra-thin films, which is demonstrated in Figs. 7.4A–E. The first two images show morphology of an ultra-thin layer of ultra-high molecular weight, UHMW, PE formed by dipping a piece of silicon wafer into a dilute polymer solution in xylene. Concentric patterns of crystalline material result from the interplay between dewetting, crystallization and solvent evaporation. A flat-on lying lamella with grainy surface seen in the center of Fig. 7.4B is surrounded by short fibrils with a block substructure, which are attached to each other. The width of the fibrils is between 10 and 15 nm and the height of the flat-on lamella is approx. 20 nm, which is in the range of a typical lamellar thickness.

Fig. 7.4. (A)–(B) Height images of ultrathin UHMW PE film obtained by dipping a Si wafer into diluted xylene solution at high temperature, $5 \times 5 \, \mu m^2$ and $600 \times 600 nm^2$ scans.

It is interesting to note that the presence of granular morphology is not necessarily related to the lamellar organization of the semicrystalline structure. Thus, in contrast to UHMW PE, a metallocene-based polyethylene sample with high concentration of octene branches does not reveal well-defined lamellar structures (Figs. 7.5A–B). The image, which was recorded in light tapping, exhibits mostly grainy surface features. When imaged in hard tapping (Fig. 7.5B), the same location looks different and short fibrils are seen everywhere. The transition between the two morphologies is fully reversible and seems consistent with the fibrillar structure schematically depicted in Fig. 7.3F.

It is instructive to compare the semicrystalline morphology and the structure of the top surface layer of PE with that of another important member of the polyolefin family, polypropylene. Morphologies of a thin film of fractionated syndiotactic polypropylene, sPP, and an ultrathin film of highly regular iPP are

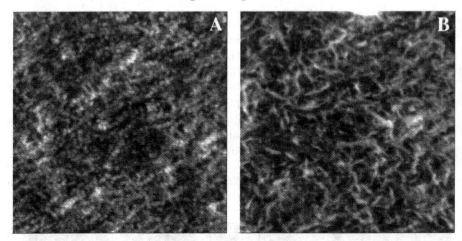

Fig. 7.5. (A)–(B) Height images of the top surface of ethylene-octene copolymer ($\rho = 0.86\mathrm{g/cm^3}$) sample prepared by molding between mica sheets. The images were obtained in light and hard tapping on the same area of $760 \times 760\,\mathrm{nm^2}$.

seen in AFM images in Figs. 7.6A–F. In both cases, one can notice single crystal-like objects surrounded by smaller crystalline entities. Rectangular platelets of sPP exhibit well-defined sectors reflecting different orientations of the crystalline lattice [32].

Numerous fibrillar structures extending from the crystal edges (Fig. 7.6A–B) are believed to be due to growth in conditions of material depletion. The high-resolution image in Fig. 7.6C presents a part of the crystal, where the surface is covered with numerous grains. The straight fibrils grown from the crystal edge also reveal grains and tiny overgrown fibrils oriented in a perpendicular direction, which is most likely the result of the homoepitaxy of polypropylene. AFM images of an ultrathin iPP film (Figs. 7.6D E) show single crystals of slightly different habit than those of sPP [33]. Nevertheless, they also present different sectors and grainy surface morphology (Fig. 7.6F). Stacks of edge-on standing lamellae and web-like patches formed by tiny fibrils are seen between the single crystals. A more detailed characterization of these morphologic features will be given elsewhere [33].

Additional information on the semicrystalline structure and thermal transitions of LDPE and iPP can be obtained from AFM imaging at elevated temperatures (Fig. 7.7A H). The first series of images shows changes in morphology of an ultrathin film of LDPE recorded during a step-like heating of this sample. Initial spherulitic morphology of the crystalline LDPE layer has been formed by dipping the substrate into a hot polymer solution [34]. It can be seen that heating of the layer to 80 °C does not bring any substantial modifications to the morphology. By contrast, further heating to 90–100 °C induces drastic changes: at 100 °C, instead of spherulitic structures linear lamellar aggregates having several microns in length and up to 1 micron in width grew in the same sur-

Fig. 7.6. (A)–(B) Height and phase images ($12 \times 12\,\mu\mathrm{m}^2$) of the surface of fractionated syndiotactic polypropylene. (C) Phase image of a part of the area shown in (A)–(B), $2\,\mu\mathrm{m} \times 2\,\mu\mathrm{m}$ scan. (D)–(E) Height and phase images of the surface of a spin-cast film of isotactic polypropylene, image size is $20 \times 20\,\mu\mathrm{m}^2$. (F) Height image ($3 \times 3\,\mu\mathrm{m}^2$) presenting a detailed view of the region shown in (D)–(E).

face region. Heating to 115 °C, which is in the main melting region of the bulk LDPE sample, leads to a featureless surface indicative of complete melting of the layer. These observations present direct evidence for recrystallization processes occurring during heating of a LDPE layer. AFM visualization of the structural transformation during recrystallization helps understanding changes in polymer properties upon annealing at these temperatures.

Melting of an ultrathin film of iPP demonstrates another kind of structural change in the pre-melting temperature region (Fig. 7.7E–H) [33]. The network of thin lamellar and fibrillar structures visible in the first image of the series melts almost completely at 120 °C (Fig. 7.7F), whereas melting of single crystals of iPP becomes noticeable at 150 °C (Fig. 7.7G). Finally, only small remnants of the initial crystalline structure can be found in the image at 170 °C (Fig. 7.7H). However, crystallization of the molten sample at 160 °C restores the initial morphology consisting of crystalline platelets embedded in a network of tiny lamellae.

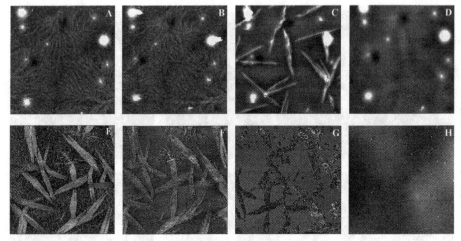

Fig. 7.7. (A)–(D) Height images ($10 \times 10 \, \mu m^2$) of ultrathin film of LDPE on Si substrate at $T = 25$, 80, 100 and 115 °C, respectively. (E)–(H) Height images of ultrathin film of isotactic polypropylene on Si substrate at $T = 25$, 120, 150 and 170 °C, respectively. Image size is $20 \times 20 \, \mu m^2$.

7.3.1.3   Blends of Poly(ε-caprolactone) and Poly(Vinyl Chloride), PCL/PVC

Binary semicrystalline/amorphous blends of PCL and PVC extensively studied in the past [35] present a difficult object for SAXS studies due to their linear crystallinity ($\varphi_{c,lin} \equiv \frac{L_c}{L_B}$) varying around 0.5 and the electron density contrast becoming faint at certain extents of interlamellar inclusion of PVC. Therefore, high temperature AFM is probably the only technique offering the possibility of in-situ crystallization studies of these systems at the lamellar scale [36].

Samples for AFM studies [37] were prepared by casting blend solutions in THF on freshly cleaved mica to obtain films of approx. $10 \, \mu m$ thickness. The reproducibility of resulting semicrystalline morphologies (e.g., the primary nucleation density) was improved by applying the self-seeding technique. Typical topography and phase images corresponding to the crystallization of PCL/PVC 75/25 wt/wt blend at 40 °C are shown in Fig. 7.8.

The whole image sequence (not shown here) corresponding to about 1 hour melt-crystallization was recorded at the same surface spot. At the selected temperature linear growth of PCL crystals occurring during the primary stage of crystallization (i.e., radial propagation of the crystallization front) is too rapid to be observed with AFM. By contrast, the processes of secondary crystallization operating at a slower pace inside the already grown spherulites could be examined with some scrutiny. Since the crystalline lamellae are better defined in the AFM phase images, the latter were used to compute the degree of crystallinity by volume and the morphological parameters of the semicrystalline structure in direct space. The phase images were firstly converted to binary format by choosing a threshold value of the phase signal above which pixels were attributed to

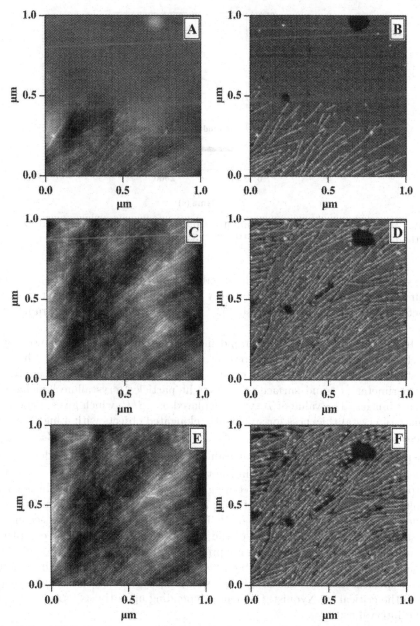

Fig. 7.8. AFM topography and phase images recorded during isothermal melt-crystallization of a PCL/PVC 75/25 wt./wt blend at 40 °C. Elapsed times are 0 (A,B), 541 s (C,D), and 2931 s (E,F). All images of this sequence were successively recorded on the same surface area: The film defect noticeable in the upper right part of the images was used as a marker. The second image of the series (C,D) roughly corresponds to completion of the primary crystallization stage at the image location. The full gray-scale is 12 nm and 16° for topography and phase images, respectively.

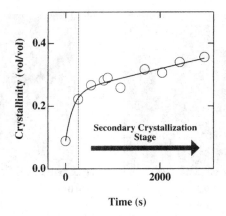

Time (s)

Fig. 7.9. Crystallinity by volume computed from the AFM phase images corresponding to the melt-crystallization of PCL/PVC 75/25 wt./wt. blend at 40 °C.

the crystalline phase. This threshold value was computed by minimizing contour line fuzziness [38], and crystallinity by volume was then calculated as the fraction of pixels above the threshold. The time evolution of this parameter is shown in Fig. 7.9.

It can be seen that the first rapid increase of crystallinity is followed by a much slower growth, which is, however, still detectable after about 1 hour of crystallization. The evolution of crystal thickness was computed by analyzing the perimeter (P) and surface (S) of bright particles (crystalline lamellae) in binary images. The value of L_c was calculated as $2S/P$, which gives a good estimate of crystal thickness for sufficiently elongated objects with a high contour persistence length. A small fraction of more roundish particles with $P_{circ} \leq \frac{1+\varepsilon}{\pi\varepsilon}$, where $P_{circ} \equiv \frac{P^2}{4\pi S}$ and ε – a small value, were excluded from consideration in order to keep a certain precision in the thickness determination ($\frac{\Delta L_c}{L_c} \leq \varepsilon$). The output of this analysis was a surface-weighted lamellar thickness distribution as a function of crystallization time (cf. Fig. 7.10). The results presented in Fig. 7.10 indicate that a slight thickening of PCL crystals occurs during the secondary crystallization stage. The results of the direct-space analysis can be completed by analysis of the same images in reciprocal space, which is performed similarly to the classical treatment of SAXS curves. In the first step, the two-dimensional power spectral density function ($P_2(\underline{s})$) was computed from AFM images ($u(\underline{r})$) up to the critical, or Nyquist, frequency depending upon the experimental sampling interval as:

$$P_2(\underline{s}) \equiv \frac{1}{A} \left| \int_A u(\underline{r}) W(\underline{r}) \exp\left(2\pi i \underline{s} \cdot \underline{r}\right) d^2\underline{r} \right|^2 , \qquad (7.1)$$

where A denotes the image area, $W(\underline{r})$ window function [39] and $\underline{s}$ the reciprocal space vector. The $P_2(\underline{s})$ function was then transformed into the one-dimensional

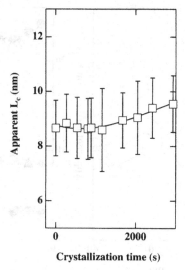

Fig. 7.10. Surface-weighted distribution of lamellar thickness computed from the AFM phase images measured during the melt-crystallization of PCL/PVC 75/25 wt./wt. blend at $40\,^{\circ}\mathrm{C}$. The error bars correspond to the standard deviation of the crystal thickness distribution.

PSD $(P_1(s))$ according to:

$$P_1(s) = (2\pi s)^{-1} \int P_2(\underline{s}')\delta(|\underline{s}'| - s)d\underline{s}', \tag{7.2}$$

and, finally, the one-dimensional correlation function was computed as a real part of the Fourier transform of $P_1(s)$:

$$\gamma_1(l) = Re\left\{ 2\pi \int_0^\infty P_1(s)s\exp(2\pi isl)\exp(4\pi^2\sigma^2 s^2)ds \right\}, \tag{7.3}$$

where the $P_1(s)$ function was preliminarily corrected for the presence of sigmoidal gradient crystal/amorphous transition layers having thickness σ [40]. Calculation of the morphological parameters of the semicrystalline structure was performed with a standard approximate relationship [41]:

$$\varphi_{\mathrm{c,lin}}\left(1 - \varphi_{\mathrm{c,lin}}\right) = r_0/L_B \ . \tag{7.4}$$

In Eq. (7.4), r_0 stands for the first intercept with the abscissa of the tangent to the linear part of the correlation function in the self-correlation triangle. The lamellar thickness L_c is found from Eq. (7.4) as $\varphi_{\mathrm{c,lin}}L_B$, where L_B is determined from the location of the first subsidiary maximum of $\gamma_1(l)$. The resulting correlation functions and the morphological parameters of the semicrystalline structure of the blend are given in Fig. 7.11.

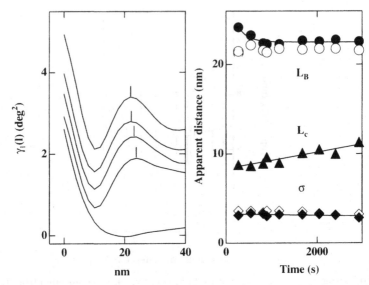

Fig. 7.11. One-dimensional SAXS-type correlation functions and the morphological parameters of the semicrystalline structure computed from the AFM images given in Fig. 7.8. Open symbols correspond to topography and filled to phase images.

It can be seen that the evolution of lamellar thickness determined by the reciprocal space analysis agrees with the results of the direct space analysis reported before, i.e., the crystal thickness shows a monotonic increasing trend. By contrast, the long period probably decreases slightly with time, which is a rather typical observation for isothermal polymer crystallization. In our case, this slight decrease can be related to the in-filling growth occurring between the primary lamellae formed during the rapid stage of crystallization. Examination of the images shows, however, that insertion takes place only in the amorphous pockets that are large enough to accommodate growth of new crystals, and, therefore, the total variation of L_B remains very limited.

7.3.2 Semi-rigid Chain Polymers

While most of AFM work to date has been performed on flexible chain polymers, some important questions related to semicrystalline morphology of semi-rigid chain polymers such as poly(ether-ether ketone), PEEK, and various aromatic polyesters remain open. Generally, despite the fact that the structure of these polymers has been in focus of numerous studies, very little is still known of their lamellar organization. This can be explained by the fact that, on the one hand, these polymers typically present difficult objects for TEM studies [42], and, on the other hand, SAXS data analysis could be ambiguous leading in some cases to very different micro-structural models of these polymers. For example, the uncertainty in L_c as determined from the SAXS correlation function analysis could be as high as 100%. Such uncertainty clearly expresses the need, in studies of the semicrystalline structure of these polymers, for new approaches that would simplify the interpretation of SAXS data. In this section, the problems of SAXS data analysis common to the majority of semirigid chain polymers will be briefly reviewed, and the role of AFM in clarifying these controversial issues will be critically analyzed.

7.3.2.1   Poly(Ether Ether Ketone), PEEK, and Its Blends with Poly (Ether Imide), PEI

PEEK is a typical member of the family of semirigid chain polymers characterized by a fully para-aromatic backbone structure. It is noted for high performance properties, among others excellent thermal and chemical stability, absence of polymorphic transitions, and an extremely large temperature interval of crystallization. The lamellar arrangement of PEEK has been a subject of heated debate due to a difference between authors in the interpretation of the SAXS results for this polymer [43-45]. Briefly, the proposed microstructural models can be divided into two groups. The first model is essentially a three-phase model, wherein thick lamellar crystals (about 70–120 Å) separated by thinner amorphous interlayers (about 30-40 Å) pack into stacks of finite size, these stacks being separated by large purely amorphous PEEK regions. The large amorphous gaps represent a significant fraction of the polymer (up to 50% according to this model). The second model is essentially a two-phase model, wherein the sample is homogeneously filled with stacks of thin lamellae alternating with thicker amorphous interlayers. Different arguments in favor of one or the other model were discussed in some detail elsewhere [45]. In short, the main reason for the difference of opinion between authors is due to the Babinet reciprocity theorem, which states that, in a two-phase system the electron densities of the phases may be interchanged, without affecting the scattering curve or, consequently, the correlation function [46]. In other words, the solution that one finds for example from the classical one-dimensional correlation function analysis is not unique, as the attribution of the two distances obtained from this analysis to the crystalline and amorphous layer thicknesses (i.e., L_c and L_a) can be inverted. Thus, clearly some additional structural information would be required to choose between the two models described above, and simple accumulation of SAXS data treated with the help of the correlation function will not provide any new insight.

    In one of the attempts to solve this problem [47], a direct study of the semicrystalline structure of PEEK was carried out with AFM. In this case, imaging was performed on miscible blends of PEEK with poly(ether imide) (PEI), an amorphous polymer fully miscible with PEEK in the amorphous state. The use of PEI as an inert macromolecular diluent made it possible to open up the dense semicrystalline structure of PEEK and visualize the details of the subspherulitic organization in the course of its evolution (cf. Fig. 7.12). Since the heating stage accessory for AFM was not readily available at the time of the measurements, the imaging had to be performed ex situ. In this experiment, the sample was quenched to room temperature after different times of crystallization at T_c or upon heating to different annealing temperatures T_a ($T_a > T_c$) and imaged

118   Dimitri A. Ivanov and Sergei N. Magonov

at room temperature. To achieve the visualization of the time- or temperature-evolution of a single crystalline entity (e.g., single spherulite or lamellar stack), a special nanopositioning technique was employed to repeatedly retrieve the same spot on the sample surface with a precision of several tens of nanometers. The results of this rather tedious approach show (Fig. 7.12) that AFM can indeed provide information on the evolution of single elements of the semicrystalline structure of PEEK. More specifically, the obtained image sequences gave access to the growth rates of individual lamellar bundles [47], which is impossible with Polarized Optical Microscopy, and to the spatial localization of the reorganization process occurring on heating. However, the imaging conditions employed in the experiments did not allow to obtain lamellar resolution, which somewhat limited the interest of such studies for structural analysis and, consequently, confined the potential of AFM in this field to only qualitative comparison with X-ray results. It became clear that high-resolution in situ AFM imaging was needed to address the issues of lamellar organization of semirigid chain polymers. Such morphological AFM work has still to be done on PEEK [48].

7.3.2.2 Poly(Ethylene Terephthalate), PET

PET is a high-performance aromatic polyester currently used for many industrial applications. The first studies on the semicrystalline structure of PET were carried out more than 40 years ago [49]. Since then, a large amount of data has been obtained on the structure and crystallization behavior of this polymer using different experimental techniques ranging from X-ray and electron diffraction to small-angle light scattering, and from differential scanning calorimetry to IR/Raman spectroscopy. Generally, PET is a low crystallinity polymer sharing many similarities with PEEK, introduced about two decades later. In particular, quite similarly to the situation with PEEK, the information available on the microstructure of PET is very scarce and, therefore, the same problem with the SAXS data analysis (i.e., the possibility of interchanging the distances L_c and L_a) is typically encountered [50].

It is worth noting that the problem with SAXS data analysis not only has direct consequences upon the choice of the structural model for lamellar ar-

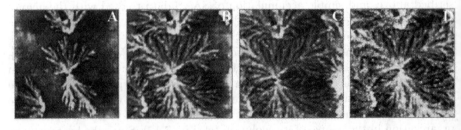

Fig. 7.12. Square $25\,\mu m^2$ topographical AFM images corresponding to the growth of a single PEEK spherulite in a PEEK/PEI 70/30 wt./wt. blend cold-crystallized at $T_c = 180\,^{\circ}C$. Elapsed times correspond to 4 (A), 8 (B), 12 (C) and 60 min (D). Imaging in contact mode was performed ex situ (see text for more details).

Fig. 7.13. Schematic diagrams of evolution of lamellar microstructure during isothermal crystallization: Dual lamellar population model (cf. [43] and references therein) containing separate stacks of thick and thin lamellae (A) and mixed stacks comprising both types of lamellae (B). In this model linear crystallinity in the stack is significantly higher than bulk crystallinity. To account for this difference, large amorphous gaps are supposed to exist in the structure.

rangement but also affects the interpretation of the multiple melting behavior well-documented for PEEK, PET and other semirigid chain polymers [44,51]. Thus, the proponents of the first structural model (see previous section for details) explain the decrease of long period during isothermal annealing either by the formation of separate stacks of thinner crystals (Fig. 7.13A), or by insertion of thinner, more defective, crystals in-between the primary (dominant) ones (Fig. 7.13B). In both cases, the low-temperature endothermic peaks present in DSC scans of the annealed polymer would correspond to melting of these later grown, ill-formed crystals. By contrast, the proponents of the second model (i.e., homogeneous, or infinite stack model) attribute multiple melting behavior to the melting-reorganization process in which the majority of crystals undergo a continuous transformation, i.e., melting followed by immediate recrystallization, to form thicker, more perfect and thermodynamically stable crystals. Since the technique of DSC is unable, despite its high sensitivity, to provide any information regarding the spatial localization of the melting processes, further studies are required to check the validity of each of the crystallization models.

The development of AFM made it possible to visualize the details of PET semicrystalline morphology [52]. However, imaging of PET in tapping mode at ambient temperature normally does not yield sufficient phase contrast between the crystalline and amorphous regions. To improve the contrast, one would need to use tapping mode at temperatures higher than the glass transition (T_g) of the amorphous interlamellar regions, at which the difference in the mechanical properties of both phases becomes much more pronounced. In our recent work [53], the melt-crystallization of PET was monitored in situ with high temperature AFM. Typical images of crystallization kinetics acquired at 233 °C are shown in Fig. 7.14. It can be seen that the PET lamellae display a marked tendency to form stacks. The arrows in the figure point to locations where crystallization mainly proceeds by growth of crystals parallel to the borders of the already existing stacks, which gives rise to the overall increase of the number of crystals per stack. This mechanism termed "stack thickening" seems to play an important role at

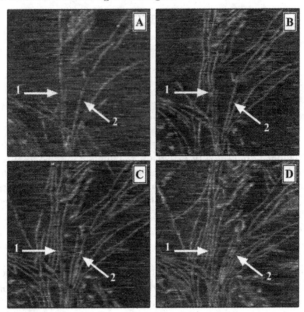

Fig. 7.14. AFM phase images $(1\,\mu m^2)$ recorded in hard tapping during crystallization of PET from the melt at 233 °C; elapsed times are 72 (A), 80 (B), 96 (C), and 112 min (D). Thin white stripes correspond to the crystalline lamellae in almost edge-on orientation. The arrows help the reader to identify the same lamellar stacks and locations on the sample surface where crystallization mainly proceeds via the stack thickening mechanism.

the secondary crystallization stage. By contrast, only very limited occurrence of insertion growth was detected upon thorough examination of the images. In addition, these rare insertion events were found to occur only in those amorphous zones which were much larger than the "normal" interlamellar regions, which is very different from the scheme depicted in Fig. 7.13B.

The time-evolution of the semicrystalline structure of PET was studied by direct space AFM image analysis [53]. The average surface-weighted crystal thickness shown as a function of crystallization time in Fig. 7.15 does not reveal any evolution, at least in the time scale of the experiment. This fact is at variance with the first micro-structural model described earlier (cf. Fig. 7.13).

It should be mentioned, however, that the fraction of the subsidiary crystals is a priori unknown. Thus, to exclude the possibility that this lamellar population has been overlooked in the treatment due to insufficiency of the image statistics, further analyses were performed. To detect this possibly missing lamellar population, it was decided, first, to increase the number of analyzed crystals (i.e., perform the same analysis on several images) and, second, to inspect the semicrystalline morphologies formed after long-time crystallizations. Typical images of the PET morphology formed after completion of crystallization are presented in Fig. 7.16. It is clear that no amorphous gaps of large size (which was

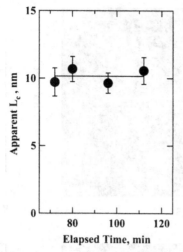

Fig. 7.15. Evolution of the average crystal thickness computed from the AFM images given in Fig. 7.14. The error bars indicate the standard deviations of the distributions.

earlier estimated to be on the order of 200 nm [43]) could be visualized in the images: the semicrystalline structure appears to be completely space-filling. This observation necessarily implies that the linear crystallinity determined by SAXS should be close to the bulk crystallinity of the sample. In addition, the apparent crystal thickness distribution at the end of crystallization remains monomodal as it was during the crystallization kinetics reported in Fig. 7.14. The average crystal thickness decreases with temperature (cf. Fig. 7.16C), which is in agreement with kinetic theories of crystallization. Thus, the hypothesis of secondary crystallization of PET producing more imperfect (thinner) crystals is not supported in our experiments.

The periodicity of lamellar stacking determined from the images obtained at $T_c = 233\,^\circ C$ was quantitatively compared to the SAXS results obtained on the same sample (cf. Fig. 7.16D). The AFM and SAXS values of L_B found from the corresponding correlation functions $\gamma_1(r)$ are rather close (19.2 and 15.5 nm, respectively), which ensures that the morphological parameters of the semicrystalline structure determined by these two very different techniques are comparable. The AFM linear crystallinity calculated using the results of the direct and reciprocal space analysis of the images is given as $\varphi_{c,lin} = 10/19.2 \sim 0.5$ (cf. Figs. 7.15 and 7.16D); the corresponding SAXS values equal either 0.4 or 0.6, depending upon the attribution of the two distances to the sizes of one or the other phase. The value of the AFM linear crystallinity is thus intermediate between the two possible SAXS values, which would formally render the choice difficult. However, it is clear that the bulk of high temperature AFM results can be reconciled only with the choice of the SAXS crystal thickness as the smallest distance, i.e., $L_c < L_a$. Indeed, in the opposite case, the fraction of large non-crystallized inter-stack regions amounting to about $1 - \frac{\varphi_c}{\varphi_{c,lin}} \propto 0.33\text{–}0.42$, where

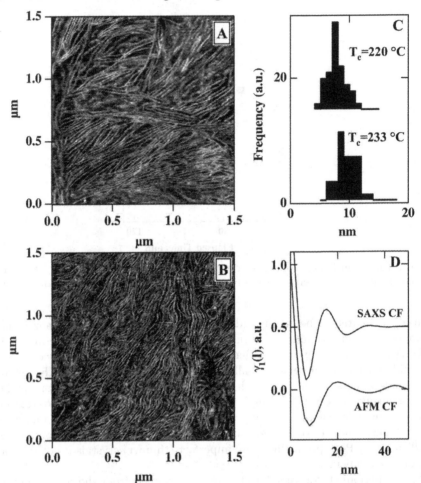

Fig. 7.16. (A, B) Semicrystalline morphology of PET after completion of melt-crystallization at 233 and 220 °C, respectively. AFM phase images were obtained at the temperature of crystallization. (C, D) Quantitative analysis of semicrystalline morphology of PET. (C) Distributions of the apparent lamellar thickness corresponding to images (A) (lower histogram) and (B) (upper histogram). (D) One-dimensional correlation functions computed from the AFM image shown in (A) and from SAXS data obtained on the same sample at room temperature; curves are vertically offset for clarity.

φ_c is bulk crystallinity of the sample, should have been distinctly observed in the images of a completely crystallized sample of PET. Another argument in favor of the homogeneous model is that the AFM value of $\varphi_{c,lin}$ could only be higher than the SAXS linear crystallinity because of the previously mentioned effect of dilation of surface features by the tip geometry. The dilation increases the apparent thickness of crystals slightly protruding the sample surface, without

Fig. 7.17. Model of the stack thickening process operating during the secondary stage of the melt-crystallization of PET. Visualized with high temperature AFM, this process consists in progressive increase of the size of lamellar stacks containing thin crystals separated by thick interlamellar amorphous regions.

modifying the intercrystalline distance, and thereby increases the apparent linear crystallinity.

To sum up, a direct assessment of melt-crystallization of PET by in situ AFM shows that the evolution of the semicrystalline structure can be mainly accounted for by the stack thickening process depicted in Fig. 7.17. In this process, new crystals grow parallel to the borders of the already existing stacks described by the homogeneous micro-structural model, i.e., containing thin crystals separated by thicker amorphous interlayers.

It is important to note that the stack thickening model can successfully account for a decrease of the SAXS long period upon annealing [49] if lamellar stacks contain second-order defects in the structure, or paracrystallinity. This effect is described in some detail elsewhere [54].

7.3.2.3 Poly(Trimethylene Terephthalate), PTT

PTT is a recently commercialized aromatic polyester, with a chemical structure intermediate between that of PET and poly(butylene terephthalate), PBT. This polymer has already been the subject of several publications focusing on its crystal structure, as well as crystallization and melting behavior [55,56]. A detailed study of the organization of PTT at the nanometer scale and its evolution during crystallization has recently been carried out by high temperature AFM and time-resolved SAXS [57]. The AFM images given in Fig. 7.18 are typical of crystallization kinetics observed at 210 °C. They show the propagation of the crystallization front in the form of large leaf-like lamellar crystals. Interestingly enough, in these images one can note the appearance of a spherulitic band formed by the constituent S-shaped lamellar stacks marked with white arrows.

Generally, melt-crystallized PTT displays banded spherulitic texture, the spacing of which was found to vary with crystallization conditions [56]. Although not expected at such high crystallization temperature [56], pronounced banding of large PTT spherulites grown at 210 °C was observed optically and inspected at

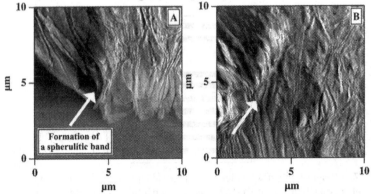

Fig. 7.18. Melt-crystallization of PTT, as visualized with in situ high temperature AFM at 210 °C. The white arrows point at S-shaped lamellar stacks forming a spherulitic band, which extends from the top center toward the left bottom of the images. Elapsed times: (A)–0, (B)–8 min.

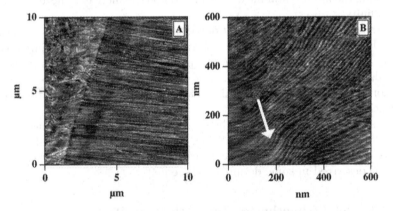

Fig. 7.19. High temperature AFM images of the semicrystalline structure of PTT: crystallization and imaging at 210 °C. Structure of a spherulitic band, as observed at the micrometer (A) and sub-micrometer (B) scales. The arrow indicates the onset of a concerted lamellar twist involving many crystals.

different spatial scales with AFM (Fig. 7.19). At the scale of several microns only a discontinuous morphological change related to spherulitic bands (Fig. 7.19A) can be observed, whereas images at the sub-micrometer scale reveal a neatly resolved concerted lamellar twist involving many lamellae (Fig. 7.19B) [58]. The onset of the twist can be identified in the images by the curvature of otherwise straight lamellae pointed with an arrow in Fig. 7.19B. However, it is technically difficult to follow the propagation of individual lamellae throughout the entire twist period because AFM provides only a two-dimensional projection of the lamellar structure. Therefore, the absence of certain elements of the structure, e.g., parts of lamellae buried underneath the surface, complicates the reconstitution of the three-dimensional structure.

The presence of a regular lamellar twist (rhythmic growth) and the dominance of lamellar stacks containing a large number of crystals make the assessment of the semicrystalline structure of PTT with AFM easier than that of other aromatic polyesters such as PET or PBT. The sub-micrometer AFM phase images of PTT obtained above the T_g of interlamellar amorphous regions normally reveal well defined lamellar crystals (cf. Fig. 7.19B). The parameters of the semicrystalline structure can then be evaluated for the regions where the crystals are grown close to edge-on orientation. Typical distributions of the crystal thickness and long period for the semicrystalline morphology of PTT formed at 210 °C are given in Fig. 7.20.

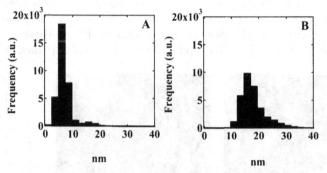

Fig. 7.20. Distributions of crystal thickness (A) and long period (B) corresponding to the semicrystalline morphology of PTT formed at 210 °C. The distribution functions were calculated from the direct space analysis of AFM images using home-built object recognition routines [57].

The comparison of the distribution functions shows that the width of the long period distribution is higher than that of crystals, which is logical since the former is the convolution of the latter with the corresponding distribution of amorphous interlamellar regions. The quantitative analysis of the distributions shows that crystals are slightly thinner than the interlamellar amorphous regions, which is similar to the results on melt-crystallized PET reported earlier. The position and shape of these distributions will be further directly compared [57] with those extracted from the synchrotron SAXS data analyzed with the generalized one-dimensional paracrystalline model of Hosemann [59,54].

7.3.2.4   Syndiotactic Polystyrene, sPS

A broad use of metallocene catalysts led to synthesis of a variety of polymers with well-controlled macromolecular structures. In addition to polyolefins, other polymers such as syndiotactic polystyrene are being produced in large quantities. sPS exists in several crystal modifications, which can be recognized not only by their X-ray diffraction patterns but also by the resulting lamellar morphologies [60,61]. Similar to the case of PCL/PVC blends, a relatively small difference in electron density between amorphous and crystalline regions of sPS limit the use of X-ray scattering techniques for the analysis of its semicrystalline structure [62]. Therefore, AFM can be particularly useful for morphological characterization of sPS in the sub-micrometer range. A series of images corresponding to the α- and β-forms of melt-crystallized sPS are given in Fig. 7.21 [63]. At a large scale, randomly oriented fibrillar entities represent the semicrystalline morphology of α-sPS, and it is best seen in the 5-micron image in Fig. 7.21A. It was specifically checked that this morphology is also representative for the bulk of the α-sPS samples. Quite loose packing is seen not only for the fibrils at the scale of microns but also for lamellar arrangement at the sub-micron scale. In some locations stacks of edge-standing lamellae can be found, which is exemplified in Fig. 7.21B. The calculation of the SAXS-type interface distribution function for this image (Fig. 7.21C) gives 8.5 nm width for both crystalline and amorphous parts of the structure. By contrast, the β-form of sPS is characterized by well-defined spherulites (Fig. 7.21D) up to 100 micron in size. A stack of the edge-on

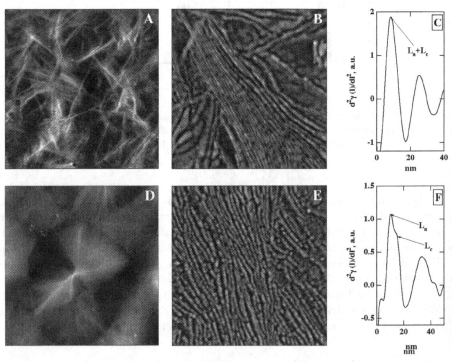

Fig. 7.21. AFM images of melt-crystallized syndiotactic polystyrene. (A)–(B) Semicrystalline morphology of the α-form of sPS showing fibrillar and lamellar organization at the scale of 5 and 1 μm, respectively. (C) SAXS-type interface distribution function computed from the image in (B). (D)–(E) Semicrystalline morphology of the β-form of sPS presenting large spherulites with tightly packed lamellar stacks, image size 100 μm (D) and 1 μm (E). (F) Interface distribution function corresponding to the image in (E). Images (B) and (E) are recorded at 220 °C.

standing lamellae (Fig. 7.21E) was chosen to obtain a quantitative estimate of lamellae packing (Fig. 7.21F). It was found that the width of the crystalline and amorphous regions is 15.5 nm and 9.5 nm, respectively. Consequently, linear crystallinity and long period of β-sPS are larger than those of α-sPS.

7.4   Conclusions

The potential of variable temperature AFM in studies of semicrystalline polymers is described. The advantages and drawbacks of this technique with respect to classical optical and electron microscopies are briefly reviewed. It is shown how the advantageous characteristics of AFM such as its non-destructive character, high precision height measurements and possibility to perform in situ studies at different temperatures can bring new insights into the structure and thermal behavior of semicrystalline polymers. Several examples including flexible and semi-rigid chain polymers illustrate how AFM is complementary to conventional scattering techniques providing quantitative direct-space structural information that can be used to interpret the results of scattering experiments.

Acknowledgements

The authors appreciate the contribution of Natalya A. Yerina (Digital Instruments/Veeco Metrology Group), who was involved in AFM studies described in this work. D.A.I. is grateful to M. Buscemi (Shell Coordination Center, Belgium) for providing a sample of PTT. The authors would like to thank Dr N. Heymans (Université Libre de Bruxelles, Belgium) for reviewing the manuscript.

References

1. Binnig, G.; Rohrer, H.; Gerber, Ch.; Weibel, E. *Phys. Rev. Lett.* **1982**, *49*, 57.
2. Binnig, G.; Quate, C.; Gerber, Ch. *Phys. Rev. Lett.* **1986**, *56*, 930.
3. Aleksander, S.; Hellemans, L.; Marti, O.; Schneir, J.; Elings, V., Hansma, P. K.; Longmire, M.; Gurley, J. *J. Appl. Phys.* **1989**, *65*, 164.
4. Zhong, Q.; Innis, D.; Kjoller, K.; Elings, V. *Surf. Sci. Lett.* **1993**, *290*, L688.
5. "Exploring the High-Temperature AFM and Its Use for Studies of Polymers" Ivanov, D.A.; Daniels, R.; Magonov, S. Application Note published by Digital Instruments/Veeco Metrology Group (2001), pp. 1–12. Available on line at URL: http://di.com/APPNotes_PDFs/AN45%20HeatingStage.pdf
6. Most experimental work was done with custom-made accessories whose temperature range was limited below 100–130 °C.
7. Magonov, S. N.; Whangbo, M.-H. "Surface Analysis with Scanning Tunneling and Atomic Force Microscopy", VCH, Weinheim, **1996**; *Scanning Probe Microscopy of Polymers*, Ratner, B.; Tsukruk, V.V., Eds., ACS Symposium Series, ACS Washington, DC, **1998**, 694.
8. Magonov, S. N. "AFM in Analysis of Polymers" *Encyclopedia of Analytical Chemistry*, (R. A. Meyers, Ed.), pp. 7432–7491, John Willey & Sons Ltd, Chichester, **2000**.

128   Dimitri A. Ivanov and Sergei N. Magonov

9. Jandt, K.D. *Mat. Sci. Eng.*, **1998**, *R21*, 221; Morris, V. J.; Kirby, A. R.; Gunning, A. P. "Atomic force microscopy for biologists", World Scientific Publishing, Singapore, **1999**.

10. Vertical displacements of a probe or a sample are obtained with a feed-back controlled Z piezo-actuator, whereas independent lateral XY-motion is performed by another piezo-actuator.

11. Magonov, S. N.; Elings, V.; Whangbo, M.-H. *Surf. Sci. Lett.* **1997**, *375*, L385.

12. Cleveland, J. P.; Anczykowski, B.; Schmid, A. E.; Elings, V. B. *Appl. Phys. Lett.* **1998**, *72*, 2613.

13. In fact, the height difference between surface locations with different density is largest when the set-point amplitude is lowered below 20% of its initial value. Yet imaging at such amplitudes is performed practically in quasi-contact mode and sample damage could be pronounced.

14. Magonov, S. N.; Elings, V.; Cleveland, J.; Denley, D.; Whangbo, M.-H. *Surface Science* **1997**, *389*, 201.

15. Ponomarenko, S. A.; Boiko, N. I.; Zhu, H.-M.; Agina, E. V.; Shibaev, V. P.; Magonov, S. N. *Polymer Science* **2001**, *43*, 419.

16. Prokhorova, S.A.; Sheiko, S.S.; Ahn, C.-H.; Percec, V.; Möller, M. *Macromolecules* **1999**, *32*, 2653.

17. Percec, V.; Holerca, M. N.; Magonov S. N.; Yeardley, D. J. P.; Ungar, G.; Duan, H.; Hudson, S. D. *Biomacromolecules* **2001**, *2*, 706.

18. Bassett, D. C., *Principles of Polymer Morphology*, Cambridge University Press, New York, **1981**.

19. Geil, P. H. *Polymer Single Crystals*, **1963**, John Willey & Sons.

20. Patil, R.; Kim, S.-J.; Smith, E.; Reneker, D. H.; Weisenhorn, A. L. *Polym. Comm.* **1990**, *31*, 455.

21. Patil, R.; Reneker, D. H. *Polymer* **1994**, *35*, 1909; Stocker, W.; Magonov, S. N.; Cantow, H.-J.; Wittmann, J.-C.; Lotz, B. *Macromolecules* **1993**, *26*, 5915; Nie, H.-Y.; Motomatsu, M.; Mizutani, W.; Tokumoto, H. *Polymer* **1996**, *36*, 183.

22. Tian, M.; Loos, J. *J. Polym. Sci. Phys.* **2001**, *39*, 763.

23. Recent attempts are directed towards modeling of this particular solid state transition, as well as that of the polymer crystal growth in general using computer simulations: Welch, P.; Muthukumar, M. *Physical Review Letters* **2001**, *87*, 218302; Sommer, J.-U.; Reiter, G. *Europhys. Lett.* **2001**, *56*, 755; Anderson, K.L.; Goldbeck-Wood, G. *Polymer* **2000**, *41*, 8849.

24. Ungar, G.; Zeng, K.B. *Chem. Rev.* **2001**, *101*, 4157; Hobbs, J.K.; Hill, M.J.; Barham, P.J. *Polymer* **2000**, *41*, 8761; Hobbs, J.K.; Hill, M.J.; Barham, P.J. *Polymer* **2001**, *42*, 2167.

25. Although the stacking periodicity of these crystals (19 nm) is approximately equal to that in the bulk, their habit is closer to fibrils since the width and thickness of these crystals are comparable.

26. Magonov, S. N.; Yerina, N. A.; Ungar, G.; Reneker, D.; Ivanov, D., in preparation.

27. The thickness of the microtomed slice is supposed to be constant.

28. Magonov, S. N.; Godovsky, Yu. K. *Amer. Lab.* **1999**, *31*, 52.

29. Kanig, G. *Colloid Polym. Sci.* **1991**, *269*, 1118.

30. Michler, G. H. *Kunstoff-Mikromechanik*, **1992**, p. 187, Carl Hanser Verlag.

31. Hugel, T.; Strobl, G.; Thomann, R. *Acta Polym.* **1999**, *50*, 214.

32. Zhu, W.; Cheng, S. Z. D.; Putthanarat, S.; Eby, R. K.; Reneker, D. H.; Lotz, B.; Magonov, S.; Hsieh, E. T.; R. G. Geerts; Palackal, S. J.; Hawley, G. R.; Welch, M. B. *Macromolecules* **2000**, *33*, 6861.

33. Yerina, N. A.; Magonov, S. N.; Jaaskalainen, P., in preparation.
34. Godovsky, Yu. K.; Magonov, S. N. *Langmuir* **2000**, *16*, 3549; Godovsky, Yu. K.; Magonov, S. N. *Polymer Science* **2001**, *43*, 1035.
35. Khambatta, F. B.; *et al. J. Polym. Sci. Polym. Phys.* **1976**, *14*, 1391; Russell, T. P.; Stein, R. S. *J. Polym. Sci. Polym. Symp.* **1983**, *21*, 999; Chen, H.-L.; Li, L.-J.; Lin, T.-L. *Macromolecules* **1998**, *31*, 2255.
36. Basire, C.; Ivanov, D.A. *Physical Review Letters* **2000**, *85*, 5587; Basire, C.; Ivanov, D.A. (unpublished data).
37. The samples of PCL and PVC were obtained from SOLVAY S.A. (Solvic ® grades CAPA ® 650 and 258RD). The molecular weights M_n and polydispersity indices, as determined by gel permeation chromatography (GPC), are 85 400, 1.55 and 34 600, 1.87, respectively.
38. Russ, J. C. *The Image Processing Handbook,* CRC Press, Boca Raton, FL, 1998.
39. Press, W. H.; et al. Numerical Recipes in C, The Art of Scientific Computing (Plenum Press, New York 1988).
40. Koberstein, J. T.; Morra, B.; Stein, R. S. *J. Appl. Crystallogr.* **1980**, *13*, 34.
41. Strobl, G.R.; Schneider, M. J. *Polym Sci: Part B* **1980**,*18*,1343.
42. Watkins, N.C.; Hansen, D. *Textile Res. J.* **1968**, *32(4)*, 388; Ivanov, D.A.; Pop, T.; Yoon, D.Y.; Jonas, A.M. *Polym. Mater. Sci. Eng.* **1999**, *81*, 335; Ivanov, D.A.; Pop, T.; Yoon, D.Y.; Jonas, A.M. (*Macromolecules,* submitted)
43. Verma, R. K.; Hsiao, B. S. *Trends Polym. Sci.* **1996**, *4*, 312.
44. Blundell, D. J. *Polymer* **1987**, *28*, 2248.
45. Jonas, A.M.; Ivanov, D.A.; Yoon, D.Y. *Macromolecules,* **1998**, *31*, 5352; Ivanov, D.A.; Jonas, A.M.; Legras, R. *Polymer,* **2000**, *41*, 3719.
46. Balta-Calleja, F.J.; Vonk, C.G. "X-ray scattering of synthetic polymers", Amsterdam: Elsevier, 1989.
47. Ivanov, D.A.; Jonas, A.M. *Macromolecules,* **1999**, *32*, 1582; Ivanov, D.A.; Nysten, B.; Jonas, A.M. *Polymer* **1999**, *40*, 5899.
48. A systematic AFM study has been recently started in our laboratory using narrow molecular weight fractions of PEEK synthesized by Dr. J. Roovers.
49. Zachmann, H. G.; Schmidt, G. F. *Makromol. Chem.* **1962**, *52*, 23; Elsner, G.; Zachmann, H. G.; Milch, J. R. *Makromol.Chem.* **1981**, *182*, 657; Elsner, G.; Koch, M. H. J.; Bordas, J.; Zachmann, H. G. *Makromol. Chem.* **1981**, *182*, 1262.
50. Wang, Z. G.; Hsiao, B. S.; Fu, B. X.; Liu, L.; Yeh, F.; Sauer, B. B.; Chang, H.; Schultz, J. M. *Polymer* **2000**, *41*, 1791.
51. Lee, Y.; Porter, R. S. *Macromolecules* **1987**, *20*, 1336; Lee, Y.; Porter, R. S.; Lin, S. J. *Macromolecules* **1989**, *22*, 1756; Holdsworth, P. J.; Turner-Jones, A. *Polymer* **1971**, *12*, 195; Medellin-Rodriguez, F. J.; Phillips, P. J.; Lin, J. S.; Campos, R. *J. Polym. Sci., Polym. Phys.* **1997**, *35*, 1757.
52. Dinelli, F.; Assender, H. E.; Kirov, K.; Kolosov, O. V. *Polymer* **2000**, *41*, 4285.
53. Ivanov, D.A.; Amalou, Z.; Magonov, S.N. *Macromolecules* **2001**, *34*, 8944.
54. Ivanov, D.A.; Jonas, A.M. *Macromolecules,* **1998**, *31*, 4546.
55. Yang, J.; Sidoti, G.; Liu, J.; Geil, P.H.; Li, C.Y.; Cheng, S.Z.D. *Polymer* **2001**, *42*, 7181; Chung, W.T.; Yeh, W.J.; Hong, P.D. *J. Appl. Polym. Sci.* **2002**, *83*, 2426.
56. Wang, B.J.; Li, C.Y.; Hanzlicek, J.; Cheng, S.Z.D.; Geil, P.H.; Grebowicz, J.; Ho, R.M. *Polymer* **2001**, *42*, 7171.
57. Amalou, Z.; Hocquet, S., Dosiere, M.; Ivanov, D.A. (in preparation).
58. Such a twist has been usually invoked to explain the formation of banded spherulites.
59. Hosemann, R.; Bagchi, S.N. (1962), " Direct Analysis of Diffraction by Matter ", North Holland, Amsterdam.

60. Greis, O.; Xu, Y.; Asano, T.; Petermann, J., *Polymer* **1989**, *30*, 590.
61. Lopez, L. C.; Cieslinski, R. C.; Putzig, C. L.; Wesson, R. D. *Polymer* **1995**, *36*, 2331.
62. Barnes, J. D.; McKenna, G. B.; Landes, B. G.; Bubeck, R. A.; Bank, D. *Polym. Eng. Sci.* **1997**, *37*, 1480.
63. Yerina, N. A., et al., in preparation.

8 Crystallization of Polymers in Thin Films: Model Experiments

Günter Reiter, Gilles Castelein, and Jens-Uwe Sommer

Institut de Chimie des Surfaces et Interfaces CNRS et Université de Haute Alsace
Mulhouse, 15 rue Jean Starcky, F–68057 Mulhouse, France, g.reiter@uha.fr

Abstract. Real-space and time-resolved observations of morphology and pattern formation resulting from crystallization of ultrathin films of low molecular weight polyethyleneoxide (PEO) or diblock copolymers containing PEO shed light on the mechanisms of how polymers order. Differences in viscoelastic properties, as detected by tapping-mode atomic force microscopy, allow distinguishing crystalline and molten (amorphous) areas with a nanometer resolution. In our approach, we use simple but restricted geometries like thin films of controlled thickness or confinement resulting from block copolymer mesostructures. Nucleation in ultrathin films is suppressed or at least strongly hindered and thus less restricted areas (thicker regions) will act as nucleation sites. This enables us to separate nucleation from growth and to follow growth exclusively. The resulting morphology can be directly related to the kinetics of crystal growth. Changes in the morphology with time and due to different thermal histories are the consequence of the mestable nature of polymer crystals. In addition, information about the nucleation process is obtained by studying crystal formation in 12 nm small spherical cells of a block copolymer mesostructure. There, growth is extremely limited and the main event is nucleation. We discuss the advantages of thin film studies for a better understanding of polymer crystallization.

8.1 Introduction

One of the most intriguing phenomena in polymer physics is ordering and crystallization of long chain-like macromolecules. The transition from an amorphous, randomly coiled state to a crystalline, ordered lamellar state is far from being fully understood [1–3]. In particular, questions concerning the formation of initial nuclei [4–8], the growth kinetics resulting in folded intermediate metastable states [3,9–11], and subsequent morphological evolutions during annealing [11,12] are still intensively debated. In this context, it is necessary to study and understand the molecular pathways polymers take when forming a crystalline state, a question which has been already discussed for more than 50 years.

Typically, polymer crystals are not in thermal equilibrium since their formation is strongly complicated and hindered due to the connectivity of the segments, involving co-operative movements of a large number of connected monomers [2,3,13]. Although the free energy would be lowest for fully extended crystalline molecules, long polymers usually form much thinner lamellae ("chainfolding"). Therefore, polymer crystals represent meta-stable states [3] with a significant degree of disorder, mainly characterized by the degree of chain folding. As a consequence, such imperfect crystals are thermodynamically driven towards states of higher degrees of order [14–23]. However, the corresponding relaxations processes in the crystalline state are usually extremely slow and rather localized because the long sequence of connected segments is not very mobile. Increasing the chain mobility by annealing polymer crystals at elevated temperatures (but below the melting point) may result in faster improvement of the crystalline order. Consequently, the morphology o f a polymer crystal does not only depend on the rate at which it was created (i.e. the crystallization temperature) but also, and sometimes most importantly, on the thermal treatment (e.g. the storage conditions) after the "primary" crystallization process.

The behavior of polymers close to interfaces and within submicrometer ultrathin films has been extensively investigated in recent years in order to understand, control and ultimately predict the properties of polymer surfaces and thin films. Specific molecular architectures or restricted geometries impose constraints on crystallization. The presence of a solid surface may modify or even prevent the ordering process. At present, for the crystallization of polymers in thin films the influence of the solid substrate or of the free surface is hardly understood and controversial observations have been reported. On the other hand, studying the crystallization behavior in thin films at a molecular resolution allows to obtain structural information on ordering, orientation and growth patterns.

Experimental observations in real space (in contrast to the reciprocal space of scattering experiments) for a simple and well-defined sample geometry, under well-controlled conditions and at a molecular resolution present promising approaches which should allow to get more direct insight in the ways how polymers crystallize. For the understanding of polymer crystallization, it is highly desirable to follow the ordering processes on a molecular scale. Atomic force microscopy (AFM) real-space observations provide information on the molecular pathways polymers take when forming a crystalline state[1]. The observed morphologies can be related to the kinetics of chain folding and, based on these real-space observations with molecular resolution, indicate possible molecular pathways polymer crystallization takes. One can find fundamental analogies between crystallization of polymers and nucleation and growth processes of other simpler systems like e.g. metals showing that the same mechanism of diffusion limited aggregation (DLA) is at work [24,25].

In our experimental work presented here, we want to observe molecular scale details of ordering and relaxation processes in a simple geometry. To this end, we use well-defined starting conditions and a simple geometry to relate morphological changes to the state of order in polymer crystals. We focus on crystallization close to an interface in quasi 2-dimensional geometry for strongly adsorbed polymers and on the influence of confinement induced by a block copolymer mesostructure. We employ tapping mode AFM, which allows to resolve differences in the viscoelastic properties of polymers, and so to distinguish between the molten and the crystalline state of a polymer.

Previous studies have clearly demonstrated that the thickness of crystalline polymer lamellae depends on the crystallization temperature [26,27] and can

---

[1] See also the Chaps. 6 and 7.

increase when the crystals are annealed below the melting point (lamellar thickening) [14–17,20–23]. However, the actual pathways taken by such relaxations have never been observed in real space. Only recently, increasing attention has been paid to real space investigations of polymer crystallization using AFM to follow changes in the crystalline state directly [12,28–37].

Confinement certainly affects the mechanism of nucleation [28,38] because the number of available possible conformations and the number of molecules are reduced. The influence of confinement on crystallization was studied in thin films [28–31,38] or in the mesophases of self assembled block copolymers [32,39–45]. It became clear [39,41] that the overall crystallization kinetics of highly asymmetric block copolymers forming spherical or cylindrical mesophases differs qualitatively from the kinetics in unconfined geometries. Size and shape of the restricted geometry confining the polymer crystals has to be compared with the size of a crystalline polymer lamella in the corresponding homopolymer melt, and with the maximum length of the crystalline block in the fully extended state. In the present study we use spheres with a radius of about 6 nm separated by about 24 nm. Obviously, this represents considerable confinement for polymer crystals.

8.2   Experimental Section

8.2.1   Polymers

For our studies we used always low molecular weight poly(ethylene oxide) (PEO), either as a homopolymer or attached to an amorphous polystyrene or hydrogenated polybutadiene block. PEO is a well investigated polymer [14–17,20–22,28,29].

For experiments with adorbed homopolymer monolayers, PEO was obtained from Goldschmidt AG/Essen (M_w=2000 g/mol) or was donated by O. Lambert/Mulhouse (M_w= 7600 g/mol). These polymers were terminated by a CH_3 group at one end -preferring the air surface- and an OH group at the other end -preferring the substrate surface. The polymers adsorbed onto the oxide surface of silicon wafers which were cleaned in a water-saturated UV-ozone atmosphere (the wafers were exposed to UV-light in a humid atmosphere for 24 hours). By this procedure, we created a surface with a high density of OH-groups. The maximum length of the two crystalline polymers in the fully extended state is: $L = l_u N = 12$ nm and 48 nm, respectively, with $l_u = 0.2783$ nm [15] and N is the number of monomers.

The melting temperatures of a similar polymer ($L = 49$ nm, both ends are OH-terminated) in the bulk for the twice folded, once folded and the fully extended state are about 61°C, 62°C and 64°C, respectively [15].

For our studies on diblock copolymers, we used low molecular weight poly-(styrene-block-ethyleneoxide), abbreviated by PS-PEO and hydrogenated poly-(butadiene-block-ethyleneoxide), abbreviated by PB_h-PEO .

PS-PEO(3-3) was synthesized in the research laboratories of Th. Goldschmidt AG/Essen, Germany and was supplied to us by Prof. Riess (Mulhouse/France).

PB$_h$-PEO(3.7-2.9) was obtained by hydrogenation of the polybutadiene block (about 50% 1-2 and 50% 1-4 units, statistically distributed). The polymer has also been synthesized anionically in the research laboratories of Th. Goldschmidt AG/Essen, Germany and their high quality has been checked by GPC (using THF solutions and polystyrene standards) and NMR. The percentage of PEO has been determined by NMR with a precision of about 1 % .

In PB$_h$-PEO(21.1-4.3) the PB$_h$ block contains statistically distributed 1-4 and 1-2 units, with a majority ($\approx$ 85%) of 1-2 units. The polymer was synthesized by Pierre Hoerner (Mulhouse/France) [46]. The PB$_h$ block is amorphous (its glass transition, as detected by differential scanning calorimetry (DSC), is at -18^0C), while the PEO block can crystallize. The length of the fully extended crystalline PEO block is about 27 nm. From complementary X-ray scattering measurements, we know that the microphase structure of the bulk consists of spherical micelles arranged on a body centered cubic lattice with a lattice constant of about 24 nm and a radius of 6 nm for the PEO core. Upon crystallization this structure remained largely unchanged. We call the PEO core the "cell" embedded in a PB$_h$ matrix.

Molecular details of all investigated polymers are given in Table 8.1

Table 8.1. Characteristics of the polymers used in our studies

SAMPLE	M$_w$(PS or PB$_h$)	M$_w$(PEO)	M$_w$/M$_n$
PEO-2k	–	2000	1.1
PEO-7.6k	–	7600	< 1.1
PS-PEO(3-3)	3000	3000	1.1
PB$_h$-PEO(3.7-2.9)	3700	2900	1.1
PB$_h$-PEO(21.1-4.3)	21100	4300	1.15

8.2.2 Sample Preparation

8.2.2.1 Monolayers of PEO Homopolymers and PS-PEO Diblock Copolymers. Samples were prepared from dilute toluene solutions of dried polymer by spincoating thin films (about 50-100 nm) onto silicon wafers. Annealing these films in the molten state led to pseudo–dewetting [28,29], thereby forming holes leaving an adsorbed monolayer behind. The origin of this autophobic behaviour is related to the difference in entropy (resulting from the different chain conformations) of adsorbed and free polymers. Crystallization of this monolayer within the holes was investigated [28,29], using AFM to visualize the resulting crystal morphology. We used films, which were first molten for about 10 min at temperatures above 70°C to allow for the formation of a monolayer. This monolayer was then crystallized at temperatures between room temperature and 55°C. In order to check for relaxation processes within the crystalline state, samples were sequentially annealed for 5 min at increasingly higher temperatures, but always below the equilibrium melting point (i.e. 64°C for PEO-7.6k). Temperature was increased in steps of a few degrees. After each step, the sample was quickly cooled to room temperature and analyzed by AFM.

8.2.2.2 PB$_h$-PEO Diblock Copolymer Films.

Thin films (about 100 nm thick) were prepared by spincoating dilute toluene or methyl-cyclohexane solutions onto UV-ozone cleaned silicon wafers. To erase any non-equilibrium states induced by spincoating, all samples were annealed in the molten state at temperatures up to 150°C before further experiments were performed. Thermal treatment (annealing and subsequent crystallizaton) was performed directly under the microscope in an inert atmosphere (nitrogen flow).

PB$_h$-PEO(21.1-4.3) annealed samples did *not* crystallize at ambient conditions, even after storage at room temperature for many months. Only after cooling the samples to temperatures below about -20^0C, crystallization was observed. Melting, on the other hand, occurred at temperatures above about $+40^0$C. Consequently, at room temperature the samples neither crystallized nor melted. As we show later, this fact enables AFM measurements of partially crystallized samples at room temperature, preserving the structure obtained by crystallization at low temperatures.

8.2.3 Observation Techniques

Optical microscopy and AFM were used to determine the morphology, either in the melt state or during crystallization (optical microscopy) or after crystallization (mostly AFM). All samples were crystallized at constant temperatures.

8.2.3.1 Optical Microscopy.

The samples were placed onto an enclosed hotplate, purged with nitrogen, under a Leitz-Metallux 3 optical microscope. No polarization or phase contrast was used. Contrast is due to the interference of the reflected white light at the substrate/film and film/air interface, resulting in well-defined interference colors which can be calibrated with a resolution of about 10 nm. We have followed the displacement of the crystal growth front in real time by capturing the images by a CCD camera. All data were stored with a VCR for later analysis. The crystallization temperature at the hot stage was controlled to within 0.1 degrees. However, for PB$_h$-PEO(3.7-2.9) at low undercooling nucleation of crystals was found to be very slow. Thus, in order to start crystallization, we first lowered the temperature to 30°C . Once we observed the onset of crystal growth, we jumped to the higher temperatures where we wanted to follow the growth process in real time. It should be noted that only very few nuclei were formed. Thus we were able to clearly distinguish (based on the online observation) within one sample regions crystallized at different temperatures.

8.2.3.2 Atomic Force Microscopy (AFM).

Measurements were performed with a Nanoscope IIIa/ Dimension 3000 (Digital Instruments) in the tapping mode at ambient conditions, using the electronic extender module allowing simultaneously the phase detection and height imaging. We used Si-tips (model TESP) with a resonant frequency of about 300 kHz. Scan-rates were between 0.2 and 4 Hz. The free oscillation amplitude of the oscillating cantilever was around 50 nm, the setpoint amplitude (damped amplitude, when the tip was in intermittent contact) was slightly lower. Topographic (height mode) and viscoelastic (phase-mode) data were recorded simultaneously. It should be noted that semicrystalline polymers are well suited for the use of the "phase-mode" as the differences in viscoelastic properties between crystalline and amorphous regions are large.

8.3   Results and Discussion

8.3.1   Crystallization of Adsorbed PEO Homopolymer Monolayers

In a first approach, we determined the morphology of mono-lamellar crystals obtained from monolayers of PEO homopolymer. Here, we focus exclusively on the growth process, excluding the nucleation process from the monolayer region, as we will describe in more detail below.

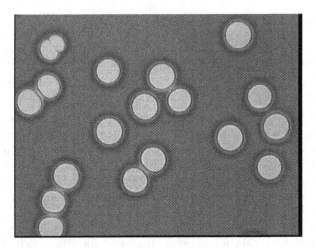

Fig. 8.1. Optical micrograph of a typical dewetting pattern. The white circles represent the holes containing a molten monolayer resulting from pseudo-dewetting. The rings around these holes are rims containing the material removed from the holes. The homogeneous gray area in between represents the unperturbed film.

We first discuss the properties of the adsorbed amorphous molecules of the monolayer within the dewetted holes (see Fig. 8.1). Each polymer segment in contact with the substrate gains a certain adsorption energy and therefore the molecules are attracted to the substrate. Consequently, their conformations are flattened resulting from a balance of energetic (adsorption strength) and entropic (conformation) terms. To a first approximation, the adsorption energy is not much changed during lateral transport on the surface as it occurs during diffusion towards the crystal.

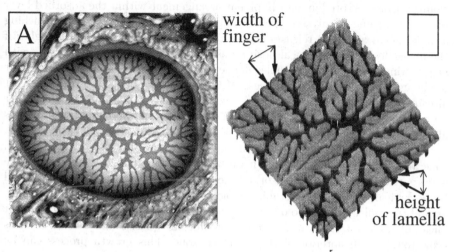

Fig. 8.2. Typical result obtained by tapping mode AFM (topography) showing the fingerlike patterns obtained by crystallization for 880 min at 44°C of an adsorbed PEO-7.6k mono-layer in a pseudo-dewetted hole. A): Fingers inside the hole (image flattened to enhance the visibility of the fingers). Note that all fingers start at the periphery of the hole. B) 3D representation of the fingers in A). The height of the mono-lamellar fingers is 8 ± 1 nm. The size of the images is 20 × 20 μm^2 and 10 × 10 μm^2 for A) and B), respectively.

The AFM results (see Fig. 8.2) now prove the existance of a monolayer inside the holes and show that this monolayer could crystallize. Our results can be reconciled with the contradictory observations by Franck et al. [38] if we assume that crystallization for films thinner than about 15nm is strongly suppressed. The formation of a nucleus in such thin films is probably not possible at the temperatures used in their and our experiments. Now, it has to be noted that our experiments indicate that crystallization of the monolayer started in the less confined thicker parts of the film at the edge (outside) of the dewetted region (i.e. at the rim) where the material from this dewetted area had been collected. No nucleation sites were found within the monolayer, except if there was a defect like a dust particle.

If we now assume that the rim represented a nucleus, which induced crystallization of the monolayer, then the crystal would start at the rim and grow towards the center of the hole. Complementary experiments [29] (not presented here) showed that crystallization of the thicker parts of the film was much faster (about one order of magnitude) than for the monolayer. Thus, the whole surrounding of a hole became crystalline before significant crystallization of the monolayer could take place. This crystalline boundary around the monolayer ensured that the number of molecules within the monolayer was a conserved quantity. Molecules from the monolayer arriving at the rim (=nucleus) got attached to the crystal. As it was energetically more favorable, the molecules changed from the predominantly planar arrangement within the adsorbed layer to a vertical and highly stretched configuration within the crystal. Attaching a molecule to a crystal releases crystallization energy. This energy is higher if the molecules are in a vertical position as each segment has more contacts with neighboring segments. At a molecular level one may suppose that attachment to the crystal is a multistep process. If for kinetic reasons other molecules get already attached before the molecule is completely stretched upright less perfect crystals are formed, which try to obtain better order in a second stage process afterwards (re-arrangement within the crystal - see later: Sect. 8.3.2).

The change to a vertical conformation leads to an increase of the depleted zone ahead of the crystal front. This process creates "free" or "empty" surface as the area occupied by an adsorbed molecule was larger than the one needed by a vertically oriented crystalline molecule. In order to get attached to the crystal the molecules had to diffuse to the crystal across this depletion zone just ahead of the crystal. So, in the course of time the quasi 2-dimensional crystals grow towards the interior of the dewetted zone. This growth process can be viewed as a random deposition of polymers at an initially smooth interface. At low temperatures the probability for desorption from the crystal surface is rather small. Thus, the arriving molecules will get attached to the crystal at the sites which are closest to the diffusing molecules. Molecules having the shortest diffusion path have the highest probability for reaching the crystal surface and getting attached to it. As a consequence of the predominance of the crystal sites closest to the "reservoir" of amorphous adsorbed polymers, the crystal-front becomes unstable, and eventually finger-like or fractal patterns are formed.

Molecules sitting at the crystal surface have also a significant probability for desorption (detachment), especially at temperatures close to the melting point. In a simplified picture, one may say that as a consequence of temperatures closer to the melting point the polymers need more time to get attached permanently to the crystal (they have a higher probability to leave the crystal surface again) and therefore, during this longer time, the effective distance increases over which molecules can diffuse. The increase of the diffusion paths is responsible for the coarsening (widening) of the fingers with increasing crystallization temperature (see Fig. 8.3).

Extensive computer simulations based on the above mentioned ideas of a diffusion-limited aggregation (DLA) process have been performed in order to test the basic concepts [28,30,47]. Their results are in remarkable agreement with the experimental observations. Even features like the relaxation processes shown the following section or the mixed patterns caused by sequential crystallization at two different temperatures (see Fig. 8.4) could be reproduced. A detailed description of these simulations, the model used and the fundamental processes involved in polymer crystallization can be found in the accompanying contribution (Chap. 9 of this book).

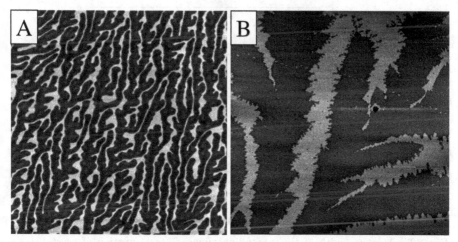

Fig. 8.3. Analogous results as in Fig. 8.2, obtained after crystallization for A) 12 min at 35°C and B) 5430 min at 53°C. The height of the mono-lamellar fingers is 6±1 *nm* and 15±1 *nm*, respectively. The size of the images is 5 × 5 *μm²* and 84 × 84 *μm²* for A) and B), respectively. One should note that the width of the fingers varies from about 150 *nm* to about 15 *μm* within a temperature range of 18°C.

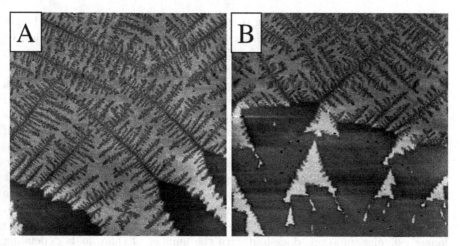

Fig. 8.4. Patterns obtained by crystallization of a PEO-7.6k monolayer at 53°C for 3782 min followed by a quench to 35°C. After 66 min the sample was finally brought to room temperature and measured directly after that by AFM. The size of the images is A) 70 × 70 *μm²* and B) 97 × 97 *μm²*. The small fingers resulted from crystallization of the polymers which at the time of the quench to 35°C were not yet crystalline. The average height of the small and the large fingers is about 6±1 nm and 15±1 nm, respectively. The triangular shape of the fingers should be noted.

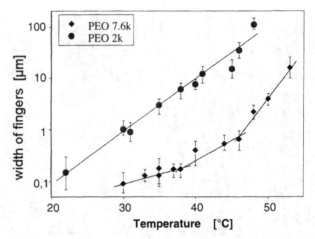

Fig. 8.5. Average width of the crystalline "fingers" as a function of crystallization temperature for two different molecular weights (PEO-2k and PEO-7.6k, respectively) on semi-logarithmic scales. The dotted lines are guides to the eye. The different slopes indicate different numbers of folds per chain, as supported by the different heights of the fingers.

Finally, it is worth emphasizing that the thickness of the lamellae (the height of the fingers) at a given crystallization temperature was rather constant (the thickness fluctuations are much smaller than the mean thickness, see e.g. Fig. 8.2B). However, this thickness varied with temperature. This indicates that the kinetics of the deposition process determines besides the width of the fingers also the height of the lamellae. At a high deposition rate (large undercooling) the molecules attached to the crystal have less time to stretch out and are thus in a state of relatively low internal chain order (folded state). In contrast, if the crystals grow slowly the attached molecules have more time to relax towards the fully extended form of lowest free energy by re-arrangements at the crystal surface. Figure 8.5 demonstrates the coupling between width and height of the lamellae via the different slopes one can fit to the PEO-7.6k data. The smaller PEO-2k polymer does not fold in the temperature range investigated. Accordingly, a single straight line is sufficient to approximate the data. The longer polymer, however, does fold with the number of chain folds depending on crystallization temperature. The influence can clearly be seen in our AFM images. E.g. in Fig. 8.4, one finds 6±1 nm and 15±1 nm thick lamellae resulting from crystallization at 35°C and 53°C, respectively. These values correspond to lamellae containing chains which are about 8 and 2 times folded, respectively. Figure 8.5 now shows that the width of the fingers does change much with crystallization temperature if the chains are highly folded while it increases more significantly for the less folded states.

In particular, for highly folded chains one has to expect that after the primary DLA-type growth process a secondary much slower relaxation process will follow. Relaxations at the crystal boundary or even within the crystal are responsible for morphological transformations of the finger patterns as we will discuss in more detail in the following section.

8.3.2   Relaxations of Polymer Crystals AFTER Formation

In Fig. 8.6, we show a typical example of what happens to rather rapidly grown finger-like crystals if they are briefly annealed, even at temperatures about 20°C *below* the melting point. Exactly the same spot, which can be identified by the defect visible in the right corner, is measured before and after the annealing procedure. One can clearly see that the finger-like pattern (Fig. 8.6A), created by a DLA process [30], changed its morphology over large distances, up to more than 1000 times the size of a monomer. We emphasize the ostensible analogy to dewetting processes in liquids.

At all temperatures below the melting point, all crystalline domains show elevations (rims) at their edges (boundaries). The height of the domains was quite uniform and changed in rather discrete steps with temperature [11]. The higher the temperature the higher were the domains. For our polymer (PEO-7.6k), we expect, based on the molecular weight, that lamellae consisting of chains containing 8, 6, 4, and 2 folds have thicknesses of 5.4 nm, 6.9 nm, 9.6 nm and 16 nm, respectively. This is in striking coincidence with the heights we actually measured (6, 8 10, and 16 nm, with an error of about 0.5 nm and a variance of about 1 nm). Corresponding to the increase in lamellar thickness we observed a decrease in the area covered by the crystalline domains. This is

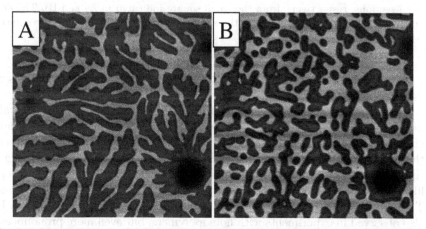

Fig. 8.6. Morphological transformations visible in a monolamellar crystal of PEO-7.6k grown at 35°C for 10 min A) before and B) after the sample was annealed for 5 min at 45°C. The size of the images is 5×5 μm^2. Note the tendency for transformation of the tree-like shape of the fingers into an arrangement of circular (droplet-like) structures with a slightly higher region at the periphery. The defect zone at the right lower corner served as a marker for an unambiguous identification of the investigated area even after annealing. This allows to follow the pathways the system has chosen for the relaxation of the metastble states (folded chains) within the "fingers".

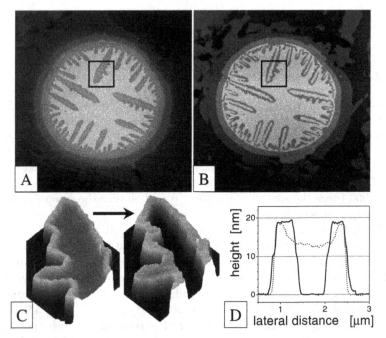

Fig. 8.7. Tapping mode AFM images showing the morphological changes induced by annealing fingers obtained by crystallization of monolayers of PS-PEO(3-3) resulting from pseudo-dewetting. A) after crystallization at 45°C, B) after subsequent annealing for 1 min at 54°C. The size of the images is $40\times40\ \mu m^2$. C) and D) show 3D-plots and cross-sections, respectively, from the small square indicated in A) and B). The dotted and the full lines in D) represent the states A) and B) before and after annealing, respectively.

a direct consequence of the conserved number of polymers within the observed area.

The observed relaxations can be interpreted as the tendency of the crystalline polymers to remove chain folds and to establish better ordered and more compact crystals. We also note that several distinct morphologies can be observed in the course of such crystal improvement. Besides the pattern shown in Fig. 8.6 (mostly indicating lateral contraction of fingers and some coalescence of droplet-like domains) one may also observe the formation of "holes" within the fingers. This phenomenon has been predicted by computer simulations [2] and has been also observed in experiments with homopolymers, but even more pronounced for PS-PEO(3-3) (see Fig. 8.7A-D). The advantage of block copolymers is that the piling up of lamellae is extremely unlikely due to the incompatible amorphous block. Thus, one can unambiguously attribute the observed height to the polymer architecture (chain folding and eventual chain tilting). On the other hand, the

[2] see Chap. 9, in particular Sect. 9.5 therein.

amorphous block may also affect the crystallization process and thus may cause complications in the interpretation of the results.

The 3D representation in Fig.8.7C and in particular the cross sections shown in Fig.8.7D cleary reveal the elevations (rims) at the boundaries of the crystals. Comparing the maximum measured height (about 20 nm) with the expected thickness (about 26 nm) of a lamella containing fully extended polymers, which are orientied exactly perpendicular to the substrate, indicates that the steric and entropic constraints induced by amorphous block cause the molecules to tilt. The degree of tilt can change continuously, even within one finger. Consequently no discrete steps in lamellar thickness are visible. Annealing such crystals of non-ideally ordered polymers allows to reduce the degree of tilt (the lamellar thickness increases locally). As a consequence, less surface area will be covered by an individual molecule, giving rise to the empty space within the fingers.

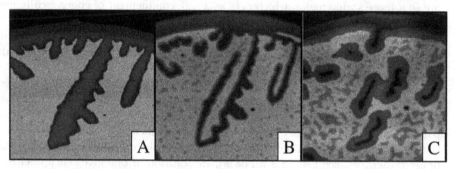

Fig. 8.8. Tapping mode AFM images showing the morphological changes induced by annealing fingers of Fig. 8.7A at comparatively higher temperatures. A) after crystallization at 45°C, B) after annealing for 1 min at 55°C and C) after annealing for 1 min at 56°C. The size of the images is $10 \times 10 \ \mu m^2$.

In Fig. 8.8 we demonstrate that annealing at even higher temperatures leads (in accordance with theory and computer simulations [47]) to a higher probablity for desorption of polymers from the edges of the crystalline domains. This, in turn, creates a "sea" of non-crystalline molecules around the remaining (now smaller) crystalline domains. Upon cooling down to room temperature, these "free" polymers (or at least most of them) will re-crystallize, however, in more folded states. The remaining crystalline domains containing highly extended polymers serve as seeds for the crystallization of the free polymers. Thus, one can find small but rather high domains surrounded by much thinner lamellae. This can be clearly seen in Fig. 8.8C.

Our data prove unambiguously that the reduction of the number of chain folds in crystalline polymer lamella is accompanied by changes in the morphology on a mesoscopic scale, large compared to the size of the molecules. This is surprising because polymer mobility within the lamellae is highly restricted. These changes, which set in at the border of the crystalline domains and propagate at a progressively slower rate towards their interior, are driven, similar to dewetting, by the tendency to minimize the total circumference of the domains thereby minimizing line and surface tensions.

The following interpretation is strongly supported by computer simulations [30] which were able to reproduce our experimental findings and provided additional information for the understanding of the underlying basic processes. Further details can be found in Chap. 9 and in [47].

After the growth process, relaxations within the crystalline domains start to dominate. Since chains at the edge of the crystal are much more mobile compared to the interior, relaxations start there. Thus, as a second step of morphogenesis, a pronounced rim is formed around the crystalline domain. This rim, in turn, prohibits further relaxation of the interior on the same time scale. Note the difference to equilibrium crystals of small molecules where the melting process sets in at the surface, i.e. the weakest part of the crystal. In contrast, non-equilibrium polymer crystals are surrounded by a rim of better ordered molecules corresponding to a more stable region which acts as a self–confinement of energetically less favorable states.

However, also this self-confined state will continue to relax since it is still not in thermal equilibrium. On much longer time scales, a third process leads to further growth of the thickened rim zone. For large areas surrounded by such rims, or for block copolymers lamellae consisting of tilted fully extended molecules, even holes may appear in the interior of these crystalline domains. Note that, as a result of the kinetic pathway the system has taken, the central part of a crystalline polymer fingers is thermodynamically weakest. This implies that re-arrangements and melting will start most likely from the interior of the finger structures (see Figs. 8.7 and 8.8). In addition, if one gives such enclosed areas sufficient time for relaxations, morphological changes in the interior will happen, even at temperatures about 40 degrees below the melting point [11].

8.3.3 Discrete Variation of Lamellar Spacings with Temperature in Block Copolymer Systems

We now turn to crystallization within the confining mesostructures found in diblock copolymer systems. The crystallization of polymers involves a hierarchy of ordered structures on several length scales starting from the crystal unit cell, over nano-sized crystallites up to hundreds of microns for spherulites. In block copolymer systems, these length-scales are superposed to, and sometimes in competition with, the mesoscopic structures due to the interaction (incompatability) of the blocks. Consequently, crystallisable block copolymers in thin films or at surfaces present interesting systems for the creation of patterned substrates exhibiting multiple lengthscales.

The representative AFM image shown in Fig. 8.9 for the almost symmetric PB_h-PEO(3.7-2.9) demonstrates the possibility to distinguish between the crystalline PEO domains and amorphous PB_h layers via the "phase" image of tapping mode AFM. The most spectacular result is represented by the line patterns visible in the phase-image. These lines are separated by a well defined distance of 22 ± 1 nm, a value of the order of the size of the molecules. We have

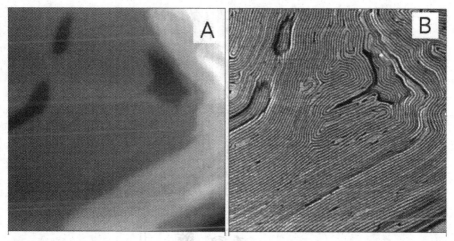

Fig. 8.9. Tapping mode AFM images from a thin film of PB$_h$-PEO(3.7-2.9), crystallized at 38°C. The film thickness is varying between one and three lamellae (indicated by the different gray levels in the topography image A). The substrate can be seen in the right lower corner (see e.g. the "phase" image in B). The phase shows a clear molecular spacing indicated by the periodic light and dark lines representing vertical lamellae. A FFT analysis yields a characteristic distance of 22±1 nm. The contact line to the substrate has a strong influence on the orientation of the lines. These lines are not neccessarily interrupted, even if they cross different height levels (see upper left part of the images).

to conclude that the molecules formed lamellae oriented vertically with respect to the substrate. The difference in viscoelastic properties (the crystalline PEO domains are much harder than the liquid PB$_h$ domains) is reponsible for the contrast in the AFM phase image. Interestingly, the vertical lamellae resulting from the crystallization process were preferentially aligned along the borders of dewetted regions. The perfectness of this alignment could be even further improved by reducing the crystal growth rate, i.e. by crystallizing at higher temperatures. In Fig. 8.9 one can see that the spacing between these lamellae is highly constant and that the phenomenon is not depending on the film thickness.

The morphologies of polymer crystals are generally meta-stable structures as they are always significantly affected by the kinetics. We investigated this influence by varying the crystallization temperature and thus the crystal growth rate. In Fig. 8.10 we have compared typical results for four crystallization temperatures T_c differing only by a few degrees. A small difference in T_c, from 38°C (Fig. 8.10A) to 45°C (Fig. 8.10C), resulted in a doubling of the characteristic spacing. For T_c between 20°C and about 40°C we observed a spacing of about 22 ± 1nm. Increasing T_c to 48°C led to the complete loss of a characteristic separation distance between lines of low and high stiffness, as demonstrated in Fig. 8.10D. It should be noted that the topography and the phase image are now clearly correlated. In particular, depressions between stripe-like features appear

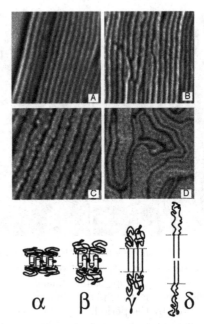

Fig. 8.10. Comparison of AFM phase-images for thin films of PB_h-PEO(3.7-2.9) crystallized at A) 38°C, B) 41°C, C) 45°C, and D) 48°C. The size of each image is $400 \times 400\ nm^2$. The characteristic spacing increases from A) 22±1nm to B) 34±1nm and C) 44±1nm and is lost for D). These spacing can be related to different levels of chain folding as indicated by the sketches $\alpha, \beta, \gamma, \delta$. While α, β and γ can explain the spacing in A, B) and C), respectively, δ can probably not be realized due to unfavorable or even impossible strong streching of the amorphous block. Thus, D) is probably due to the re-orientation of the vertical lamellae of C into a horizontal alignment. This would allow for lateral spacings of any width, not related to the size of the molecules

as dark lines in the phase image while the stripes (independent of their width!) are represented by uniformly light parts.

Based on simple geometric arguments (we know the number of the monomers and their size in the crystalline state as well as the lamellar spacing observed by AFM) we conclude that for temperatures up to about 40°C we have once folded interdigitated PEO blocks (sandwiched between PB_h layers) resulting in a repeat period of about 22 nm. This value has to be related to the maximum length of the fully extended crystalline block alone which is 18.4 nm. Assuming that we have one fold, we would obtain a length of 9.2 nm, only for the PEO block. Taking into account that the lamellar period L consists of amorphous and crystalline layers (ABA sequence), and further assuming that the PEO blocks are interdigitating ("zipper" we obtain a value between 20-25 nm for L, depending on the conformation of the PB_h block.

At higher temperatures the system starts to remove the fold and eventually we end up in a state of fully extended but interdigitated PEO blocks with a length of 18.4 nm at around 45°C (this is only about 9 degrees below the melting point and growth proceeds only at a speed of less than 1 nm/sec). In Fig. 8.10 we have sketched these ideas.

In contrast to molten systems, the resulting lamellar spacing in the crystalline state cannot simply be described by thermodynamic arguments. It also depends strongly on the kinetics of the crystallization process. As long as the equilibrium between the loss of entropy due to the stretching of the PB_h blocks and the gain in crystallization energy of the PEO blocks is not obtained, the PB_h blocks have no possibility to control the number of folds of the PEO blocks. In such a case, the PEO blocks crystallize more or less unaffected by the PB_h blocks. Thus, crystallization of the PEO blocks sets the lateral separation of PB_h-PEO(3.7-2.9) junction points and the degree of stretching of the PB_h blocks. The evolution of the patterns with temperature (Fig. 8.10) proves that the once folded state at low temperatures is NOT an equilibrium state but is the result of the kinetics of the crystallization process. The equilibrium situation is most likely obtained if crystallization takes place very slowly at a temperature only a few degrees below the equilibrium melting temperature.

8.3.4   Individual Crystallization and Melting of Polymer Nanocrystals

In the last set of experiments we used the asymmetric block copolymer PB_h-PEO(21.1-4.3), where PEO presents the minority phase and forms spherical cells with a diameter of about 12 nm. A typical AFM image is shown in Fig. 8.11. The first point to realize is the ability of the AFM phase mode to distinguish even between two liquid polymers, the liquid PEO cells embedded within the liquid PB_h matrix. There is no correlation between topography and phase image.

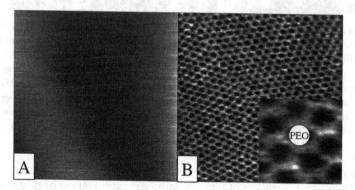

Fig. 8.11. AFM topography (A) and phase (B) images showing the surface of a thin PB_h-PEO(21.1-4.3) film. While the film is rather smooth, the different viscoelastic properties of PB_h and PEO allow to identify the distribution of the two block in the hexagonally packed mesostructure of spherical PEO cells in an amophous PB_h matrix. The size of the images is $500 \times 500\ nm^2$. The inset in B) shows a magnification of a $65 \times 65\ nm^2$ section of B where one PEO cells is indicated schematically.

The key question we aim to answer is how crystallization proceeds under such extreme confinement. One has to realize that within such small spherical cells with a diameter of about 12 nm about 145 PEO blocks each having about 100 monomers are contained. In addition, each block is attached to a much longer amorphous block forming the confining matrix around the PEO cells. Previous experiments [39,41] already showed that the overall growth kinetics of highly asymmetric block copolymers forming spherical or cylindrical mesophases differs qualitatively from the kinetics in unconfined geometries. As crystallization did not destroy the mesophases, it was concluded [39,41] that crystallization was initiated by homogeneous nucleation, separately in each compartment. However until now it was not possible to verify by direct visualization of the crystallization process if nucleation was really independently occuring in each compartment. Employing again tapping mode AFM phase contrast to distinguish now between crystalline and amorphous cells, we observed in direct space that crystallization occurred in a random manner, cell-by-cell (see Fig. 8.12). Systematic AFM experiments using samples held at -23°C for increasing times showed that the fraction of crystalline cells increased with crystallization time.

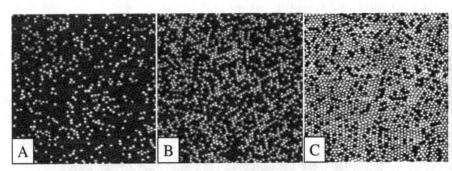

Fig. 8.12. AFM phase images showing the variation in the number and the distribution of crystalline cells after crystallization at -23°C for A: 5 min, B: 15 min and C: 120 min. The percentage of crystalline cells as a function of time follows a relation which can be approximated by $n = n_\infty(1 - e^{-t/\tau})$. Here, n_∞ and τ are the maximum fraction of crystallizable cells and the characteristic time of the the process, respectively. Fitting this equation to the data at early times yields $\tau = 35$ min. Note, however, that even after 120 min a measurable fraction of about 5% was still not crystalline, indicating deviations from above equation at long times. The size of the images is $1 \times 1\mu m^2$.

An obvious question arises: Is there a correlation between the crystalline cells? Does crystallization of one cell induce crystallization of neighboring cells? Our experiments [48] show that neither at low nor at high densities of crystalline cells, any indications for a coupling between cells can be found. Our results demonstrate further that the probability for nucleating a crystal in a cell is not visibly affected by the state (either molten or crystalline) of the neighboring cells. Thus, at all densities of crystalline cells, the PEO spheres crystallize independently, with a random spatial distribution of the cells.

As shown by the AFM images of Fig. 8.12, the mesophase structure was not changed upon crystallization and the shape of the crystalline spheres as detected by AFM did not indicate large scale deformations. Thus, the polymer crystals seemed to keep the spherical shape of the cell. A spherical shape of the crystalline domain can only be achieved by imperfect and thus comparatively unstable crystals. The fact that the crystals were formed at extremely large undercoolings supports the assumption that the crystals contain many defects. A strong tendency to improve crystalline order via relaxations, even at temperatures well below the melting point, has to be expected. Such improvements of crystalline order were clearly detected by DSC [49].

Just after crystallization and some relaxations, we should not expect that the properties of all the indiviual crystals are exactly identical. An indirect proof for heterogeneities of the crystals comes from direct observations of the distribution of crystalline cells after partial melting. AFM clearly shows a distribution of melting temperatures of the individual spheres (see Fig. 8.13). Different melting temperatures are attributed to different degrees of perfection of the crystals. Images of the partially molten samples after progressively increasing the temperature towards $+45^{0}$C do not show any signs of correlation between cells during melting (see Fig. 8.13). Like the distribution of the crystalline cells upon crystallization, the molten cells are also distributed randomly [48,49].

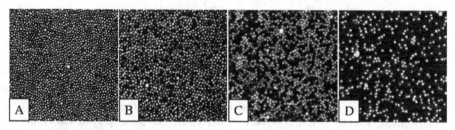

Fig. 8.13. The sample of Fig. 8.12 after 240min at $-23°$C (A) was split into several pieces which were separately annealed for 2 min at B: $39°$C, C: $42°$C, and D: $44°$C, respectively.

We note that all cells which were unstable at a given temperature melted already within less than 2 min, and that no further cells melted upon prolonged annealing. In contrast to techniques averaging over large volumes, AFM allows to distinguish which cell is melting at which temperature. Already molten cells *cannot* re-crystallize as nucleation is only possible at low temperatures. Therefore, the fact that a certain number of spheres remained crystalline, independent of the duration of annealing (melting), indicates a heterogeneity in stability. The perfectness of crystalline order of the individual spheres as it results from crystallization and reorganization during heating is indeed of statistical nature.

8.4 Conclusions

We have demonstrated that AFM in combination with well-defined samples like thin films or even polymer monolayers allowed a direct relation between the morphology and the crystal nucleation and growth processes.

Due to a strongly reduced nucleation probability of polymers confined in the two-dimensional adsorbed monolayers, crystallization starts always in the thicker regions, e.g. at the rims collecting the material from the dewetted areas. Thus, in such experiments, we can clearly separate between nucleation and growth phenomena, putting the focus on the growth processes. In the dewetted holes, the resulting growth patterns are analogous to structures resulting from diffusion limited aggregation (DLA), frequently found in nature (e.g. snowflakes). Moreover, using the basic principles of DLA but adding polymer specific features like the possibility for chain folding, we succeeded in reproducing even fine details of the experimentally found features of the monolamellar crystals with a generic computer simulation model [28,30,47] presented in Chap. 9. In particular, the increase of the characteristic width of the fingers with crystallization temperature and the relaxation behaviour upon annealing of the non-equilibrium polymer crystals has been understood based on these simulations. The complex nature of polymers is introduced via the possibility of chain folding, leading to a sequence of metastable states.

Studying crystallization of adsorbed polymer monolayers allows the direct observations of molecular parameters like the width and the thickness of the fingers (average stem length). The latter is a direct measure of the degree of chain folding and thus gives access to the basic steps of the organization of segments within a polymer crystal. In spite of the complexity of this process, our results clearly showed that the primary step of polymer crystallization is related to a DLA process. The kinetics of this process results in highly non-equilibrated structures which then, in a second step, will relax towards a thermodynamically favored state of more extended molecules. Such morphological changes in meta-stable polymer crystals may occour on large mesoscopic scales even at temperatures well below the melting point. This increase of internal chain order (removal of chain folds) is directly detectable for the quasi-2 dimensional crystals. The resulting morphology is strongly depending on the time-temperature pathway taken by the thermal treatment. Based on the well-suited conditions of our experiments (direct visualization with molecular resolution, a conserved number of molecules in well-oriented crystals, and control of temperature) we are able to test theoretical approaches of polymer crystallization and melting. In combination with results from computer simulation, this represents a promising way for solving the question of how polymers crystallize.

A key parameter in controlling the crystalline morphology is the number of chain folds (selected by the kinetics of crystallization) which ultimately determines the lamellar spacing L. For diblock copolymers, the strong incompatability of the two blocks prevents the exchange of molecules between layers on the timescale of the crystallization process. In contrast to amorphous copolymers, where the lamellar thickness is controlled by thermodynamic aspects, in semicrystalline polymers the morphology is dictated by the kinetics of the crystallization process. It has to be noted that the crystallization energy is usually much larger than interfacial energies or configurational entropy of the polymers. At different temperatures, the kinetics of the crystallization process leads to different but well-defined metastable states which are reflected in different lamellar spacing, also for block copolymer systems.

An approach to investigate questions of how polymer crystals are nucleated is based on highly confining geometries provided by block copolymer mesostructures. Time-resolved real-space AFM measurements allow to observe statistical nucleation and melting via the corresponding distributions of crystalline domains within an array of small confined spherical cells. Homogeneous nucleation at large undercoolings results in crystals of different degrees of order, prone to internal reorganization processes and reflected in a range of melting temperatures for the individual cells. Most remarkably, all cells crystallize or melt independently, without any visible correlation between cells. We expect that further systematic studies of this kind using polymers differing in size and/or nature of the confining matrix will shed even more light on basic problems like how polymer crystals are nucleated.

Certainly, many question have been left open but our results demonstrate the advantages of thin film studies for a better understanding of polymer crystallization. Several possible and obvious extensions of our work are planned for the future.

Acknowledgements

We are indebted to G. Riess, P. Dumas, P. Hoerner, Olivier Lambert and R. Krikorian (Mulhouse/France) for providing us with the polymers. Fruitful discussions with T. Thurn-Albrecht and partial financial support by the EC under Contract No HPRN-1999-00151 are acknowledged.

References

1. E.W. Fischer, Kolloid Z. u. Z. Polym., **231**, 458 (1968).
2. G. Strobl, The Physics of Polymers, Springer, Berlin, Heidelberg, N.Y., 1997.
3. A. Keller, S.Z.D. Cheng, Polymer **39**, 4461 (1998).
4. P.R. ten Wolde, D. Frenkel, Science **277**, 1975 (1997).
5. C. Liu, M. Muthukumar, J. Chem. Phys. **109**, 2536 (1998).
6. P.D. Olmsted et al. Phys. Rev. Lett. **81**, 373 (1998).
7. D.W. Oxtoby, Nature **406**, 464 (2000).
8. L.Zhu et al., Macromolecules **34**, 1244 (2001).
9. J.P.K Doye, D. Frenkel, Phys. Rev. Lett. **81**, 2160 (1998).
10. G. Strobl, Eur.Phys.J.E, **3**, 165 (2000).
11. G. Reiter, G. Castelein, J.-U. Sommer, Phys. Rev. Lett. **86**, 5918 (2001).
12. W. Zhou et al., Macromolecules **33**, 6861 (2000).
13. K. Armistead and G. Goldbeck-Wood, Adv. Polym. Sci., **100**, 219 (1992).

14. A. Kovacs, A. Gonthier, and C. Straupe, J. Polym. Sci.: Polym. Symp. **50**, 283 (1975).
15. A. Kovacs, C. Straupe, and A. Gonthier, J. Polym. Sci.: Polym. Symp. **59**, 31 (1977).
16. A. Kovacs and C. Straupe, J. Cryst. Growth **48**, 210 (1980).
17. S. Cheng, A. Zhang, J. Chen, and D. Heberer, J. Polym. Sci. B, Polym. Phys. **29**, 287 (1991).
18. S. Cheng, J. Chen, A. Zhang, and D. Heberer, J. Polym. Sci. B, Polym. Phys. **29**, 299 (1991).
19. S. Cheng and J. Chen, J. Polym. Sci. B, Polym. Phys. **29**, 311 (1991).
20. S. Cheng, J. Chen, A. Zhang, J. Barley, A. Habenschuss, and P. Zschak, Polymer **33**, 1140 (1992).
21. S. Cheng, J. Chen, A. Zhang, J. Barley, A. Habenschuss, and P. Zschak, Macromolecules **25**, 1453 (1992).
22. S. Cheng, S. Wu, J. Chen, Q. Zhuo, R. Quirk, E. von Meerwall, B. Hsiao, A. Habenschuss, and P. Zschak, Macromolecules **25**, 5105 (1993).
23. Z.-G. Wang, B. Hsiao, B. Sauer, and W. Kampert, Polymer **40**, 4615 (1999).
24. J. Langer, Rev. Mod. Phys. **52**, 1 (1980).
25. P. Meakin, *Fractals. scaling and growth far from equilibrium*, vol. 5 of *Cambridge nonlinear science series* (Cambridge University Press, Cambridge, 1998).
26. P. Barham, R. Chivers, D. Jarvis, J. Matinez-Salazar, and A. Keller, J. Polym. Sci.: Polym. Lett. Ed. **19**, 539 (1981).
27. G. Hauser, J. Schmidtke, and G. Strobl, Macromolecules **31**, 6250 (1998).
28. G. Reiter, J.-U. Sommer, Phys. Rev. Lett. **80**, 3771 (1998).
29. G. Reiter, J.-U. Sommer, J. Chem. Phys. **112**, 4376 (2000).
30. J.-U. Sommer, G. Reiter, J. Chem. Phys. **112**, 4384 (2000).
31. K. Taguchi et al., Polymer **42**, 7443 (2001).
32. G. Reiter, et al., Phys. Rev. Lett. **83**, 3844 (1999).
33. Y. Sakai, M. Imai, K. Kaji, M. Tsuji, J. Crystal Growth **203**, 244 (1999).
34. A. Winkel, J. Hobbs, M. Miles, Polymer **41**, 8791 (2000).
35. C. Basire, D.A. Ivanov, Phys. Rev. Lett. **85**, 5587 (2000).
36. Y.K. Godovsky, S. Magonov, Langmuir **16**, 3549 (2000).
37. L. Li, C.-M. Chan, K.L. Yeung, J.-X. Li, K.-M. Ng, Y. Lei, Macromolecules **34**, 316 (2001).
38. C.W. Frank et al., Science **273**, 912 (1996).
39. B.Lotz, A.J. Kovacs, ACS Polym. Prepr. **10**, No. 2, 820 (1969).
40. A.J. Ryan, I.W. Hamley, W. Bras, F.S. Bates, Macromolecules **28**, 3860 (1995).
41. Y.-L. Loo, R.A. Register, A.J. Ryan, Phys. Rev. Lett. **84**, 4120 (2000).
42. C. De Rosa, C. Park, E.L. Thomas, B. Lotz, Nature **405**, 433 (2000).
43. Y.-L. Loo, R.A. Register, D.H. Adamson, Macromolecules **33**, 8361 (2000); J. Polym. Sci. Polym. Phys. **B38**, 2564 (2000).
44. H.-L. Chen et al., Macromolecules **34**, 671 (2001).
45. L.Zhu et al., Polymer **42**, 5829 (2001).
46. P. Hoerner, G. Riess, F. Rittig, G. Fleischer, Macromol. Chem. Phys. **199**, 343 (1998).
47. J.-U. Sommer, G. Reiter, Europhys. Lett. **56**, 755 (2001).
48. G. Reiter, G. Castelein, J.-U. Sommer, A. Röttele, T. Thurn-Albrecht, Phys. Rev. Lett. **87**, 226101 (2001).
49. T. Thurn-Albrecht, A. Röttele, J.-U. Sommer, G. Reiter, submitted.

9 A Generic Model for Growth and Morphogenesis of Polymer Crystals in Two Dimensions

Jens-Uwe Sommer and Günter Reiter

Institut de Chimie des Surfaces et Interfaces CNRS et Université de Haute Alsace
Mulhouse, 15 rue Jean Starcky, F–68057 Mulhouse, France
ju.sommer@uha.fr

Abstract. We discuss a generic algorithm to simulate the growth of chain crystals from adsorbed monolayers. Chains are considered as elementary statistical units which can exist in states of different internal order. Non-equilibrium growth is combined with internal reorganization processes with a tendency to improve the crystalline order towards the fully stretched chain state. The thickness of the grown crystals is self-organized as a result of the interplay between a barrier to increase local chain order and the gain of enthalpy by accessing higher degrees of order (longer stems). When the reservoir of liquid chains is exhausted relaxation processes dominate. Since chains located at the crystal rims have a higher mobility they are prone to spontaneous reorganization into higher ordered state. This results in striking and very stable morphological phases such as overgrown rims or hole-rim patterns. Increasing the temperature yields to further morphogenesis. In particular droplet-like patterns can be obtained which show liquid-like features on large scales but are made up of highly ordered crystalline chains. We argue that final melting of polymer crystals is not related to the structure which is crystallized but only to the descendants in the morphogenesis of the chain crystal.

9.1 Introduction

Polymer crystals generally represent non-equilibrium states usually referred to as folded chain crystals [1–3]. Only crystals containing fully stretched chains can be regarded as an equilibrium thermodynamic state.[1] The occurrence of non-equilibrium folded states has its origin in the high internal conformational entropy of individual chains in solution or in the melt and is sketched in Fig. 9.1.

To illustrate the basic idea, we assume ideal Gaussian conformations for the chains in the liquid state. Then, the averaged extension of the randomly coiled chain R in the amorphous state is proportional to the square root of the degree of polymerization N, i.e. $R = lN^{1/2}$, where l denotes the statistical segment length. This state represents the thermodynamic equilibrium in the liquid phase. On the other hand, the thermodynamic equilibrium in the crystalline phase is given by the fully stretched chain crystal having a thickness of $L = lN$. Now, let us

---

[1] This might be different for copolymers, where an equilibrium between the stretching of the amorphous block at the expense of folding or tilting of the crystalline block is conceivable

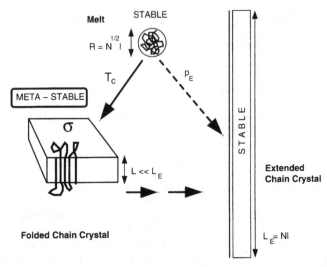

Fig. 9.1. Sketch of the transformation from the coiled amorphous state to the stretched crystalline state via intermediate metastable states.

consider a relatively short chain made of 100 statistical monomers. Assuming further that an individual statistical monomer has only $z = 2$ times more microscopic states (bond orientation) in the liquid state then in the crystalline state, an estimate which is certainly much too low. Than, the probability for this chain to access the crystalline conformation due to statistical fluctuations is given by $p = \exp\{-100\ln 2\} \simeq \exp\{-70\}$, which is an inconceivably small number. The time needed to produce such a conformation t_c is of the order of $t_c = \tau/p$, where τ is in the pico-second range (segmental relaxation time). Thus, $t_c \simeq 10^{58}s$, much longer than the age of the universe. Therefore, for a given under-cooling chain crystals are formed by much shorter stretched sequences (stems) at the expense of additional surface energy at the folding surfaces. However, these states can spontaneously gain free energy by transforming into less folded states. Three remarks have to be made at this point

- The transformation into more stable states can involve complex morphological transformations including spatially distributed lamellar thicknesses and breaking-up of the original lamellae. Only the trivial possibility of homogeneous lamellar thickening has been considered in some detail in the past.
- The non-equilibrium character of polymer crystals is not necessarily manifested by the lamellar thickness L alone, but can correspond to non-equilibrium growth morphologies as well. This is well known also for ordinary liquid-solid-transitions. Snow-flakes being a commonly known example.
- Also the stem lengths in newly born polymer crystals are not due to equilibrium fluctuations between the liquid and the crystal phase but are obtained as the result of a kinetical pathway. This is most obvious for polyethylene under high pressure, where lamellar thicknesses in the range of μm are obtained,

see Chap. 3 on page 17. However, this requires pre-cursor-states where the chain favors the stretched state.

In short: Polymer crystals are out-of equilibrium structures since equilibrium fluctuations are virtually impossible between the liquid and fully extended crystalline state.

In the past, attempts to understand and predict the properties of chain crystals have been focused on kinetic models for chain attachment onto the crystal growth front assuming the resulting structure as frozen and quasi-stable [3,4]. This represents a one-step crystallization process. Here, we want to put forward a more general point of view by considering polymer crystallization in the context of non-equilibrium growth and reorganization phenomena. We show that a simplified lattice model constructed from basic assumptions about the polymer crystallization process is capable to predict a variety a morphological features and transformations which can be observed under the AFM (atomic force microscope) on crystallized polymer films. Furthermore, we scrutinize basic thermodynamic concepts such as "melting" and "recrystallization" for systems far from equilibrium. We want to put forward a more general concept of "morphogenesis" . During the life cycle from growth out of a super-cooled melt until final melting a polymer crystal can go through various intermediate stages of different stability.

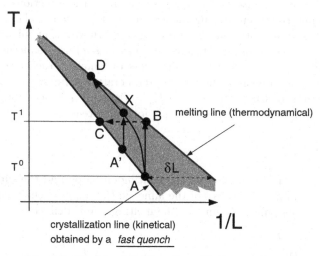

Fig. 9.2. Sketch of the Lauritzen-Hoffmann diagram.

Before we describe our model assumptions in more details, we would like to discuss the use of thermodynamic terms in the context of non-equilibrium states, in particular in case of the widely used Hoffman-Lauritzen (HL) diagram as displayed in Fig. 9.2. Let us consider temperature to be the only thermodynamic variable, i.e. in equilibrium thermodynamics (ETD) all states are uniquely defined for a given temperature, the phase diagram of ETD has only one axis. Now, a polymer crystal is obtained by a *fast quench* from the stable liquid state.[2] The thickness $L(T_c)$ of the polymer lamellae obtained under such non-equilibrium conditions at the crystallization Temperature T_c shows a pronounced dependence from T_c. Plotted in a $(1/L, T)$–diagram this can be approximated by straight line which we call the crystallization line, see also Chap. 4 on page 48 in this book. The crystallization line can be described by the *empirical* relation

$$L = \frac{a}{\Delta_c} + \delta L \ , \tag{9.1}$$

where a and δL are empirical constants. Usually, the *small* correction δL is only introduced for theoretical reasons (see below). The dimensionless under-cooling is defined as

$$\Delta_c = \frac{T_c^0 - T_c}{T_c^0} \ , \tag{9.2}$$

where T_c^0 should be considered again as an empirical parameter.

On the other hand, the crystallization temperature $T_c(L)$ does not coincide with the melting temperature T_m where the transformation of the crystallite, born at T_c, back to the liquid phase takes place. For a given thermal treatment, for instance a constant heating rate, this defines an array of maps $T_c \rightarrow T_m$, which, by continuity can be expressed as a function $T_m(T_c)$, and because of the previous observation also as $T_m(1/L)$. Generally, these functions must depend on the thermal treatment during the annealing stage such as the heating rate.

So far, the argumentation is still very general. However, at this point, usually arguments borrowed from ETD come into play. Assuming the picture of geometrically regular and very thin (but rather large) lamellae, the finite thickness L gives rise to a depression of the melting temperature compared to a (hypothetical) equilibrium melting temperature $T_m^0 = T_m(1/L \rightarrow 0)$ according to the well known Gibbs-Thomson rule. In a simple form this reads

$$\frac{1}{L} = b \cdot \Delta \ , \tag{9.3}$$

where b is constant depending on the melting enthalpy and the surface tension of the bottom and top surfaces of the lamellae (folding and chain end surfaces). The dimensionless under-cooling at the melting-line is defined as

$$\Delta_m = \frac{T_m^0 - T_m}{T_m^0} \ . \tag{9.4}$$

This would provide another straight line on top of the crystallization line as sketched in the figure. By this definition, the Gibbs-Thomson-melting-line should

---

[2] The crystalline phase can also be reached from a glassy state (cold crystallization). Then a fast annealing step to the crystallization temperature is required. Also here, the system is brought far from equilibrium to an anticipated equilibrium melting temperature.

be the upper envelope (stability limit) for all observed melting temperatures in the HL-diagram provided that the lamellar thickness L is well defined in all experiments. In models for the crystallization kinetics [3–5] often the following identifications are made: $T_m^0 = T_c^0$ and $a = b$. Then δL refers to the distance between the melting and the crystallization line at T_c, see Fig. 9.2.

A few general problems in this context should be highlighted. First, the Gibbs-Thomson rule is valid for an equilibrium state, where L can be *fixed* as an *independent* parameter (For instance melting of a material squeezed within a thin slab). In the present case, L is a result of the crystallization process, in particular of the cooling history of the sample. It cannot be varied or fixed by independent means. When the crystalline lamella approaches its stability limit T_m sufficiently slowly, several kinds of reorganization processes can take place which can change the morphology including the apparent thickness L, instead of direct melting. This can be easily seen from the HL-diagram in Fig. 9.2. If we start from the state of crystallization A in the figure and anneal the system just up to the melting line, state B at temperature T_1, there exist a state C on the crystallization line at the same temperature, which corresponds to a crystal which just appear at this temperature. So, instead of melting the crystal can "decide" to "recrystallize" with a thicker lamella.

Secondly, the theoretically established melting line cannot be independently checked, since the surface tension (which controls the constant a) is not measured independently. By contrast, it is usually calculated on the basis of the Gibbs-Thomson rule taking the actually measured values of T_m.

Thirdly, and most importantly, the HL-diagram is often considered as an analogue to a thermodynamic phase diagram. This implies that the non-equilibrium valiable L together with the temperature defines the state of the crystals uniquely and no further knowledge of the sample is required. First of all, we note that only states between the crystallization and the melting line are defined, because L cannot be defined outside this region (gray shading in Fig. 9.2). Let us now consider a crystal born at the state A. The crystal is then *annealed sufficiently slowly* so that it does not approach the "direct" melting point B, but, due to various reorganization processes, it melts in a different state D (see also Fig. 4.7 in Chap. 4 at page 57). The annealing process can be related to a continuous path in the HL-diagram, defining a trace of states. Now, consider fast annealing of a crystal born at the state A' in Fig. 9.2, which leads to the same state X by means of a different thermal history. However, there is no way to assume that both states representing the point X show the same physical behavior and correspond to the same macroscopic state. By contrast, we will show that this is highly unlikely since the HL-diagram is an incomplete description of the non-equilibrium states of polymer crystals. For a more complete description, in addition to L additional non-equilibrium variables such as measures of the crystal morphology must be included.

9.2   A Lattice Model for Crystallization in Polymer Monolayers

Crystallization from polymer monolayers opens novel insights into the processes of growth, reorganization and melting of polymer crystals as is outlined in Chap. 8 on page 131. On the other hand, the geometric restrictions for growth in two dimension makes it possible to cast basic assumption about the polymer crystallization process into a simple and fast lattice algorithm. In this section we describe first our basic assumptions for the states and processes relevant for polymer crystallization. Then, we propose a model for the implementation of these assumptions in a Monte Carlo algorithm. More details can be found in [6].

In the introduction we emphasized the point of the large entropy reduction of individual chains during crystallization. In fact, a polymer crystal can be regarded as a super-crystal made of complex units which are the folded and frozen-in chain conformations. Traditionally, in models of the crystallization kinetics the unit for the crystallization process is chosen to be much smaller than the chains (stems [4], or small groups of statistical monomers [5]). Here, we will take the opposite point of view and consider the chain as the smallest unit of the crystallization process. In contrast to crystallization in simple liquids, the polymer chain changes its *internal* state in the crystalline phase. This is illustrated in Fig. 9.3.

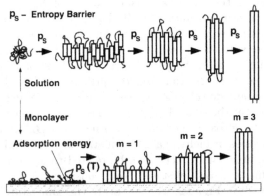

Fig. 9.3. top) Different states of internal order of a polymer chain during crystallization: In the amorphous phase the chains form random coils with a high conformational entropy. In the crystalline phase the chains have to order into a folded structure which can be characterized by the average length of stems. Also here different degrees of disorder such as loops, non-regular folding patterns must be accounted for. We assume a folding pathway of individual chains from the amorphous state into the fully ordered, stretched state via intermediate less ordered states. bottom) In addition, when polymer crystallization occurs in thin films onto substrates the ordering process involves desorption of monomers.

We denote the degree of order of the individual chains by an order parameter m. The liquid state is defined by $m = 0$ which corresponds to the disordered random conformation. When a chain enters the crystalline phase its internal state of order must change (chain folding, reduction of conformational entropy). We denote the states of increasing internal order by $m = 1, 2, ..., M$, were M represents the maximum degree of order, i.e. the fully stretched state. For all ordered states $m > 0$ we assume a binding energy to each neighboring chain in the crystal which increases with m:

$$E(m, m') = \epsilon \cdot \min(m, m') \; , \tag{9.5}$$

where ϵ denotes the binding energy per unit of the order parameter. On the other hand, the transition into a state of higher internal order is accompanied with a reduction of entropy. In our simplified model, we assume that the probability of increasing the internal degree of order from m to $m + \Delta m$ is described by a simple exponential law

$$\bar{p}(\Delta m) = p_S^{\Delta m} \; , \tag{9.6}$$

where p_S gives the ordering effort per unit of the order parameter. We call this the entropic penalty. Note that these assumptions are quite similar to those of the model of Sadler [5] but generalized to whole chains.

The interplay between both effects can qualitatively explain the appearance of folded chain crystals with an optimal value m_{opt}. If the order parameter at the growth front is low ($m < m_{opt}$), the binding energy will not be sufficient to localize new crystalline chains at the growth front. These chains can fall back into the liquid state with a probability of $p \simeq \exp\{-f\epsilon m/k_B T\}$ where f is the effective number of nearest neighbors at the growth front, T denotes the absolute temperature, k_B being the Boltzmann constant. Here, we assumed the same value of m for all chains in the crystal for the sake of simplicity. On the other hand, a high value of $m > m_{opt}$ is related to a low probability $\bar{p}(m)$ for the liquid chains to enter the crystal, which in turn favors the crystallization of weaker ordered chains in competition with these high ordered chains. As a consequence of these opposing effects an optimal value m_{opt} will dominate the population of ordered crystal chains. The value of m_{opt}, however, will depend on the incoming flux of liquid chains and the growth morphology obtained under the given conditions (via f). Most importantly, however, is that also crystalline chains can change their degree of internal order within the crystal and can move to different places. This will take place in particular at the crystal boundary where the restrictions are lowest.

In order to get a more detailed and quantitative picture, we apply these ideas to the growth of monolayer crystals in form of a lattice model. Let us start with a few facts about polymer crystallization from adsorbed polymer monolayers

- Homogeneous nucleation is highly suppressed in monolayers. Crystallization starts usually from the thicker parts surrounding the monolayer, see Fig. 8.2 and 8.7 in Chap. 8.

- Lamellae grow flat-on. Hence, the adsorbed chains have not only to stretch but also to desorb partially from the substrate, see also Fig. 9.3. As a result of this process the occupied area per chain is reduced for chains in the ordered state. This is illustrated in the left hand side of Fig. 9.4.

In particular the second point allows us to find a very simple realization of the internal order parameter m in a lattice model. For crystallization in polymer monolayers it can be directly related to the reduction of the occupied surface area. Hence, we can define m as

$$m = \frac{A_0}{A} \; , \tag{9.7}$$

where A is the surface occupied in the crystalline state (upright stems) and A_0 corresponds to the surface area the chain occupies in the flatly adsorbed liquid state. In simple words: The area of gyration of a single single adsorbed molecule is occupied by m molecules in the crystalline state. Within a lattice model this leads to the idea of *multiple occupation* of lattice sites in the crystalline state. Therefore, we can treat the polymer chains as elementary objects (points on a lattice) having an internal degree of freedom m which is represented by multiple occupation of lattice sites.

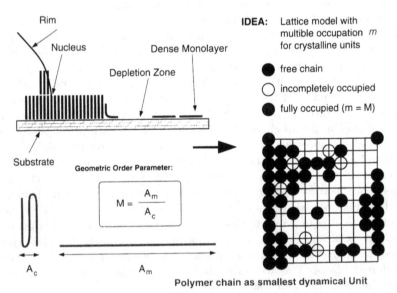

Fig. 9.4. Left: Model for crystallization of polymer mono-layers. The stems must be oriented upright using the rim of the dewetted hole as primary nucleation sites. The area per chain is reduced in the crystal phase. The area ratio between adsorbed, non-crystallized chains and crystalline chains represents a suitable order parameter m. Right: Lattice growth model with multiple occupation of sites in the crystalline phase.

Let us summarize our model assumptions for the case of a Monte Carlo algorithm:

- Chains are the elementary units of dynamics and are modeled as points on a two-dimensional lattice. In the liquid state (strongly adsorbed chains) diffusion is realized by changing the position of the chains to neighboring lattice sites randomly. Because of excluded volume, lattice sites can be occupied by one liquid chain only $(m = 0)$.
- Since homogeneous nucleation is suppressed (see first point above), no spontaneous change $m = 0 \rightarrow m > 0$ can occur. The increase in the order parameter must be mediated by an existing growth front.[3] Therefore, nucleation sites must be provided. In most of our simulations we use a straight line at the bottom and the top of a rectangular simulation box representing the rim surrounding of the monolayer region.
- In the crystal phase $m \leq M$ molecules can occupy the same lattice site. For simplicity, those m molecules are considered to be all in the same state of order (coarse graining remains on the size of the adsorbed chains). The binding energy corresponds to the number of pairs of oriented units between neighboring sites, see (9.5).
- A liquid chain can enter a crystalline site of order m when it is located next to the growth front. Then, it must increase the local order of this site by one unit, which corresponds to a probability $\bar{p}(m+1)$ as given in (9.6). If the move is rejected a new crystalline site is created at the actual position of the chain in the lowest state of order $m = 1$. This corresponds to a precursor state. We do not allow for growing further chains directly at a site with $m = 1$. During growth there will be many chains in this state next to the growth front which corresponds to an adsorbed (but not yet fully crystallized) layer.
- A chain at the growth front can leave the crystal and enter the liquid phase by breaking all bonds to the neighboring crystal sites which is given by a probability $p = \exp\{-E/k_BT\}$, where E denotes the sum of all binding energies to neighboring sites, see (9.5).
- A chain in the crystal phase can change to a neighboring site within the crystal. This requires a breaking of all crystalline bonds at the original site and the formation of new bonds at the new site. If the degree of order is increased during this move also a probability $\bar{p}(\Delta m)$ has to be multiplied to the net probability of such a move.
- All m chains in a crystal site are treated individually during the Monte-Carlo procedure.

The model assumptions and its realization on a two-dimensional square lattice are illustrated in Fig. 9.4. During a Monte Carlo step a chain is chosen randomly and a random direction for a possible move is generated. If the new position would violate the excluded volume condition (for liquid chains, only single lattice occupation is allowed, for crystalline chains $m_{newposition} \leq M$ must

---

[3] This can take place via adsorption of the liquid chain at the growth front and further change of the conformational state, which may include liquid crystalline precursor states [7], see also Chap. 10

be fulfilled) the step is rejected. Otherwise, all weights for the steps have to be taken into account. This means that the move is accepted only if a random number between 0 and 1 is smaller than the product $\bar{p}p$ as discussed above.

The two parameters

$$p_0 = \exp\{-\frac{\epsilon}{k_B T}\} \tag{9.8}$$

and p_S, see (9.6) are abstract parameters on a coarse grained scale. It is not possible to associate these parameters with microscopic parameters within our model. The parameter p_0, however, represents a strong temperature dependence. Therefore, we interprete p_0 as a measure of the temperature of our model according to

$$T = -\frac{1}{\ln p_0} \; . \tag{9.9}$$

Here, we have taken ϵ/k as the natural unit of our temperature definition. On the other hand, the entropic penalty p_S represents many complex features on the scale of the chain. The transition between differently ordered states involves highly cooperative processes which cannot be reflected on the scale of our model. Certainly, the simple relation for $\bar{p}$ in (9.6) is a rough oversimplification. Nevertheless, as we will show below, even such a simplified model reflects fine details as observed in real experiments. To get more insight into the local processes which control the entropic penalty parameter, direct simulations of the freezing polymer chains have to be carried out, see Chap. 10. The most important achievement of our algorithm is to reduce the enormous number of parameters which can influence the polymer crystallization process to only two empirical model parameters which make it feasible to check their impact on the crystallization process by direct comparison to experiments. Moreover, only simulations on a coarse grained scale can be used to model the behavior of the growth and reorganization processes on time and length scales which are experimentally relevant.

9.3   Growth Morphologies

In order to see the impact of the both model parameters we start to investigate the role of the temperature and the maximum order parameter M, which is related to the polymerization degree N, without taking into account the entropic penalty ($p_S = 1$). As a consequence a crystal site is populated by incoming free chains until M is reached. After completion the next incoming chain creates a new crystal site in the state of lowest order. Hence the crystal will be dominated by fully occupied sites, i.e. fully extended chains. However, during the growth process some partially filled sites can be surrounded by others which, when fully occupied, frustrate enclosed partially ordered crystal sites. This induces disorder in the lamellar structure in form of defects created by the 2D nature of the growth process.

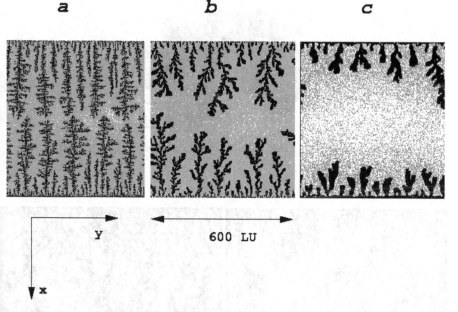

Fig. 9.5. Growth patterns obtained without entropic penalty $(p_S = 1)$ for $M = 4$. Lattice size is 600×600. a) $p_0 = 0$ $(T = 0)$, locally irreversible growth after 30 TMS. b) $p_0 = 0.3$ after 60 TMS and c) $p_0 = 0.5$ after 60 TMS.

We start the discussion with an maximum occupation number of $M = 4$ on a lattice of size 600×600. The results for three different values of p_0 are displayed in Fig. 9.5. Growth proceeds in the x-direction, see Fig. 9.5a. The first row $x = 0$ serves as the rim (primary nucleation) as described above. Because of the periodic boundary conditions growth starts simultaneously from the top $(x = 0)$ and from the bottom line $(x = 600)$. In Fig. 9.5a the temperature is zero $(p_0 = 0)$, hence locally irreversible growth takes place. The picture was made after 30 000 MCS (30 TMS). Figures 9.5b and 9.5c are obtained after 60 TMS with a value of $p_0 = 0.3$ and 0.5 respectively. In addition to the height $< m >$, two length scales can be identified from the pictures and are illustrated in Fig. 9.6

First, the correlation length ξ of the branching structure which corresponds to the characteristic distance between the fingers. For the case $T = 0$ (Fig. 9.5a) about six large growth animals. This corresponds to $\xi \simeq 100$ for this case. On the other hand there is a smaller characteristic length scale W which can be identified with the width of finger structures and strongly depends on temperature [8,9]. A simple approach to understand the temperature behavior of W was presented in [10]. It is based on the assumption that the width W of the fingers is directly related to the average path of diffusion of a chain until it gets trapped finally in the crystal phase.

The effect of a change in the order parameter is shown in Fig. 9.7, where $M = 2$ is used. For the case $T = 0$, Fig. 9.7a, densely branched morphologies

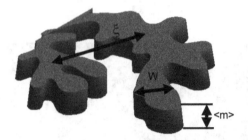

Fig. 9.6. Illustration of the relevant length scales of the growth patterns.

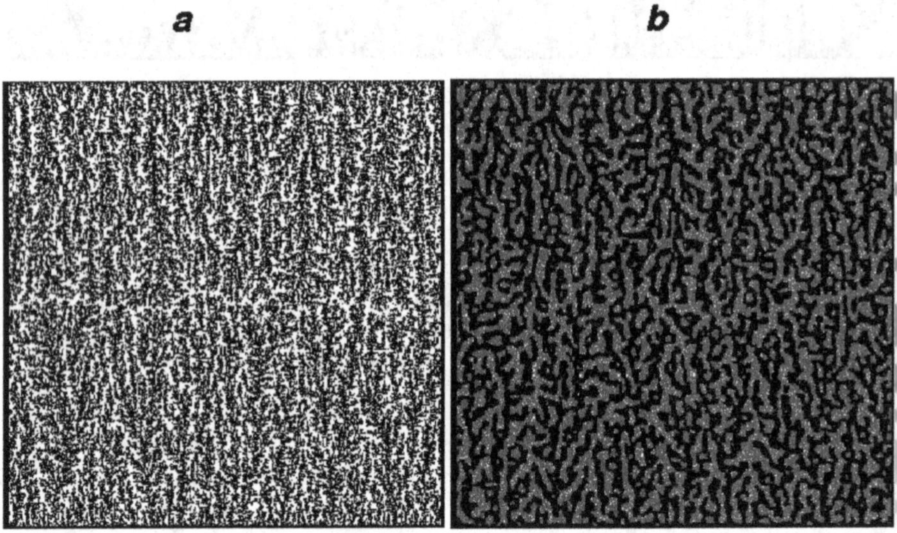

Fig. 9.7. Growth patterns obtained without entropic penalty for $M = 2$. a) $p_0 = 0$ and b) $p_0 = 0.2$

can be seen. Note the smaller correlation length ξ, i.e. the distance between neighboring branches. A qualitative change in the growth morphology emerges for higher temperatures. Fig. 9.7b displays the result after 30 TMS for $p_0 = 0.2$. Now loops and islands occur, indicating a matching of both length scales ξ and W. In order to obtain a measure of the correlation length ξ the one-dimensional structure function perpendicular to the growth front averaged over the x-range is used:

$$S(n) = \frac{1}{L_x} \sum_{x=0}^{L_x-1} \frac{1}{L_y} \sum_{y,y'=0}^{L_y-1} \exp\left\{2\pi i \frac{n}{L_y}(y - y')\right\} \sigma(x,y)\sigma(x,y') \tag{9.10}$$

where $\sigma(x,y)$ is unity if the lattice site (x,y) belongs to the crystal and zero otherwise.

The result is plotted in Fig. 9.8. For $T = 0$, as represented in Fig. 9.7a, the maximum is located at about 27 in units of the inverse overall length scale

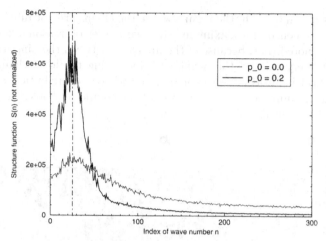

Fig. 9.8. Structure function $S(n)$ in the direction parallel to the growth front averaged over the lattice for the both cases of Fig. 9.7

L_y. Associating this value with the dominating length scale ξ of the patterns in y-direction we obtain a value of $\xi \simeq 22$ lattice units which fits well to the average period obtained directly from visual inspection of the picture. Note that this maximum is not shifted by temperature. Increasing the temperature only pronounces this characteristic length scale, see Fig. 9.7b. For $p_0 = 0.2$, most of the fractal-like details which are visible for $T = 0$ are lost. This indicates that the two length scales behave quite differently with temperature. While the correlation length ξ does not show a significant temperature dependence the width of the fingers W increases rapidly with temperature.

To conclude this section we have shown that the order parameter M controls the overall morphology of the growth patterns. The higher the surface density jump is, the larger is the correlation length ξ. For small M, matching of the strongly temperature depended lower scale W (width of fingers) with the correlation length ξ is obtained leading to multiply connected morphologies. Both trends have been observed in experiments.

9.4   Self-organized Crystal Thickness and Growth Velocity

After having considered the appearance of growth morphologies and their properties as a function of the area density jump m, we now turn to the full problem by considering a finite entropic penalty. Now, the average lamellar thickness $< m >$ can be much smaller than M and therefore also prone to further reorganization phenomena. We start by fixing the entropic penalty to an arbitrary value of $p_S = 0.6$, representing a finite effort for individual chains to increase their internal degree of order. Later, we will study the influence of different values for p_S. The maximum order parameter is taken as $M = 10$. A systematic variation for p_0 mimics temperature variation. The value of M is chosen in order to allow us to study both, the regime where the chains are not fully stretched as well as the influence of a maximum order parameter on reasonable time scales. It should be noted that, because of the interplay between the binding energy for chains in the crystal and the penalty for increasing the local chain order, the resulting effect is always due to combination of both contributions. One has to distinguish carefully between the nature of the competing forces and the mixed nature of the resulting effect [11].

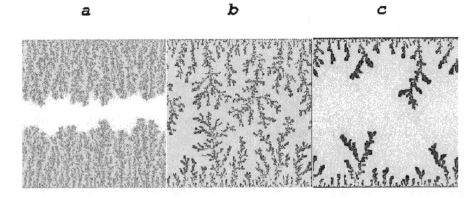

Fig. 9.9. Growth patterns obtained with an entropic penalty of $p_S = 0.6$ for different temperatures. Lattice size is 600×600 a) $T = 0.0$, b) $T = 0.72$ and c) $T = 0.95$.

Typical results of the growth morphologies are presented in Fig. 9.9 for $T = 0.0, 0.72$ and 0.95. The effect of a self-adjusting averaged order parameter $< m >$ can been seen already for the locally irreversible case $T = 0$ in Fig. 9.9a. Note that $< m >$ corresponds to thickness L (average stem length) in the crystal as used in the empirical (9.3). A distribution of stem lengths creates a fine structure of the patterns (roughness of the folding surface) indicated by different gray levels. The increase of temperature changes the morphology of the growth animals as well as the surface roughness and the average stem length.

Let us first consider the distribution of stem lengths and their dependence on temperature. The data were obtained after 30 TMS for lattice of size 200×200. In Fig. 9.10a the average stem length $< m >$ is plotted versus temperature T. As expected, the average stem length (thickness of crystalline lamellae) is an increasing function of temperature until the melting point is reached. In Fig. 9.10b the distribution of stem length is displayed for different temperatures. We can qualitatively distinguish three regions: For low temperatures fully stretched chains are highly suppressed (M is irrelevant). The increase of the stem length with temperature shows a convex behavior. Above $T \simeq 1$ the fully stretched chains dominate the ensemble, which results also in the imprisoning of non-stretched chains (for details, see the next section). In this region, the thickness of the lamellae still increases with temperature but now in a concave, i.e. gradually saturating manner. Eventually, for $T > 2$ the crystallites which were formed at the preset nucleus are rapidly disassembling which produces dynamic, cloud-like

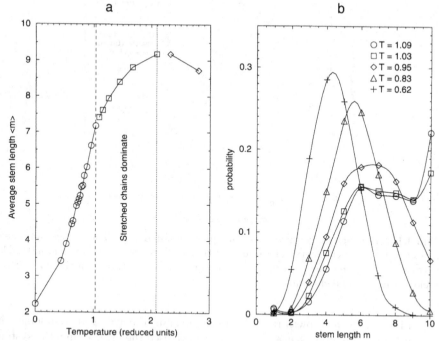

Fig. 9.10. a) The average stem length $< m >$ is plotted versus the temperature T taken in reduced units according to (9.9). Three regions of qualitative different behavior are marked with different symbols: ○ - un-stretched chains, □ - stretched chains are dominating and ◇ - no persisting growth structure. b) The distribution of stem lengths for different temperatures are displayed.

patterns. In this region, we do not have a persisting crystalline phase. These three regions are marked by using different symbols in Fig. 9.10a.

To test the empirical linearity of the crystallization line, see Fig. 9.2 the corresponding HL-representation is applied in Fig. 9.11. Here, the temperature is plotted against the inverse average stem length. Indeed, a linear fit can be obtained, using the region of un-stretched chains only.

A second parameter which has been investigated extensively is the crystal growth rate G. The following equation has been empirically tested and can be obtained from different models of the crystallization kinetics:

$$G = G_0 \exp(-A/\Delta) \ . \tag{9.11}$$

Combining this with the *empirical* equation for the crystallization line (9.1) one obtains

$$G = G_0' \exp(-L/L_0) \ , \tag{9.12}$$

with $L_0 = a/A$.

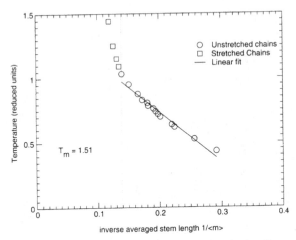

Fig. 9.11. Temperature vs. inverse stem length (HL-diagram). The vertical line separates the extended chain crystal phase from the non-extended chain crystal phase. The extrapolated melting temperature T_m is obtained as $T_m = 1.51$ from the data of the region of non-extended chains.

In Fig. 9.12a the growth rate G is plotted versus the averaged stem length in a semi-logarithmic representation. Here we have used the inverse time the system needed to display a maximum in the occupied area of the crystalline phase, see Fig. 9.12b. Because of relaxational processes, the crystal area *decreases* again after the density of not crystallized chains has decreased sufficiently, see the next section. Hence, the maximum of the occupied area serves as a measure for the time where relaxation starts to dominate crystallization. This yields the same behavior of G as taking the gradient of the crystal mass versus time for small times. The maximum method is easier to apply and can be extended to the region where stretched chains dominated, where a gradient region is difficult to obtain. Again, the relation can be well applied to our data.

To conclude this section, we have shown that a simple lattice model provides a self-organized lamellar thickness which follows the empirical relation in the temperature interval where the chains are not stretched (the maximum order parameter is not involved). Also the growth rate coincides with the empirical observation, i.e. an exponential decay with $< m >$ is obtained. This shows that these relations are very general for a growth process with internal reorganization and are therefore not sufficient to discriminate the particlar features of polymer crystallization processes.

9.5   Reorganization of the Polymer Crystal

As has been emphasized in the introduction, polymer lamellae containing unstretched chains are usually not in thermal equilibrium. In contrast to previous models [2,3] chains in the crystalline state are still mobile and can change

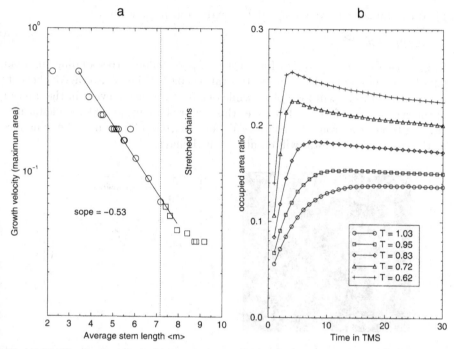

Fig. 9.12. To obtain a measure for the growth rate the maximum of the area occupied by the crystal phase has been used. a) Growth rate (inverse time for reaching the area maximum) is plotted versus the average stem length $< m >$. The dotted vertical line indicates the cross-over from non-extended to extended chain crystal growth. Within the region of non-extended chain crystals a simple exponential fit is applied. b) Area occupied by the growing crystal versus time for several temperatures.

their position as well as their internal degree of order. Therefore, relaxations towards thermodynamical equilibrium will take place. However, there is a hierarchy of metastable states which show relative persistence on increasingly long time scales so that true equilibrium is practically not reached. On the other hand, these metastable structures can be identified in experiments. It is just one of the intriguing features of quasi two-dimensional film crystallization to display several stages of meta-stability within the observable time window. Because the appearance of these structures is the consequence of the interplay of growth and relaxation processes controlled by the microscopic crystallization and diffusion events, agreement between the different phases in the simulation and in the experiment is a highly non-trivial feature.

First, we show the importance of the crystalline rim at the edge of the crystalline domains with respect to such reorganization processes. Let us consider that at the growth front typically three nearest neighbors (NN) are present (at edges or dislocations there are only two NN, new crystal units on a growth front have only one NN), while in the interior there are four NN. Using for simplicity an averaged order parameter $< m >$, the ratio of internal reorganization events

p_I to reorganization events at a flat growth front p_F is given by

$$p_I/p_F = p_0^{<m>} \ . \tag{9.13}$$

In the case of $< m >= 6$ and for a value of $p_0 = 0.35$ (corresponding to about 0.95 kT per bond of ordered units) this ratio is 0.0018, i.e. on average 540 events at the growth front will take place while one reorganization event in the interior of the crystal is performed. For edges this factor is about 300,000, while for new sites on top of the front the factor is already 160,000,000. Hence, the dynamics at the front of the crystal will dominate the behavior.

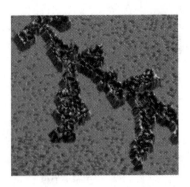 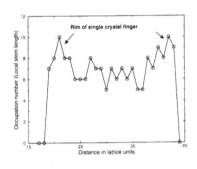

Fig. 9.13. Three dimensional representation of a growth animal for $M = 10$, $p_S = 0.6$ and $p_0 = 0.35$. The right part shows a height profile taken from an individual finger.

When the reservoir of free chains gets exhausted, therefore, relaxation processes at the crystal border will dominate the dynamics. This can be clearly seen from Fig. 9.13, where part of the growth pattern for $T = 0.95$ is displayed in a topographical representation. It can be clearly seen, that the stems near the crystal edges are higher compared to the enclosed inner regions. This effect is easily understood by considering our numerical example above. The outermost chains are in a kinetically favored position because of the lower number of nearest neighbor bonds. As a consequence, relaxation towards a higher ordered state of the chains takes place first at the edge of the crystal. This, in turn, restricts the mobility of the enclosed chains further. The resulting islands enclosed by rims of more extended polymers are therefore *more stable* compared to the situation directly after their formation by the growth mode. Hence, such patterns should be observable in experiments too and represent a second morphological phase after the formation of single polymer crystals. Indeed, the appearance of higher rims at the edges of the fingers is clearly shown in experiments, see Fig. 8.7c and d in Chap. 8.

The special role of crystal edges for the relaxational processes becomes even more pronounced if we switch to smaller p_S (high ordering penalty). In the experimental system this would correspond to stronger adsorption. In Fig. 9.14 we show a system for $M = 4$, $p_S = 0.1$ and $p_0 = 0.2$ at different times. Here, we can observe the formation of holes within the growth animals. After 100 TMS only

a   **b**   **c**

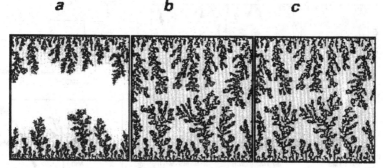

Fig. 9.14. Structures formed for $p_S = 0.1$ and $p_0 = 0.2$ at different times. a) 5 TMS, b) 30 TMS and c) 100 TMS.

the rims remain which are, however, in a locally highly ordered state. Further relaxation takes place on a very long time scale, since the binding energy in the self-assembled rim-structure is high. This morphology has also been observed experimentally, see Fig. 8.6B) in Chap. 8.

At this point one has to note that for crystallization from strongly adsorbed thin films the parameter p_S also depends on temperature since desorption of chains occurs. Therefore, a shift in both p_S and p_0 towards smaller values should correspond to a decrease of temperature in the real system. One would expect a next step in the relaxation of the crystalline phase where now the rims form again more compact objects driven by the line tension of the patterns. This, however, requires extremely long time scales – or an increase of temperature.

9.6   Annealing and Morphogenesis – The "Crystalline Liquid State"

We expect that the relaxation effects under isothermal conditions are even more pronounced if temperature is increased. In contrast to crystals in equilibrium, a temperature increase (annealing) not only causes melting but also faster access to better ordered states due to higher mobility. To investigate this behavior we applied a temperature jump after the initial growth process [12]

In Fig. 9.15a, the growth of a system with $M = 10$, $p_S = 0.6$ and $T = 0.95$ is shown for three different times, see also Figs. 9.9 and 9.13. After 100 TMS a temperature jump is applied, as shown in Fig. 9.15c. The evolution of the crystal morphology (morphogenesis) is shown for different times in Fig. 9.15b. In Fig. 9.15d we display the internal energy (negative number of interacting units per lattice place) as a function of time after the temperature jump has been applied.

Immediately after the temperature jump, a strong increase of internal energy U is observed in Fig. 9.15d which corresponds to detachment of chains at the edge of the crystal. However, after a rather short period, U displays a maximum (Max I) and than turns to decrease. The second picture in Fig. 9.15b at $t = 0.7$

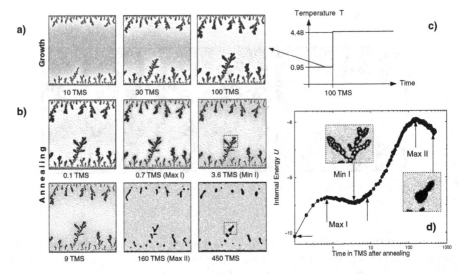

Fig. 9.15. a) Three stages during isothermal growth for $M = 10$, $p_S = 0.6$ and $T = 0.95$ ($p_0 = 0.35$). This corresponds to the parameters used for Fig. 9.13. b) Morphogenesis for different times after the temperature jump displayed in c). d) Internal energy (negative number of interacting units per lattice place) as a function of time after the annealing step. The arrows indicate the times where the morphology snap-shots of b) have been taken. Two characteristic stages of morphogenesis are magnified.

TMS shows the reason for this behavior: the temperature jump also accelerated processes in the interior of the finger structures, which locally improved the internal order (local increase of the height of the lamella) leading, in turn, to the formation of holes. The edges of the crystal which were already in a state of higher order prove to be more stable and act as nucleation sites for the reorganization of molecules in the interior of the finger structures.

After about 3.6 TMS the process of improving chain order can no longer compensate for the increased chain detachment at the periphery of the crystal and the internal energy starts to increase again. The third picture in Fig. 9.15b displays the state of the morphogenesis where the internal energy shows a minimum (Min I). At later times, the fingers are broken up and the structure starts to decompose as shown by the fourth picture at 9 TMS after annealing. At even longer times (note the logarithmic time scale in Fig. 9.15d), a third process sets in which stops the structural decay again. Now, parts of the remaining crystal structure take the advantage of the locally higher concentration of chains, a consequence of the transformation of the original morphology, and start to grow in size by forming compacted patches, thereby decreasing the total length of the perimeter of crystalline structures (energetically unfavorable sites). The transition to more compact structures is reflected by a second maximum (Max II) in the internal energy after about 160 TMS, fifth picture in Fig. 9.15b. Several dynamical features characterize this transition to more compact patches (droplets and are responsible for a renewed decrease of the internal energy: Incorporation of additional free chains, diffusion via shape fluctuations and, coalescence. Note that our initial nuclei (first row at the bottom and at the top of the simulation box) now also favor growth of compact line-shaped crystals.

The droplets in Fig. 9.15b resemble a liquid-like structure on large scales. However, on small scales high crystalline order is realized. Almost all molecules in the droplet phase are fully stretched out. Because of the common history (same nucleation site) the droplets are also in crystal registry, i.e. display long-range crystalline order even beyond their own size. Can we apply terms established by equilibrium thermodynamics such as *solid* and *liquid* in polymer crystallization? The structures observed should be rather denoted as a *"Crystalline Liquid"*.

9.7   Heating with a Constant Rate and the Role of Morphogenesis

The capability of non-equilibrium crystals to obtain different morphological stages during annealing leads to a distortion of thermodynamic response functions such as the heat transfer under heating with a constant rate as used in DSC (Differential Scanning Calorimetry). DSC is a standard method for investigation of polymer crystallization. More details can be found in Chap. 14 and 16.

To investigate the behavior at constant heating rates, we used the same system as in the previous section, see Fig. 9.15b after 100 TMS. Then, we applied tiny temperature jumps of 0.01, repeated after a period of Δt. The values used for Δt are 0.1, 0.5 and 5.0 TMS respectively. This results in heating rates of 0.1, 0.02 and 0.002 ϵ/k/TMS) respectively. Most remarkably is the *initial decrease* of the internal energy which is most pronounced for slow heating rates ($\Delta t = 5$ TMS). This corresponds to an extended period of relaxation of the non-equilibrium structure per step in temperature and can be associated with the minimum in the isothermal relaxation as shown in Fig. 9.15d. For the lowest heating rate also the appearance and subsequent destruction of a transient intermediate morphology (first maximum in Fig. 9.15d) is observable as an additional peak of C_V (indicated by the arrow in Fig. 9.16). Without consideration of the morphogenesis as shown in the previous section the first peak in the DSC curve may erroneously be taken as a sign of "melting-recrystallization". Increasing the heating rate reduces the possibility for relaxations. Thus, the minimum of U becomes shallower and is shifted to higher temperatures. Note that in experiments the heat capacity is not negative because it is superposed by many other contributions which are not directly related to the internal energy of the crystal.

Independent of the heating rate, the internal energy must reach a common value at high temperatures when all polymers are in the liquid phase. Convergence is well achieved for the two slower heating rates. However, complete destruction of the crystalline phase, which leads to a zero-slope of the internal energy, is only reached at longer times outside the time frame of the simulation displayed here. Of course, the heating rates and relaxation times are still much faster compared to experiments. Assuming a diffusion constant of about $10^{-10} m^2/s$ for the chains the elementary time step in our simulation corresponds

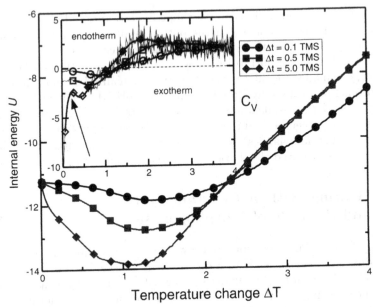

Fig. 9.16. Internal energy as a function of annealing temperature for heating at a constant rate. The starting conditions after 100 TMS are exactly the same as for the one step annealing shown in Fig. 9.15. At the beginning of each time interval consisting of Δt TMS the temperature is increased by 0.01 units. The internal energy is taken at the end of Δt and plotted vs. the total difference to the temperature at which the structure has been grown initially. The inset shows the formally derived heat capacity C_V which takes negative values at the beginning as a consequence of relaxation processes. For the lowest heating rate an additional maximum occurs as indicated by the arrow.

to one μs. The relaxation phenomena take than place on a time scale of ms. To observe such phenomena under high heating rates in experiments a similar time resolution would be necessary. In addition, the magnitude of the first peak associated with reorganization increases with decreasing heating rates but may be difficult to observe at high heating rates.

One of the main conclusion of this section can be obtained in the following way: Let us suppose for a moment that in an experiment integration of U is performed only over the area under the second maximum, which disregards the relaxation precesses leading to the negative contribution of the heat capacity. Then, the apparent heat absorption during melting would be too high by an amount proportional to the neglected exothermic region. In addition, an apparent shift in melting temperature may be deduced. Most importantly, relating such an apparent melting temperature to the state of the structure immediately after growth is clearly a mistake and physical interpretation of a relation between crystallization and melting temperature can be completely wrong. In simple words: The crystalline structure which melts is *not* identical with the originally grown crystal but represents the last stage of the morphogenetic pathway.

9.8   Conclusions

Describing the features of non-equilibrium crystals requires a simple and numerically tractable model which allows us to relate the macroscopic observations to microscopic phenomena in order to predict the behavior of various experimental situations. This seems to be possible for the crystallization of quasi-two dimensional adsorbed monolayers. The main assumption here is to consider the polymer crystal as made of units (chains) having various degrees of internal order. During the growth process at a given under-cooling a temporary stable lamella is formed which represents the optimal degree of internal chain order which is possible under the condition of steady growth. At the same time a growth morphology is selected which can be characterized by two length scales: the thickness of compact parts (fingers) and the correlation length between these fingers.

Our model allows a simple understanding of morphological changes which take place spontaneously in time or are forced by annealing processes. Because of the higher mobility of the molecules there, the borders of the lamellae play an important role. Relaxation effects yield higher order in this boundary region which self-confines the lamellae and gives rise to the formation of holes within the enclosed regions at later times or at elevated temperatures. Also other morphological phases can be explained on the basis diffusion, relaxation and growth processes. An example is the formation of droplet phases which contains highly ordered molecules but shows liquid-like features on larger length scales.

Despite of the fact that our model is restricted to polymer monolayers, several general remarks can be made:

- How can we define a unique state of polymer crystal without considering the history and morphological features of the system? These properties might not be easy observable in crystallizing melts, but, can be nevertheless as important as in thin films. Thus the classical HL-represention of the state of polymer crystal is insufficient and must be extended.
- What is the generalization of phases and phase transformation known from ETD to system far from equilibrium? We have seen that annealing can yield an increase of order and melting proceeds via a morphogenesis. The originally grown lamellae will not melt but only their descendants. This calls in serious question not only the classical one-step crystallization models but also the interpretation of the melting line.

The model which we have outlined in this chapter together with the experimental observations presented in several other chapters of this book[4] makes clear that a new and complete theory of polymer crystallization must involve many

---

[4] See for instance Chaps. 8, 6, 7.

general aspects of non-equilibrium processes. It also shows how computer simulations, i.e. simple algorithms, are useful in particular for a basic understanding of complex phenomena.

References

1. A. Keller, Phil. Mag. **2**, 1171 (1957)
2. E. Fischer, Kolloid-Zeitschrift und Zeitschrift für Polymere **1-2**, 458 (1968)
3. K.A. Armistead, G. Goldbeck-Wood, Advances in Polymer Science **100**, 219 (1992)
4. J.D. Hoffman, L.J. Frolen, G.S. Ross, J.I. Lauritzen, J. Res. Nat. Bur. Stand. A: Phys and Chem. **79A**, 671 (1975)
5. D.M. Sadler, Nature **326**, 174 (1987)
6. J.-U. Sommer and G. Reiter, J. Chem. Phys. **112**, 4384 (2000)
7. P.D. Olmsted, W.C.K. Poon, T.C.B. McLeish, N.J. Terrill, A.J. Ryan, Phys. Rev. Lett. **81**, 373 (1998)
8. T. Irisawa, M. Uwaha, and Y. Saito, Fractals **4**, 251 (1996)
9. T. Irisawa, M. Uwaha, and Y. Saito, Europhys. Lett. **30**, 139 (1995)
10. G. Reiter and J.-U. Sommer, Phys. Rev. Lett. **80**, 3771 (1998)
11. J. P. K. Doye and D. Frenkel, J. Chem. Phys. **110**, 7073 (1999)
12. J.-U. Sommer and G. Reiter, Europhys. Lett. **56**, 755 (2001)

10   Structure Formation and Chain-Folding in Supercooled Polymer Melts. Some Ideas from MD Simulations with a Coarse-Grained Model

Hendrik Meyer

Institut Charles Sadron, CNRS,
6, rue Boussingault, 67083 Strasbourg, France
hmeyer@ics.u-strasbg.fr

Abstract. This chapter describes molecular dynamics (MD) simulations with a coarse-grained (CG) model. Starting from the disordered melt, the formation of ordered structures is observed which reproduce essential features of polymer crystallization. Short chains form extended chain crystals. As a function of chain length, homogeneous crystallization and melting temperatures show a good agreement with droplet experiments on alkanes. For longer chains, chain-folded lamellar-like structures are formed. For isothermal relaxation runs, well defined crystallization and melting lines as a function of inverse stem length are recovered. The model consists of excluded volume interactions, connectivity and an angle bending potential reflecting the torsional states of the underlying atomistic backbone. The simulations demonstrate that chain stiffness alone without an attractive inter-particle potential is a sufficient driving force for the formation of chain-folded structures. The growth front of these structures extends over several chain diameters and hairpins appear to contribute by an easy growth mode in this model.

10.1   Introduction

In this chapter, structure formation in dense melts is discussed by doing computer experiments with a realistic coarse-grained polymer model [1,2]. Studying polymer crystallization from the melt by chemically realistic computer simulations is a difficult task since large systems need to be simulated over a long time. Consequently, early MD simulation studies used e.g. an isolated chain to analyze chain folding during the collapse into a crystalline globule [3–5], the properties of crystals were examined using predefined clusters [6], or chain-folding was examined under specific boundary conditions like a predefined growth front [7]. More recently, the structure formation from a disordered state has also been studied by MD for a melt of short chains [8] and for ensembles of chains in solution [9,10].

One possibility to access larger length and time scales is to use so-called coarse-grained (CG) models and Monte-Carlo schemes to accelerate the simulation by nonphysical moves [11,12]. The present simulation study is motivated by the debate concerning the early stage in polymer crystallization and the possibility of intermediate phases [13]. We are interested in the dynamics of structure formation from the amorphous melt and in conformational details of each chain.

178   Hendrik Meyer

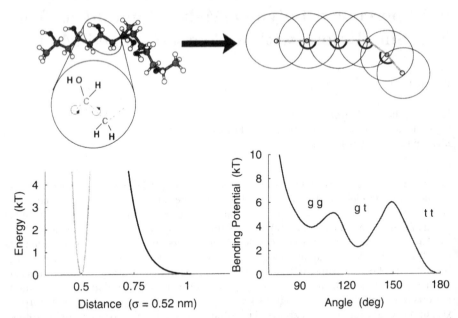

Fig. 10.1. Illustration of the mapping of the atomistic chain of poly(vinyl alcohol) onto the coarse-grained (CG) model. One monomer is represented by one sphere. The spheres are connected by harmonic springs and an angle-bending potential. On the bottom from left to right: the bond potential, the excluded volume interaction, and the angle bending potential between successive CG beads. A CG-angle corresponds to two successive torsions of the atomistic backbone; "gg", "gt", and "tt" stand for gauche-gauche, gauche-trans, and trans-trans.

So we use classical molecular dynamics simulations with a realistic coarse-grained model. Such a model was recently introduced which makes it possible to access large systems of chains long enough to examine the formation of chain-folded lamellar-like structures. The approximations of the model give also hints which parameters are essential for explaining these structures. The properties of this model are reviewed in the following.

10.1.1 The Model CG-PVA: Excluded Volume, Connectivity and Rotational Isomeric States

The model used in this chapter is a bead–spring model with an additional angle bending potential [1,2]. Many simulation studies using chemically realistic polymer models already use the "united-atom" approximation: hydrogen atoms are absorbed with their neighboring carbon into one so-called united atom. We go one step further and map a whole monomer onto one bead as shown in Fig. 10.1. The interaction parameters have initially been determined by a systematic coarse-graining procedure based on all-atom simulation data of poly(vinyl alcohol) (PVA) [14]. However, approximations make it a rather generic physical model which fits also very well to polyethylene (PE). The effective excluded-volume interaction and the bond length have been determined such that the radial distribution function of the monomers in the CG model coincides with the corresponding distribution of the atomistic simulation data. The angle bending potential was determined by a Boltzmann inversion of the angle distribution in the atomistic melt. Since one CG monomer contains two carbon atoms of the backbone, one CG angle corresponds to two successive torsions on the atomistic level, see Fig. 10.1. Due to this approximation, one looses the zig-zag structure of PE or PVA chains. However, conformations are considered at the level of the rotational isomeric states.

10.1.2   Simulation Details

Classical molecular dynamics simulations are combined with a Langevin thermostat to maintain constant temperature [15]. Constant pressure is maintained by Berendsen's manostat [16]. In this chapter simulation units are used: length scales are given in $\sigma = 0.52$ nm, time is reported in τ ($1\tau = 200$ integration steps). The mapping onto the coarse-grained model gives a precise length scale (through bond length and density). The time scale, however, is approximate since the elimination of degrees of freedom and the stochastic thermostat remove local metastable states and thus accelerate the dynamics. A comparison with the diffusion constant for alkanes gives the estimate that τ is of the order of picoseconds. Note that the nonbonded potential obtained after the CG procedure is softer than a "standard" Lennard-Jones 6-12 potential. This and the overlap of the monomers make that the units σ and τ (and thus the density σ^{-3}) are not directly comparable with standard bead-spring model simulations [17], though they are in the same order of magnitude. Temperatures are given in dimensionless units with respect to the melt state where the model was parameterized. $T = 1.0$ corresponds to 550 K for poly(vinyl alcohol). When comparing with poly(ethylene), $T = 1.0$ should be rescaled to 440 K (see Fig. 10.10 in Sect. 10.3).

Start configurations are generated by random walks according to the angular distribution in the melt. The amorphous structure is equilibrated at $T = 1.0$. The end-to-end-vector time-autocorrelation function was calculated to determine a typical relaxation time of the chains. This time is quite short for the shorter chains (260 τ for $N = 20$), it starts to matter for chains of length $N = 100$ (10000 τ). With the Langevin thermostat, temperature relaxes almost immediately ($<$ 2τ) to a new average value.

10.2   Structure Formation in a Melt of Short Chains

We start by describing a short chain system to introduce a typical simulation protocol and different quantities measurable in computer simulations. The short chains of $N = 10$ monomers behave nicely in the sense that they form extended chain crystals. The first thing we want to know is the crystallization and melting

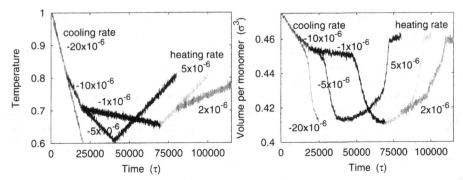

Fig. 10.2. Crystallization of 960 chains of length $N = 10$: The simulation is started with an amorphous melt at temperature $T = 1.0$. It is cooled down at different rates, slowing down at $T = 0.8$ and again at $T = 0.7$. Left: temperature protocol, right: volume per monomer as a function of simulation time (instantaneous values sampled every 50 τ).

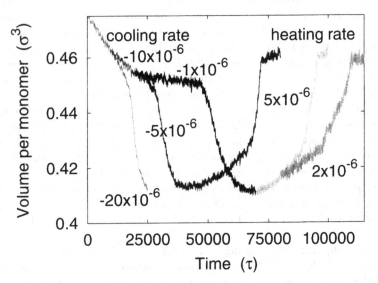

Fig. 10.3. Crystallization of 960 chains of length $N = 10$: Volume per monomer as a function of temperature at different cooling and heating rates.

temperature. They are most easily determined by continuous cooling and heating experiments. Since the crystal phase has higher density, the volume is a good indicator of crystallization [18]. Figure 10.2 shows the temperature protocol and the volume per monomer as a function of time. One observes a slow decrease of volume with temperature. At the onset of crystallization, the volume drops drastically. Figure 10.3 shows the volume as a function of temperature in these cooling and heating cycles. They exhibit a huge hysteresis (see next paragraph). Figure 10.4 shows snapshots of the initial melt and the final crystal obtained

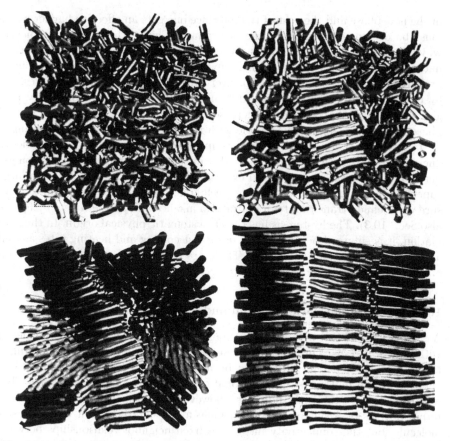

Fig. 10.4. Crystallization of 960 chains of length $N = 10$. (a) top left: initial configuration of the melt at $T = 0.9$, (b) top right: the first nucleus at $T = 0.67$ during cooling at the rate $10^{-6}\tau^{-1}$, (c) bottom left: final configuration at $T = 0.65$ which consists of several ordered domains, (d) bottom right: perfect crystal set up at $T = 0.60$ with three layers of 16x18 chains (periodic boundary conditions apply for all Fig.s).

at the slowest cooling rate $10^{-6}\tau^{-1}$. Note that by continuous cooling from the melt, we usually get polycrystalline samples. The higher the cooling rate, the smaller are the different crystal domains. For well defined growth, one must work at higher T with heterogeneous nuclei. The perfect configuration in Fig. 10.4d was set up by hand.

10.2.1   Homogeneous Nucleation

Suppose T_m^0 being the temperature at which disordered and ordered phases are in equilibrium. When cooling the disordered melt below T_m^0 the ordered phase would have a lower free energy. To initiate the phase transformation, interfaces must be introduced and this costs energy. Balancing the gain in volume energy of the new phase and the cost of the interface one gets an effective barrier which must be overcome to nucleate the new phase. This is the qualitative picture why a hysteresis is observed in first order phase transformations. The barrier to nucleation decreases with increasing supercooling [19] making nucleation more probable at lower temperatures. For ever slower cooling and heating rates, the hysteresis should become smaller and smaller. However, the connectivity may give rise to another systematic increase of the hysteresis because the crystal phase may remain highly metastable with respect to the melt during heating.

In practice, there are often many impurities which serve as heterogeneous nuclei so that little supercooling is needed in condensation or crystallization experiments. However, under careful experimental conditions, samples with large supercooling can be prepared. Droplet techniques for example were used to measure well defined transition temperatures in alkanes and poly(ethylene) [20] (see also Sect. 10.3). The hysteresis has thus a systematic physical origin. In the simulation, it is enlarged because of the quite fast cooling and heating rates as well as due to the boundary conditions: There are no free surfaces where nucleation of the new phase can be easily initiated.

10.2.2 Positional, Orientational and Conformational Order

A crystal is associated with long-range positional ordering which can be achieved with hard sphere-like entities such as metal atoms or small molecules. For anisotropic molecules, partial order may be achieved by ordering the relative orientation of the molecules. Orientational order gives rise to the rich phase behavior of liquid crystals. In polymers, however, the connectivity introduces internal degrees of freedom. The chain molecule is not rigid as a liquid crystal molecule, but the connectivity introduces fixed neighbor relationships between certain monomers. To fit into some (partially) crystalline structure, each chain has to undergo a conformational ordering prior to or during the crystallization process. In the following, we quantify these kinds of order for the four configurations shown in Fig. 10.4.

The positional order is characterized in real space by the radial distribution function $g(r)$ and in reciprocal space by the static structure factor

$$S(q) = \left\langle \frac{1}{m} \sum_{j=1}^{m} \exp(i\mathbf{R}_j \mathbf{q}) \right\rangle_{|\mathbf{q}|=q \pm dq} \tag{10.1}$$

as shown in Fig.s 10.5 and 10.6. In the melt, the typical liquid structure is obtained. $g(r)$ shows a peak of the second neighbor shell, and uniform density beyond. For the crystal, marked structuring is seen on $g(r)$. The structure factor reveals the peaks corresponding to a hexagonal lattice. These peaks are quite sharp for the perfect crystal and broadened for the polycrystalline sample (Fig. 10.4c). At small q, the structure factor of the perfect crystal shows peaks according to the smectic layering.

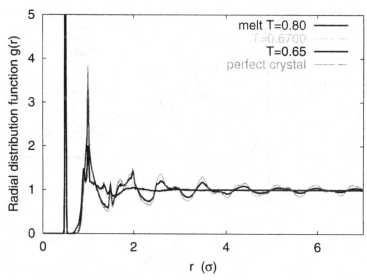

Fig. 10.5. Characterization of positional order: radial distribution function $g(r)$. The peak at $r = 0.5$ corresponds to the nearest neighbors along each chain. In the melt, no structure is observed beyond the second neighbor shell ($r > 2\sigma$). At $T = 0.67$ (Fig. 10.4b), where a small ordered nucleus is formed, $g(r)$ is still almost unchanged from the liquid structure. In the crystal at $T = 0.65$, long-range oscillations occur which are still more pronounced for the perfect crystal.

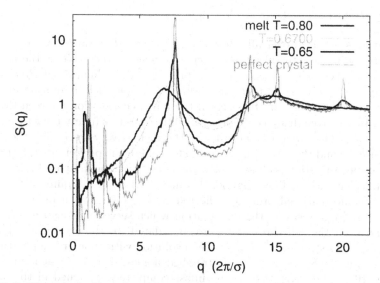

Fig. 10.6. Characterization of positional order in reciprocal space: the structure factor $S(q)$ of the crystal exhibits Bragg peaks of the hexagonal lattice. For the perfect crystal, the peaks are sharper and one obtains also peaks at small q corresponding to the layer-distance. (The box length corresponds to $q_L = 2\pi/L = 0.39$.)

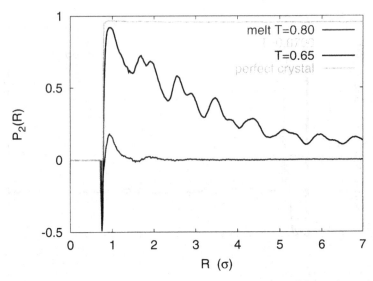

Fig. 10.7. Characterization of orientational order: Average 2nd Legendre polynomial of bond vectors at distance R. Intra-chain correlations are excluded in this figure.

The orientational order is described by the following orientational distribution function

$$P_2(R) = \langle 1.5(\boldsymbol{U}_i \cdot \boldsymbol{U}_j)^2 - 0.5 \rangle_{R=|\boldsymbol{R}_i - \boldsymbol{R}_j|} \qquad (10.2)$$

The average is taken over all pairs of bond vectors $\boldsymbol{U}_i$ and $\boldsymbol{U}_j$ whose distance $\boldsymbol{R}_i - \boldsymbol{R}_j$ is in the interval $[R, R + \Delta r]$ (the position vector $\boldsymbol{R}_i$ is the midpoint of the bond vector $\boldsymbol{U}_i$, this leads to a pronounced modulation of $P_2(R)$). This 2nd Legendre polynomial takes the value $+1$ for parallel orientation and -0.5 for orthogonal orientation. It is zero if there is no correlation between the bond vectors in the given distance. Figure 10.7 illustrates this function for the systems of Fig. 10.4. The orientational order in the melt is very restricted, there is no signature beyond the second neighbor distance. For the perfect crystal, the bond orientation correlation is almost 1. For the polycrystalline sample, this high value is only reached at the first neighbor distance, it then decays quite fast.

The conformational order is reflected by the distribution of bond angles. At $T = 1.0$, one recovers the distribution which served to construct the angle bending potential, all three states are populated. In the crystal, most angles are completely stretched. To follow the evolution with time or temperature, we trace in Fig. 10.8 the fraction of stretched angles, $\alpha > 150°$, denominated tt in Fig. 10.1. The short chains are almost completely stretched in the ordered state and almost no difference would be seen between the angle distributions of the polycrystalline and the perfect crystal configuration. A few defects persist in the ordered state which are mainly localized at the chain ends. This will be completely different for long chains which are folded in the ordered state.

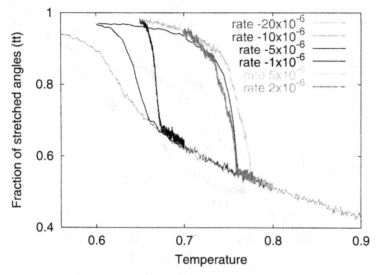

Fig. 10.8. Characterization of conformational order: fraction of stretched angles (tt-state, $\alpha > 150°$, compare Fig. 10.1) as a function of temperature. The behavior is analogous to the volume shown in Fig. 10.3.

10.3 Chain-Length Dependence of Crystallization and Melting Temperatures

In Sect. 10.2.1 the origin of a hysteresis between (homogeneous) crystallization temperature T_x and melting temperature T_m has already been discussed. We are now interested in the variation of this hysteresis with chain length. For short chains, the crystallization and melting temperatures vary considerably. This is well known from alkanes and for T_m a quantitative description was already given in the early days of polymer theory [21]. For longer chains, the values saturate and converge to some value characteristic of high molecular weight polymers. This behavior is well reproduced by the present model.

Figure 10.9 shows the volume per monomer as a function of temperature for different chain lengths. One sees the hysteresis which is shifted to higher temperatures with increasing chain length. If $N \geq 20$, the effect is less pronounced and the drop in volume is much smaller. This indicates that crystallization is incomplete. As a matter of fact, all cycles shown in Fig. 10.9 were obtained with the same cooling rate $(-5 \times 10^{-6} \tau^{-1})$ and a faster heating rate $(10^{-5} \tau^{-1})$. Longer chains need more time to crystallize and the ordering is thus less complete. But the order is also qualitatively different: longer chains crystallize in folded conformations.

In Fig. 10.10, the dependence of T_x and T_m as a function of chain length is shown [22]. The simulations discussed here are performed with periodic boundary conditions. There is no perturbation for initiating nucleation different than the decrease of temperature. We thus reproduce the situation of homogeneous nucle-

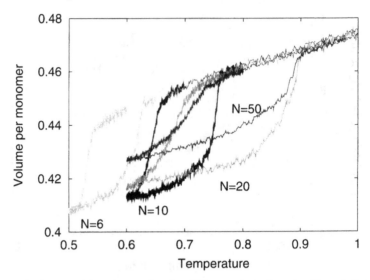

Fig. 10.9. Cooling and heating cycles for different chain lengths (from left to right: N=6, 10, 20, 50). For $N = 50$, little difference is seen compared to $N = 20$. At this chain length, folding sets in.

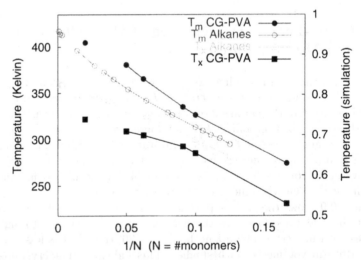

Fig. 10.10. Chain-length dependence of melting temperature T_m and homogeneous nucleation temperatures T_x. Simulation data are plotted with filled symbols, simulation units on the right scale. Open symbols and the left scale correspond to experimental values of droplet crystallization with alkanes taken from Kraak et al. [20]. The temperature axis is scaled to $T = 1.0 \equiv 440$ K; the lines are just a guide to the eye. The simulation shows a larger difference between T_m and T_x which can be explained by finite size effects and large cooling and heating rates. However, the slope as well as the widening of the hysteresis as a function of chain length are in good qualitative agreement with the experimental data.

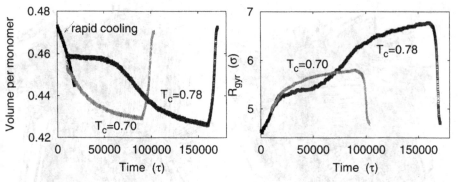

Fig. 10.11. $N = 100$: Evolution of the volume per monomer and of the gyration radius. A melt system is cooled rapidly from $T = 1.0$ at the rate $-2 \times 10^{-5}\tau^{-1}$. From $T = 0.80$ the system is quenched to $T_c = 0.78$ and $T_c = 0.70$ from where isothermal relaxation is started. At the deeper quench to $T_c = 0.70$, structure formation starts immediately, the volume and R_g then saturate rapidly. At $T_c = 0.78$, there is first an induction time before the volume goes down. The final melting at rate $2 \times 10^{-5}\tau^{-1}$ is also shown (starting at $t = 90000$ for $T_c = 0.70$ and $t = 170000$ for $T_c = 0.78$).

ation which can be measured experimentally in droplet experiments. Figure 10.10 also contains data from such droplet experiments with alkanes taken from [20]. The simulation results agree very well with the experimental findings on alkanes. The only adjustable parameter in the figure is the prefactor in the temperature scale. This fixes the temperature scale of the simulation to 440 ± 10K.

10.4   Chain-Folding in Isothermal Relaxation of Long Chains

In this section, the behavior of longer chains forming chain-folded structures is now discussed. We also consider a different type of temperature protocol: the preceding sections discussed structure formation during continuous cooling at a constant rate. The results in this section are obtained by isothermal relaxation after a sudden quench from the melt. A detailed discussion of the case $N = 100$ was recently given in [2]. In the first subsection, the most important results are illustrated by data on a larger system of 1000 chains of length $N = 100$ which confirm the results previously obtained on a smaller system. Then some results of ongoing work on highly entangled chains of length $N = 400$ are discussed.

10.4.1   Slightly Entangled Chains ($N = 100$)

Figure 10.11 shows the evolution of the volume per monomer and of the gyration radius at two different relaxation temperatures. At the higher temperature, first an induction period is observed. It takes some time until a nucleus is formed which then grows further. When the volume starts to drop, R_g starts to increase considerably. At the lower temperature, the structure formation starts

Fig. 10.12. Snapshots after isothermal relaxation of chains $N = 100$: $T_c = 0.70$ (left) and $T_c = 0.78$ (right); cut through the simulation box of 1000 chains, periodic boundary conditions apply laterally. At the lower temperature, the stems are shorter and the crystal domains are smaller.

immediately. However, the volume as well as R_g saturate rapidly. The values in Fig. 10.11 are not completely saturated, yet. Further relaxation, however, would lead to little improvement. The structure will mainly relax locally and thus improve the order in the interfaces between the domains.

Snapshots of the final configurations after these isothermal relaxation runs are shown in Fig. 10.12. One sees that at $T_c = 0.70$ the stems are much shorter than at $T_c = 0.78$. The difference in R_g reflects this difference in the average conformation of each chain. Also laterally, the ordered domains are much larger at the higher relaxation temperature. This can be seen on the structure factor in Fig. 10.13 where the peaks at the higher temperature are sharper than at $T_c = 0.70$. Figure 10.14 shows the time evolution of some peaks of the structure factor by plotting the average value obtained over corresponding q-intervals. The small-q region corresponds to SAXS experiments, the two Bragg peaks would be measured in WAXS experiments. Actually, measurement of the onset of the scattering intensity for the different regions triggered the present debate on the polymer crystallization process. In some cases, the SAXS intensity is seen to appear before the WAXS intensity [23]. This indicates that some large scale structure appears before the local crystal structure. However, the delay may also be due to experimental difficulties in resolving very low crystallinities [24].

Comparing the two temperatures in Fig. 10.14, a behavior similar to the volume or R_g is seen. The intensity grows immediately at $T_c = 0.70$ and only after some induction time at $T_c = 0.78$. On the linear scale, no significant delay of the onset of the small angle and the wide angle signal can be observed as stated also in [2]. For the higher temperature, however, a delay of the Bragg-peak signal can be seen on the logarithmic scale on the bottom part of Fig. 10.14. However, at small q, the statistics is intrinsically poor because of the limitation

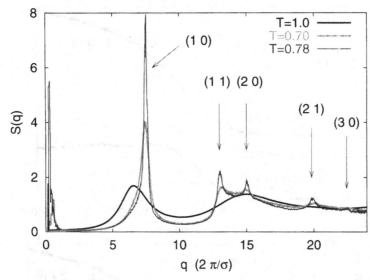

Fig. 10.13. $N = 100$: Structure factor $S(q)$ of the initial melt ($T = 1.0$) and of the final configurations after isothermal relaxation at $T_c = 0.70$ and $T_c = 0.78$ shown in Fig. 10.12. The peaks of the hexagonal lattice are labeled. They are sharper at the higher crystallization temperature.

of the box size (the smallest possible scattering vector is $q_L \approx 0.18$). Further studies are thus needed to clarify a possible delay and its origin. Nevertheless, it is clear that the small$-q$ peak grows much faster in the beginning than the Bragg peaks. This evolution of the structure factor suggests the presence of two regimes: the formation of a larger structure preceeds the ordering on a local scale and the long-range ordering of the local structure evolves only slowly. Once the large-scale structure is formed (the small-q peak saturates) further densification is observed and the Bragg peaks still increase.

In [2], an average stem length was determined for a range of relaxation temperatures. This yields a crystallization line which is reproduced in Fig. 10.15. Each configuration is then heated to determine its melting point. This gives the melting line. A fast heating rate was chosen such that no further thickening occurred [18].

The melting line in Fig. 10.15 is an upper limit of the melting temperatures of the given stem length. This and the high heating rate may explain that the lines do not cross. It may surprise that we obtain such well defined lines with small systems of still quite short chains. The chains of length $N = 100$ are only weakly entangled. We mentioned above that the relaxation is not completely finished, the configurations obtained may thus be interpreted as the result of the primary crystallization. On the other hand, because of the small system size, the crystal domains rapidly grow into other domains or even into its periodic image. This rather corresponds to the relaxation of secondary crystallization. Thus, further work is necessary to make this clearer.

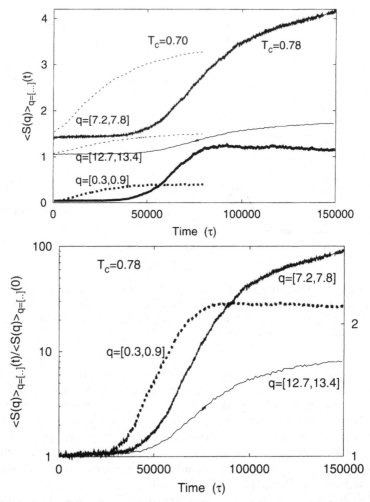

Fig. 10.14. $N = 100$: Evolution of different regions of the structure factor $S(q)$ during isothermal relaxation at $T_c = 0.70$ (dashed) and $T_c = 0.78$ (solid). The top figure has a linear scale, the bottom figure shows only the data of $T_c = 0.78$ normalized by the value at $t = 0$ on a logarithmic scale (small-q on the left, Bragg-peaks on the right scale). The interval $q = [0.3, 0.9]$ corresponds to the small angle scattering peak, the interval $q = [7.2, 7.8]$ corresponds to the first Bragg peak and $q = [12.7, 13.4]$ to the second Bragg peak (see also Fig. 10.13). [The data of the small-q and the second Bragg peak was shown in Fig. 5 of [2]. However, the curves of the high q-interval of the big system published in [2] were calculated with insufficient statistics and are very noisy. Here, the corrected data is used which is much smoother.]

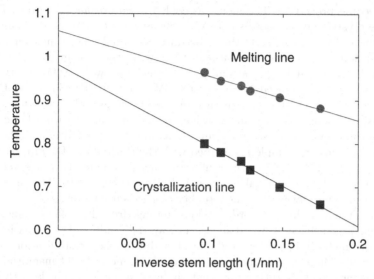

Fig. 10.15. $N = 100$: Crystallization and melting line obtained from isothermal relaxation runs. After relaxation at a given temperature, the average stem length of the final structure is determined. Then, this structure is heated continously to determine its melting point.

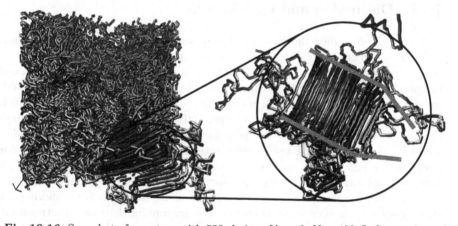

Fig. 10.16. Snapshot of a system with 500 chains of length $N = 400$. Left: cut through the simulation box (side length 45 σ), 12 chains in the crystal domain are shown completely. The zoom on the right shows these chains from the side where the lamellar structure is seen. The growth front extends over several chain diameters and has conical shape indicated by the light gray lines.

10.4.2   Entangled Chains ($N = 400$)

To approach the situation of highly entangled chains, simulations were started with chains of $N = 400$ monomers. Figure 10.16 shows a first result where the growth of an isolated lamella can be observed. The simulation was started by a self-seeding procedure: The melt was cooled rapidly from $T = 1.0$ to $T = 0.78$ where crystallization starts soon. After $20000\ \tau$ at $T_c = 0.78$, the temperature was raised to $T_c = 0.82$. By this procedure, the small nuclei melt and only the largest domain remains and continues to grow. In the time evolution, it is interesting to note that in the beginning, the lamella grows very fast by thickening (this was also recognized for the system $N = 100$ in [1]). Soon, the thickness saturates and more stems are aligned on the lateral growth front. Here, the system is big enough to have the growing lamella in contact with a large amount of disordered melt. We see that new stems join the ordered domain before the previous stems are completely aligned with the thickness of the already existing lamella. This yields a growth front which actually extends over several chain diameters and has a conical shape as indicated in Fig. 10.16. The chains in the ordered domain have still some mobility along their contour.

The stem lengths in the ordered domain range from 40 to 45 monomers. Thus, each chain can have up to 10 stems. The average radius of gyration is about 9 σ in the melt. It increases to about 11 σ during the crystallization shown in Fig. 10.16. The chain diameter is thus of the same order of magnitude as the lamella formed. It is rather rare that chains traverse the lamella with a single stem. Parts of a chain rather extend locally often forming hairpins which then reel in on the growth front (see also discussion of Fig. 10.17 below).

10.5 Discussion and Conclusions

This chapter described the structure formation in supercooled melts of a bead - spring model including an angle bending potential which reflects the stereochemistry of poly(vinyl alcohol) or poly(ethylene). Two parameters, chain length and temperature, were varied. The simulation results agree very well with experimental findings: Short chains form extended chain crystals, they order and melt at lower temperatures than longer chains. Only the hysteresis between homogeneous nucleation and melting temperatures is quite large which can be understood by the fast heating and cooling rates and the absence of heterogeneous nuclei in the simulations. One most easily observes the regime where structure formation sets in spontaneously. However, the growth of single nuclei can also be observed after careful examination of the phase diagram. For exploring the slow growth at smaller supercooling, heterogeneous nuclei can be introduced. The present systems are still quite small and the chains rather short to test the regime of highly entangled melts which is addressed by the contribution of Strobl [13]. However, several interesting observations can be made.

An important point is that the present model allows to observe the formation of chain-folded structures in a bulk system starting from the amorphous melt. For the present model, chain folding sets in at around $N = 50$ monomers. The case $N = 100$ was most systematically examined. First calculations are reported showing that bulk systems with $N = 400$ are also accessible. It is very interesting that already with the system of $N = 100$ one obtains well defined crystallization

Fig. 10.17. Schematic representation of a lamella growing into the melt. The simulations suggest that deposition of a hairpin is an easy growth mode because it acts as a wedge which is driving between the lamella and the melt.

and melting lines in the representation of temperature versus the inverse of the average stem length.

In the case $N = 100$, the gyration radius R_g is nearly doubled at the higher T_c, but the chain diameter $2R_g$ is still significantly smaller than the lamella. For these still short chains, their "global" conformation is completely changed during the crystallization. In the case $N = 400$, R_g of the melt is twice as large as for the chains $N = 100$. It increases slightly during the crystallization. The chain diameter is now in the order of the lamellar domain. This means that most chains participate mainly in one single domain. We are not yet in the limit of high polymers where it was shown that R_g remains unchanged [25]. This lead to the so-called solidification model ("Erstarrungsmodell") saying that the structure formation affects each chain only locally. We agree with this qualitative picture. Comparing the chains $N = 100$ with $N = 400$, the ordering kinetics at the same temperature is almost the same. This means that the slowing down of chain dynamics due to entanglements does not affect the ordering. It seems even that the average monomer dynamics is accellerated in the region of crystallization with respect to the melt.

Even though the global extension of a long chain changes little, locally, on the scale of the lamella, there is a significant reordering. It is frequent that chains reenter the crystal by a tight backfold. We often observed that two stems connected by a hairpin fold move together. Figure 10.17 shows schematically this possible growth mode for a single chain near the lamellar edge suggesting the following mechanical picture. The hairpin may drive like a wedge between the melt and the lamellar edge for opening free volume where the new stems can be deposited.

In the present model, the final crystal structure is a hexagonal lattice of the stems; the real crystal structure of e.g. poly(ethylene) is suppressed by the representation of one chemical monomer by a soft sphere. In this respect, one could interpret the present simulations as a modeling of the mesomorphic phase. It would now be interesting to reintroduce the atomistic details at different moments of the simulation to observe the evolution to a final crystal structure. Such a local evolution with atomistic details should be possible with the present system size contrary to a simulation from the very beginning of the amorphous melt, which would probably still be too expensive in computer ressources.

It was pointed out that different kinds of ordering can be identified in polymer systems and this distinction may be useful for further theoretical descriptions. Due to the approximations of our model, the structure formation here is mainly driven by the orientational ordering typical of liquid crystalline systems. ([26] discusses the possibility of combining density and orientational degrees of freedom as order parameters. The present model is perhaps an implementation of the hidden liquid-crystalline-like phase because the real crystal structure is suppressed.) This is mainly an entropic reason: at lower temperature, where the melt becomes denser (and thermal motion reduced), the ordered state provides more vibrational states than the disordered one and this overcomes the loss in conformational entropy. However, the ordering is triggered by an energetic component: since the stretched bond angle has a lower potential energy, the folded states will be less populated at lower temperatures, this leads to a stiffening of the chains and an increase in the persistence length. The energy gain of the stretched chains is quite large so that the crystal is also energetically more favorable. But note that the model contains no explicit attraction between monomers. Since many theoretical considerations argue with the attraction of a chain to the crystal front, it is interesting that the chain-folded structures may also be obtained without it in dense systems.

In the homogeneous nucleation regime, the system is so far supercooled that the local stiffening leads to small nuclei which then rapidly grow. We found in the simulation that these initial nuclei are always intra-chain nuclei of two or three stems connected by hairpin folds. Above the limit of homogeneous nucleation, larger multi-stem nuclei which usually involve several chains are necessary to induce growth. These nuclei can be produced by self-seeding or "dust particles" serving as heterogeneous nuclei. Above the homogeneous crystallization temperature, a surface (crystal front) is necessary to "catalyze" the ordering of the chains.

I would like to thank J. Baschnagel, B. Lotz, F. Müller-Plathe, and J.-U. Sommer for stimulating discussions. I further acknowledge the generous allocation of computer time on the Alpha-Cluster at MPI-P Mainz and NEC-SX5 at IDRIS, CNRS Orsay, and the financial support by the Laboratoire Européen Associé ICS–MPI-P.

References

1. Meyer, H.; Müller-Plathe, F. *J. Chem. Phys.* **2001**, *115*, 7807.
2. Meyer, H.; Müller-Plathe, F. *Macromolecules* **2002**, *35*, 1241.
3. Kavassalis, T. A.; Sundararajan, P. R. *Macromolecules* **1993**, *26*, 4144.
4. Liao, Q.; Jin, X. *J. Chem. Phys.* **1999**, *110*, 8835.
5. Fujiwara, S.; Sato, T. *J. Chem. Phys.* **2001**, *114*, 6455.
6. Sumpter, B.; Noid, D. W.; Liang, G. L.; Wunderlich, B. *Adv. Polym. Sci.* **1994**, *116*, 73.
7. Yamamoto, T. *J. Chem. Phys.* **1997**, *107*, 2653.
8. Takeuchi, H. *J. Chem. Phys.* **1998**, *109*, 5614.
9. Liu, C.; Muthukumar, M. *J. Chem. Phys.* **1998**, *109*, 2536.
10. Welch, P.; Muthukumar, M. *Phys. Rev. Lett.* **2001**, *87*, 218302.
11. Baschnagel, J.; Binder, K.; Doruker, P.; Gusev, A. A.; Hahn, O.; Kremer, K.; Mattice, W. L.; Müller-Plathe, F.; Murat, M.; Paul, W.; Santos, S.; Suter, U. W.; Tries, V. *Adv. Polymer Sci.* **2000**, *152*, 41.
12. Müller-Plathe, F. *ChemPhysChem* **2002**, in press, .
13. Strobl, G. *Eur. Phys. J. E* **2000**, *3*, 165.
14. Reith, D.; Meyer, H.; Müller-Plathe, F. *Macromolecules* **2001**, *34*, 2335.
15. Grest, G. S.; Kremer, K. *Phys. Rev. A* **1986**, *33*, 3628.
16. Berendsen, H. J. C.; Postma, J. P. M.; van Gunsteren, W. F.; DiNola, A.; Haak, J. R. *J. Chem. Phys.* **1984**, *81*, 3684.
17. Kremer, K.; Grest, G. S. *J. Chem. Phys.* **1990**, *92*, 5057.
18. Fischer, E. W. *Koll.-Z. Z. Polym.* **1967**, *231*, 458.
19. Oxtoby, D. W. *Adv. Chem. Phys.* **1988**, *70*, 263.
20. Kraack, H.; Deutsch, M.; Sirota, E. B. *Macromolecules* **2000**, *33*, 6174.
21. Flory, P. J. *J. Chem. Phys.* **1949**, *17*, 223. Volkenstein, M. V., *Configurational Statistics of Polymer Chains* (Interscience Wiley, chapter 5).
22. Meyer, H. *unpublished.*
23. Ryan, A. J.; Fairclough, J. P. A.; Terrill, N. J.; Olmsted, P. D.; Poon, W. C. K. *Faraday Discuss.* **1999**, *112*, 13.
24. Wang, Z.-G.; Hsiao, B. S.; Sirota, E. B.; Srinivas, S. *Polymer* **2000**, *41*, 8825.
25. Fischer, E. W. *Pure & Appl. Chem.* **1978**, *50*, 1319. Stamm, M.; Fischer, E. W.; Dettenmaier, M.; Convert, P. *Faraday Discussion* **1979**, *68*, 263.
26. Olmsted, P. D.; Poon, W. C. K.; McLeish, T. C. B.; Terrill, N. J.; Ryan, A. J. *Phys. Rev. Lett.* **1998**, *81*, 373.

11 Lamellar Ethylene Oxide–Butadiene Block Copolymer Films as Model Systems for Confined Crystallisation

Wim H. de Jeu

FOM-Institute for Atomic and Molecular Physics
Kruislaan 407, 1098 SJ Amsterdam, The Netherlands
dejeu@amolf.nl

Abstract. Semicrystalline poly(ethylene oxide–b–butadiene) diblock copolymers in uniform lamellar films are discussed as a model system for confined crystallisation. At low supercooling, crystallization of the PEO block leads to an increase in the lamellar thickness of both blocks, which is accompanied by a contraction in the lateral direction resulting in cracking of the films. A tentative model is discussed with an integer or half-integer number of folds in the vertically oriented crystalline stems. However, emphasis is also on aspects still not understood.

11.1 Introduction

In a diblock copolymer the interaction between the chemically distinct blocks leads below the order-disorder transition (ODT) to microphase separation [1]. If one of the blocks is semicrystalline, a specific situation occurs of confined crystallisation within the microphase-separated morphology. Studies of such materials have the potential to shed light on some of the fundamental aspects of polymer crystallisation. In particular, confinement provides an example of an external 'field' as a means to influence and possibly control the pathways to nucleation and chain folding. Interestingly, in a block copolymer system equilibrium chain folding can be achieved [2–4], contrary to the situation for homopolymers. This is due to the competition between the enthalpic drive to minimize the fold surface energy in the crystallisable block and the entropic penalty of the attendant stretching of chains in the amorphous one. As a result a structure of alternating amorphous and crystalline layers has been predicted, the latter with regular folding of the chain stems perpendicular to the interface. However, this does not necessarily imply that the equilibrium situation can easily be reached; kinetic effects still may play an important role. In this chapter we want to discuss critically the use of uniformly aligned lamellar films of ethylene oxide–b–butadiene (PEO b-PB) to confine the crystallisation process of the PEO-blocks. Several papers have recently been published on this system [5 8], but rather than only emphasizing what has been accomplished, we will also discuss the still outstanding questions.

Bulk investigations of diblock copolymer systems with one crystallisable block are numerous and have been reviewed [1]. In the present context especially work involving PEO is of interest. Small-angle x-ray scattering (SAXS) investigations of PEO crystallisation in PEO–PB systems have been reported by Hong et al. [9] and Li et al. [10]. Crystallisation in a series of PEO–b–PBO (polybutylene oxide) and PEO–b–PPO (polypropylene oxide) was studied by Ryan and co-workers [11–14]. Early work on the PEO–b–PS (polystyrene) system was done by Lotz and Kovacs [15], while more recently Cheng and co-workers [16,17] investigated the effects of changing the ODT. The confined surroundings have a profound influence on the crystallisation properties. The relation between the ODT, the crystallisation temperature T_{cr} of PEO, and the glass transition temperature T_g of the noncrystallisable block is of decisive importance. In the case of PEO–b–PS, PEO crystallisation from a microphase-separated melt can take place between hard glassy PS boundaries (depending on the effective T_g of PS). Alternatively, if $T_{cr} > T_g$ confined crystallisation develops between 'rubbery' boundaries. If on the temperature scale the ODT and T_{cr} are close, further complications occur. Depending on the morphology, crystallisation can lead to a 'breaking out' of the crystalline structure from the original morphology, which is destroyed. In all these cases the choice of the molecular mass of the crystallisable block is another factor of importance. Confinement will not be effective if the crystalline lamellar thickness is appreciably smaller than the domain spacing associated with the block phase separation.

Despite the extensive bulk work, far less is known about confined crystallisation in thin films. Of special interest is the situation of about equal volume fractions of the blocks, which gives below the ODT rise to a lamellar structure. These randomly oriented lamellar microdomains become macroscopic under the influence of the surfaces. Hence in the lamellar morphology thin block copolymer films provide a precise control of the boundary conditions over macroscopic areas. The approximately symmetric PEO–b–PB diblock copolymers to be discussed correspond to the limit of strong segregation with a low T_g amorphous block. The confinement of the crystallisation of the PEO blocks arises rather from the large enthalpy penalty of segmental interactions between the blocks than from the immobility of the amorphous block ('rubbery' confinement). Relatively small molecular masses have been used. The films have been investigated by optical and atomic–force microscopy (AFM) and by x-ray reflectivity (XR). While the first techniques provide surface information, the latter allows a quantitative determination of the changes in the various (sub)layers throughout the film upon crystallisation. In addition some in-plane information is available from electron diffraction.

11.2   Experimental Situation

Most of the results to be discussed are related to a PEO–b–PB$_h$ block copolymer with a block mass of 4.3 and 3.7 kg/mol, respectively (sample I). This leads to 98 PEO monomers (volume fraction 0.46) and to 66 (hydrogenated) PB$_h$ monomers [5,8]. Other results are for molecular masses of 5.0 and 5.5 kg/mol PEO and PB, respectively (sample II) [6,7]. Reiter et al. [5,18] also considered variations on sample I with the same PB_h block but a PEO one of 2.9, 3.6 and 5.0 kg/mol, respectively. The latter two cases give essentially the same results as sample I and II. However, the more asymmetric case of a PEO block of 2.9 kg/mol behaves rather differently [5,18]. Recent SAXS results indicate that this probably can be attributed to a cylindrical morphology in the melt. Crystallisation then leads to a 'breaking out' and subsequent destruction of the original morphology. Hence we will not include this case in the further discussion.

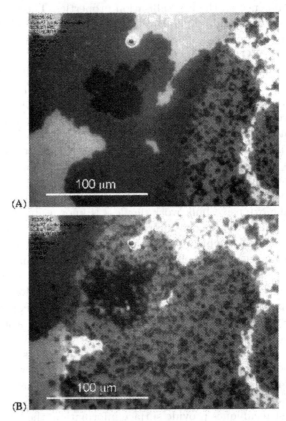

Fig. 11.1. Optical microscopy of isothermal crystallisation of sample I at 45°C. (A) Crystallisation front moving in from the right while at the left well-ordered lamellar terraces can be seen. (B) The same area 1 min later; the crystallisation front has almost crossed the full window (after Opitz et al. [8]).

For both samples I and II the typical microscopic picture of the dynamics during crystallisation at the surface of a film is as shown in Fig. 11.1. Two different regions corresponding to the molten and the crystallized structure can easily be distinguished. In the melt (here at $T = 45°C$) at the left and middle part of Fig. 11.1A, the typical terrace structure of a lamellar block copolymer film is visible [1]. Different gray scales indicate distinct levels, which differ in height by a full lamellar period. After a nucleus has been formed, the crystallization front moves laterally over the film. During this process the different height levels are preserved [5–8]. The crystallisation front is characterized by a depletion zone at its front and leaves a cracked film surface behind. An increase of the total film thickness upon crystallisation can be inferred from the lighter color of the terraces. Hence, material is pulled out locally and rearranged vertically during the process of crystal formation. In some cases distinct growth fronts with equal growth rates have been seen, indicating that the growth of the PEO crystals proceeds independently in each layer [6]. There are slight differences in the times at which the fronts reach a given position in the film [7].

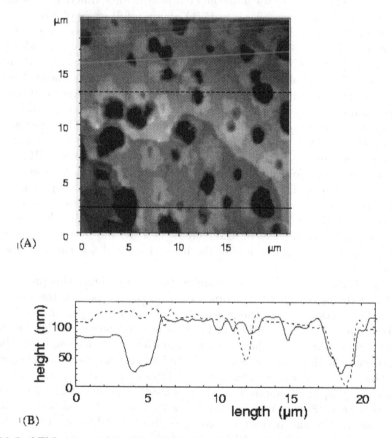

Fig. 11.2. AFM picture of the film of Fig. 1 in the crystallised state. (A) Height picture with terrace structure and deep holes formed during crystallisation. (B) Height profiles along the two lines indicated in A (after Opitz et al. [8]).

The depth of the cracks after crystallisation has been investigated by AFM [5,8]. Fig. 11.2A shows a $20 \times 20 \ \mu m^2$ detail of the surface of the same film as in Fig. 11.1; height profiles along the two lines are displayed in Fig. 11.2B.

The terrace structure with distinct steps is very clear, while also holes due to the cracking of the film can be quantified. Some of the cracks extend deep and occasionally go through the whole film down to the substrate. The terrace height, which coincides with the lamellar period L, is after crystallisation about 20 nm, while the total film thickness D amounts to approximately 110 nm (asymmetric morphology with 5.5 periods). In the depth profiles of the cracks multiples of L can clearly be distinguished; however, interestingly in some cases also values around $L/2$ appear.

X-ray reflectivity is well suited to obtain a more complete quantification of the structural properties of block copolymer films [19,20]. Data were taken for sample I in dependence of the annealing temperature and different crystallisation temperatures. As an example, Fig. 11.3A shows a full series of XR curves with their respective fittings. The upper curve depicts the measurement of a 6.5 layer film at $T = T_{ann} = 90°C$ in the molten phase-separated state. Besides the so-called Kiessig fringes, which indicate the total film thickness, several orders of Bragg peaks are visible of which the third one is most pronounced. Hence, we can conclude to a rather perfect lamellar ordering of alternating PEO and PB_h layers oriented parallel to the substrate. Precise results are obtained by fitting the experimental curves to the model sketched in the inset. The lamellar period is given by $L = 18.7 \pm 0.1$ nm and the total film thickness by $D = 122.9 \pm 0.1$ nm, which is close to 6.5 layers. Lowering the temperature to 45°C does not immediately result in crystallisation of the sample: until a nucleus is formed the system stays in the molten state. Further stretching of the polymer chains occurs [21,22], see the middle curve of Fig. 11.3A. The thickness of both the PEO and the PB_h layers have increased, leading to $L = 20.3$ nm. In addition changes are observed in the top layers: the increase of L results in a reorganisation of material in the sense that some of the upper islands disappear and new holes are created.

After formation of a nucleus crystallisation occurs. Once this process is finished and the sample remains stable over a prolonged period, the lower curve of Fig. 11.3A results. Now the general features have changed drastically. First the Bragg positions have shifted to lower values, indicating another increase in the lamellar thickness. Furthermore, the amplitude of the Kiessig fringes has decreased and the decay of the whole reflectivity curve has become stronger, due to an increased roughness both at the surfaces and at the internal interfaces. From the fitting we obtain at $T_{cr} = 45°C$ a lamellar period of 24.1 ± 0.1 nm. At the substrate a PEO layer is situated with a thickness of 5.5 nm, which coincides with a four times folded chain consisting of five vertical stems. In the stack of layers the fitting results in a PEO sublayer thickness $d_{PEO} = 10.1$ nm, slightly less than twice the value of the layer at the substrate. The (sub)layer spacings at various temperatures are displayed in Fig. 11.3B. The value of d_{PEO} can be compared with the extended chain length of 27.4 nm. The ratio $27.4/d_{PEO}$ shows discontinuous changes from 2.6 at 45°C to around 3.0 at both 40 and 35°C down to $3.5 - 3.3$ at 30 and 25°C. While these thin-film results indicate stepwise changes of d_{PEO}, SAXS investigations of bulk samples rather show a

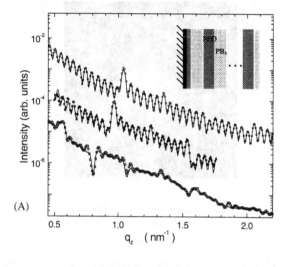

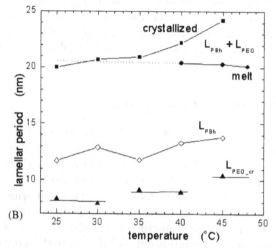

Fig. 11.3. X-ray reflectivity of a 6.5 layer film of sample I. (A) Reflectivity curves: melt at 90°C (open circles), melt at 45°C (filled dots), crystallised at 45°C (open squares). Curves have been shifted for clarity. Full lines are best fits to the model in the insert. (B) Lamellar period L and sublayer thicknesses at various temperatures as determined by fitting the x-ray reflectivity (after Opitz et al. [8]).

continuous variation with temperature [9,10]. In bulk samples the size of the microphase-separated 'grains' in the melt is of the order of $5-20$ µm, which can be compared with typical crystallites originating from a single nucleus of the order of 500 µm [9]. This difference is very unlike the situation upon crystallisation in the large block-copolymer domain of a uniform lamellar film.

Fig. 11.4. Electron diffraction image of sample II passing through multiple layers. Only a four-point pattern is observed (after Hong et al. [6]).

Finally Fig. 11.4 shows an electron diffraction image of a multilayer film of sample II. Only four spots are seen corresponding to the (120) reflection. This indicates that the PEO unit cell and chain axis are oriented normal to the layers. More importantly, as this observation is independent of the number of layers, the PEO crystals in adjacent layers are in orientational registry even though they are separated by amorphous PB blocks.

11.3   Discussion and Modelling

From the x-ray results of Fig. 11.3 the present PEO–b–PB samples evidently have an asymmetric lamellar stacking with a thickness quantified as $D = (n + 1/2)L$, confirming the AFM picture. The PEO block with the larger electron density is situated at the Si–substrate on top of the SiO_x layer, while a low-density PB layer forms at the interface to air. These outermost sublayers have about half the thickness of the corresponding interior layers. The asymmetric layer stacking contradicts some microscopic studies [5–7], in which the authors conclude to a symmetric configuration with PB at both outer interfaces. These latter results are anyhow somewhat counterintuitive, as PEO is expected to be more compatible to a SiO_x layer than the nonpolar PB. If the initial film thickness is somewhat larger than the closest quantized value, islands form at the surface; for a film somewhat short of material holes appear. The associated height c.q. depth is equal to the full period L and can be seen as different uniform gray levels in the microscopic pictures (Fig. 11.1). In fact, the situation is somewhat more complicated as for unclear reasons three or more levels were observed. This would in principle also be possible for a precisely quantized thickness.

For both sample I and sample II SAXS in the bulk [9,10] indicates a strong temperature dependence of the lamellar period. In the melt of sample I an increase is observed from $L = 15.3$ nm at 250°C to $L = 18.5$ nm at 90°C. As it is not possible to reach the ODT (which is well above 300°C), no reference value at the phase transition is available. Compared to a Gaussian coil, the molecules in the lamellae are before crystallisation already stretched by about a factor two. Upon crystallisation, the lamellar period increases further despite the decreasing volume (increasing density) of PEO. This expansion allows adjusting the thickness of the PEO sublayer to an integer (or half-integer) number of stems. Density conservation forces the PB block to follow this stretching. Evidently, the loss of entropy associated with the PB stretching is more than compensated for by the favourable packing of the PEO stems. As a consequence, a lateral contraction results through the whole film, resulting in macroscopic pits and cracks. Implicit in this interpretation is that the folded PEO chains are perpendicular to the layers. In addition to the direct observation of perpendicular chains by electron diffraction (Fig. 11.4), this is also supported by additional circumstantial evidence. First, the block segregation implies that the chains are oriented perpendicular to the interfaces in order to minimize the contact area. Second, if the stems were tilted with respect to the interface normal, a simple variation in the tilt angle could accommodate the increased length without PB stretching and lateral shrinkage.

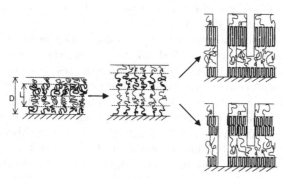

Fig. 11.5. Model for the development of the layer structure and the polymer folds during crystallisation (after Opitz et al. [8]).

Modern ideas about the early stages of crystallisation assume local 'embryos' of ordered stems. From this point of view the present situation of considerable uniform stretching of the chains provides a perfect preconditioning to crystallisation. Hence, we anticipate in the uniform films a somewhat larger degree of crystallisation than the about 85% calculated for the bulk [5]. To determine the number of stems at the various temperatures we can take the extended chain length of PEO, divide it by the number $n = 2, 3, \ldots$ and compare with the PEO sublayer thickness. For the PEO layer at the substrate this leads with decreasing temperature to $n = 5, 6$ and 7 stems. Taking for the interior PEO layers in the film the same number of folds as for the bottom layer would lead to crystalline double layers (see Fig. 11.5). Alternatively, we could assume a 'zipper' model in which stems originating from opposite interface pass along each other (co-crystallisation). This would double the stem length leading to 2.5, 3 and 3.5 stems. Whatever be the case, the number of folds increases when the crystallisation temperature is lowered. This indicates that for fast crystallisation kinetic (non-equilibrium) effects dominate. Conversely, one may assume that at the highest temperatures probably equilibrium chain-folding has been reached.

Though the two different folding models of Fig. 11.5 should lead to different melting points, in thin films it is difficult to discriminate experimentally between these situations. In the absence of conclusive evidence, various arguments can be given. (i) In the zipper-model some stems with a length of half a sublayer thickness are accompanied by stems originating from the other side. Though this seems to be a difficult process in terms of reaching optimum packing, it might be facilitated by hydrogen bonds between the hydroxy-terminated end groups. (ii) The zipper model gains a kink energy at the expense of a loose end. This point is not expected to be decisive, as both can be estimated to be of the order of $\sim kT \ln 3$/chain. (iii) Upon cooling from 45 to 40°C the original situation remains perfectly stable for at least over weeks. This is what one would expect for 5 stems that do not change into 6, but perhaps less so for 2.5 stems that could change into an integer number 3. (iv) For crystalline double layers the individual layers would decouple at the PEO–PEO interfaces [5]. This process could account for the observation of steps in the various cracks and craters corresponding to $L/2$ (c.f. the AFM picture of Fig. 11.2).

It is instructive to compare with the crystallisation of PEO homopolymers. These have been extensively studied [23–25], though the role of confinement or even thin films has been hardly addressed. Crystallites with an integral number of folds were found to be more stable than other ones, and variation in supercooling resulted in stepwise changes from one integrally folded state to another. In [24] double lamellae were reported for odd-number fold PEO molecules in contrast to the usual situation of single lamellae for even-number folds. This was attributed to hydrogen bonding between the hydroxy end groups [26]. In the case of an odd number of folds an uneven distribution of hydrogen bonds on the fold surfaces arises, which can be avoided for double lamellae. For the present block copolymers only at one side of a crystalline PEO stem is an hydroxy group present. Assuming that isolated end groups near block interfaces are again to be avoided, the double layer structure of Fig. 11.5 would be favourable for an even number of folds $2n$. For a double layer with an odd number of folds, $2n+1$ non-hydrogen bonded hydroxy groups at the block interfaces can be avoided by choosing instead a $(2n + 1)/2$ folded single layer (as pictured at the upper right part of Fig. 11.5). Such a reasoning holds only for hydroxy-terminated end groups, and becomes irrelevant in the case of methylation.

The crystallographic registry between adjacent crystalline layers is a remarkable observation, somewhat unexpected because the crystals are separated by an intervening layer of amorphous PB. Hence, some form of connectivity between the layers must exist, that allows a single nucleus as common origin. Hong et al [6,7] postulated a defect-controlled mechanism for the spreading of crystallinity between the layers. A simple edge dislocation occurring at the boundary of a hole or an island can connect crystallinity in adjacent top layers. However, a screw dislocation (see Fig. 11.6) is needed to provide a spiral that can connect

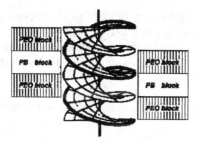

Fig. 11.6. Schematic representation of a screw dislocation in a diblock copolymer. Note that the structure is actually a double helix, one of which is indicated in bold (after Gido et al. [27]).

several layers. Transmission electron microscope pictures of such dislocations can be found at many places; they have been discussed explicitly by Gido et al. [27]. A screw dislocation can be described by analogy to the spiralling of a multilevel parking garage, with different spirals going up and down. Such an interface separates the PEO and PB domains in the grain boundary region, while minimizing the sum of interface and chain stretching energy as much as possible. When a growing crystallite encounters such a screw, its growth will carry it up or down and allow it to spread outward in many adjacent layers. The practical spatial limits of such an effect still remain to be explored.

11.4   Conclusions and Open Questions

Summarizing we come to the following tentative model for confined crystallisation of lamellar PEO–b–PB block copolymer films. Cooling down from the melt to the ODT (out of reach for the present compounds) into the phase-separated melt, one would theoretically expect to start with polymers in a Gaussian coil conformation. Upon decreasing the temperature the scaling behavior of block copolymers $L \sim \chi^{1/6}$, where χ is the Flory-Huggins interaction parameter [1], leads to stretching of the coils and thus to a parallel preorientation. Subsequently crystallisation starts and – depending on the crystallisation temperature – the polymer chains continue to stretch in order to accommodate an integer or half-integer number of perpendicular stems upon folding, leading to a further increase of the PEO sublayer thickness. In combination with the decreasing volume, this causes a strong lateral shrinkage of the PEO layers. Because of the linkage between the PEO and PB blocks, the latter must also stretch and cracks are formed through the whole thickness of the film. The results do not allow a final decision between an integer and a half-integer number of stems. Arguments based on hydrogen bonding suggest an alternation between these situations for even-number and odd-number fold stems. Crystal growth at screw dislocations can account for the observed crystallographic registry between layers at different levels.

Various open questions can be formulated. Regarding PEO–b–PB in the amorphous state the following points remain to be explored.

- The temperature dependence of the lamellar spacing should be further investigated in relation to the ODT (χ parameter).
- It is not clear why multiple domain levels occur in films. This contrasts with many other block copolymer systems in which only two levels are observed as predicted theoretically.

Regarding the PEO crystallisation in PEO–b–PB several aspects need clarification.

- A careful comparison is needed between films that indicate a stepwise variation of the crystal thickness with crystallisation temperature, and SAXS results on bulk samples that indicate a continuous thickening. Do different kinetics play a role in relation to the different size of the 'grains' in the two situations? Contrary to the situation in thin films, in the bulk a single crystallite will encounter many grain boundaries during its growth process.
- In the absence of the possibility of thermal analysis in thin films, can we differentiate between the two possibilities of Fig. 11.5: integer and half-integer folds? Or do both situations occur at the same time? Is the observation of steps of $L/2$ in 'craters' after crystallisation proof of the double-layer structure (integer folds)?
- Can experimental evidence be found for a difference between the situations of odd and even integer folds (influence of the hydrogen bonds associated with the OH–termination)? In that case also a comparison should be made with methyl-terminated systems that should differ in this respect.

Further studies are needed of the in-plane order upon crystallisation of PEO in lamellar PEO–b–PB films.

- Grazing-incidence x-ray diffraction would give an average over a much larger area that would nicely complement the electron diffraction results.
- Though the explanation of the crystallographic registry of the various layers via screw disclinations is appealing, it remains to be explained how this situation can be combined with the boundary conditions of the two blocks.

We conclude that though important progress has been made, the process of confined crystallisation still contains a considerable number of subtleties that are not well understood.

Acknowledgement

The author would like to thank Denitza Lambreva, Liangbin Li, Ricarda Opitz, (Amsterdam) and Günter Reiter (Mulhouse) for valuable discussions. This work is part of the Softlink research program of the 'Stichting voor Fundamenteel Onderzoek der Materie (FOM)', which is financially supported by the 'Nederlandse Organisatie voor Wetenschappelijk Onderzoek (NWO)'.

References

1. I.W. Hamley: *The Physics of Block Copolymers* (Oxford University Press, New York 1998)
2. E.A. DiMarzio, C.M. Guttmann, J.D. Hoffman: Macromolecules **13**, 1194 (1980)
3. M.D. Whitmore, J. Noolandi: Macromolecules **21**, 1482 (1988)
4. T. Vilgis, A. Halperin: Macromolecules **24**, 2090 (2000)
5. G. Reiter, G. Castelein, P. Hoerner, G. Riess, J.-U. Sommer, G. Floudas: Eur. Phys. J. E **2**, 319 (2000)
6. S. Hong, W.J. MacKnight, T.P. Russell, S.P. Gido: Macromolecules **34**, 2398 (2001)
7. S. Hong, W.J. MacKnight, T.P. Russell, and S.P. Gido: Macromolecules **34**, 2876 (2001)
8. R. Opitz, D.M. Lambreva, W.H. de Jeu: Macromolecules **35**, 6930 (2002)
9. S. Hong, L. Yang, W.J. MacKnight, S.P. Gido: Macromolecules **34**, 7009 (2001)
10. L. Li, D.M. Lambreva, W.H. de Jeu: Eur. Phys. J. E, *to be published*
11. S.-M. Mai, J.P.A. Fairclough, I.W. Hamley, M.W. Matsen, R.C. Denny, B.-X. Liao, C. Booth, A.J. Ryan: Macromolecules **29**, 6212 (1996)
12. S.-M. Mai, J.P.A. Fairclough, K. Viras, P.A. Gorry, I.W. Hamley, A.J. Ryan, C. Booth: Macromolecules **30**, 8392 (1997)
13. I.W. Hamley, J.P.A. Fairclough, F.S. Bates, A.J. Ryan: Polymer **39**, 1429 (1998)
14. I.W. Hamley, M.L. Wallwork, D.A. Smith, J.P.A. Fairclough, A.J. Ryan, S.-M. Mai, Y.-W. Yang, C. Booth: Polymer **39**, 3321 (2001)
15. B. Lotz, A. Kovacs: Kolloid Z. Z. Polym. **209**, 97 (1966)
16. L. Zhu, Y. Chen, A. Zhang, B.H. Calhoun, M. Chun, R.P. Quirk, S.Z.D. Cheng, B.S. Hsiao, F.J. Yeh, T. Hashimoto: Phys. Rev. B **60**, 10022 (1999)
17. L. Zhu, S.Z.D. Cheng, B.H. Calhoun, Q. Ge, R.P. Quirk, E.L. Thomas, B.S. Hsiao, F.J. Yeh, B.J. Lotz: J. Am. Chem. Soc. **122**, 5957 (1999)
18. G. Reiter, G. Castelein, P. Hoerner, G. Riess, A. Blumen, J.-U. Sommer: Phys. Rev. Lett. **83**, 3844 (1999)
19. T.P. Russel: Mater. Sci. Rep. **5**, 171 (1990)
20. M. Tolan: *X-ray Scattering from Soft-Matter Thin Films* (Springer, Heidelberg 1999)
21. K. Almdal, J.H. Rosedale, F.S. Bates, G.D. Wignall, G.H. Fredrickson: Phys. Rev. Lett. **65**, 1112 (1990)
22. M.A. Hillmyer, F.S. Bates, K. Almdal, K. Mortensen, A.J. Ryan, P.A. Fairclough: Science **271**, 976 (1996)
23. A.J. Kovacs, C. Straupe: J. Crystal Growth **48**, 210 (1980)
24. C.Z.D. Cheng, H.-S. Bu, B. Wunderlich: Polymer **29**, 5794 (1988)
25. A. Keller, S.Z.D. Cheng: Polymer **39**, 4461 (1998)
26. T. Shimada, N. Okui, T. Kawai: Makromol. Chem. **181**, 2643 (1980)
27. S.P. Gido, J. Gunther, E.L. Thomas, D. Hoffman: Macromolecules **26**, 4506 (1993)

12 Crystallization Kinetical Peculiarities in Polymer Blends

Bernd-J. Jungnickel

Deutsches Kunststoff-Institut (German Institute for Polymers), Schlossgartenstr. 6, D-64289 Darmstadt, Germany, bjungnickel@dki.tu-darmstadt.de

Abstract. Crystallization in polymer blends can proceed along a large number of so-lidification paths which results in a huge variety of supermolecular structures. These features are interesting from the underlying thermodynamics, from the rheology which governs the crystallization kinetics, as well as from their impact on the material properties. What happens in a particular case will decidingly depend on the miscibility of the components, i.e., whether they are miscible in the whole accessible composition and temperature range, whether they exhibit a miscibility gap, or whether they form separated phases. In this chapter, at first a survey is given on the several effects which can be observed and which are unknown from low molecular weight materials. Some of these features are then discussed in more detail. Particular attention is payed to the change of the nucleation conditions in phase separated blends and to the inhomogeneous composition changes which may be induced by the crystallization of one component in an initially homogeneous melt. A suitable analysis of these phenomena yields deeper insight into the nucleating efficiency of heterogeneities in the material, on the probability of homogeneous nucleation, on chain mobilities and diffusion coefficients, and on the competition between crystallization and vitrification.

12.1 Introduction

It is well known that the crystalline solidification of polymers is a complicated phase transition process with many faces. Although subject of intense research since decades, it is far from being understood. The complications base on the particular importance of the transformation kinetics, which competes with the equilibrium thermodynamics. This in its turn bases on the low mobility of the macromolecular chains and on the topological constraints, which are due to the chain character. This, at the one hand, inhibits complete volume filling crystallinity – yielding partial or semi-crystallinity – and, at the other hand, results in a large variety of crystalline and supermolecular non-equilibrium structures.

If this is true for one-component polymer systems, it must be the more for polymer blends where a number of additional kinetic and thermodynamic degrees of freedom are at disposal. This is indeed a fact. What happens in a particular system, among others, will mainly depend on the miscibility of the blend components, i.e., whether they are miscible in the whole accessible temperature and composition range, whether they exhibit a miscibility gap, or whether they demix at all conditions in the pure components. It is, consequently, of particular importance how the actual phase separation structure looks like. It is to mention, however, that, from a practical point of view, miscibility as defined by equilibrium thermodynamics is less important for the solidification than compatibility, i.e., the ability to mix the blend components by technical means down to the molecular level. Kinetic features are such considered, too.

A list of particular effects, which may be observed in blends, is given in Table 12.1. This list is clearly by far not complete. There are moreover a number of effects, which are well known from low molecular weight materials and which are not considered in that table. Most of those peculiarities, which are typical for polymers base on the comparably slow crystallization or phase separation kinetics of these substances, on the fact that the melting point depression of polymer mixtures is usually very small, or on the importance of the nucleation step. Some

Table 12.1. Survey on the most important structural or kinetic features of crystallizing polymer blends. The distribution of the several effects into the three groups of blend systems is arbitrary in some respect. All features listed for absolutely miscible blends can also take place in systems with miscibility gap in the coexisting phases, if existent. Dependent on the phase connectivity, at the other hand, also fractionated crystallization may be observed in such systems. T_c^0, T_g: equilibrium crystallization and glass transition temperatures, respectively.

	Blend system		
	absolutely immiscible	with miscibility gap	absolutely miscible
Features	Fractionated crystallization	Crystallization induced demixing	Melting point depression
	Interface induced crystallization	Demixing induced crystallization	Kinetic eutecticity
		Crystallization crossing an interface	Drastic change of the crystallization kinetics by change of the crystallization temperature window $T_c^0 - T_g$
		Interface induced polymorphism	Composition dependent variation of crystal and supermolecular morphologies, respectively
		Kinetic competition between crystallization and mixing or demixing, resulting in	
		self-retardation of crystallization kinetics	development of crystallization induced composition inhomogeneities

of these effects are well documented and understood; others are subject of current research whereas on some is still speculated. It is the purpose of this article to refer some of these features which are typical for polymers – particularly those which the author is working on , to describe their thermodynamic or kinetic background, the resulting crystalline and supermolecular structures, and, their academic or technical importance where applicable.

12.2 Immiscible Blends: Fractionated Crystallization

Upon suitable mixing, the components exist in immiscible blends as pure polymers side by side as individual phases, building a particular phase separated structure. At a first glance it was to expect that these phases crystallize like the pure components and that the overall crystallization and melting behavior is just a superposition of the contributions of the pure components weighted by their blend content. There are however several structural features existent in the melt which may alter the crystallization behavior drastically. At first, the interface between the phases or the second component itself may act as new nucleating means. If it is more efficient as the other heterogeneities in the blend component under consideration which usually induce crystallization or as homogeneous nucleation if such impurities are not present , it can induce crystallization at a higher temperature or faster, such enhancing the crystallization ability. Another yet more important feature, however, is the phase separation structure as a whole, particularly the dispersion of one component in the other. It is then occasionally observed that the crystallization of the disperse phase is distributed over two or with decreasing size of the dispersed droplets even more distinctly separated steps on the temperature scale upon cooling although it proceeds in the pure material in one step at a defined temperature range. This phenomenon is called "fractionated crystallization" [1–3].

Figure 12.1 gives some examples of that feature [3–5]. The upper figure illustrates that the crystallization of polyoxymethylene (POM) if being the minor component in a blend with polyethylene (PE) decomposes in several distinct steps where the number of these steps increases with finer dispersion. The lower left figure shows a similar phenomenon in a blend of polyamide-6 (PA-6) and poly (vinylidene fluoride) (PVDF). In the sample where the PA-6 is finer dispersed (extrusion cycle number $Z = 4$), the initial crystallization peak of the PA-6 at 190 °C disappears. A quantitative inspection of the observation reveals that the PA-6 crystallizes now together with the main part of the PVDF at 140 °C. Crystalline PVDF acts obviously as a nucleating agent for that component. The crystallization of the PVDF, in contrast splits at $Z = 1$ into two peaks since a new one arises at about 115 °C. The phase structure of that PVDF/PA-6 blend is displayed in Fig. 12.2. The low temperature peak is the only one if PVDF is the minor phase (Fig. 12.1 below right).

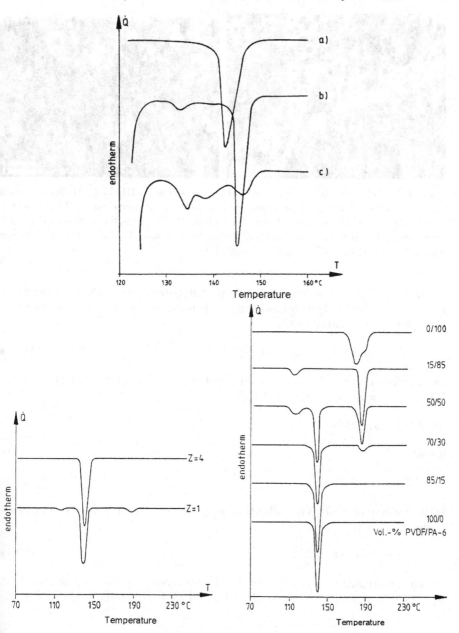

Fig. 12.1. Fractionated crystallization in some immiscible polymer blends [3–5]. (Above) extruded PE/POM = 85/15 vol-% blend after one (b) and four (c) extrusion cycles Z. Increasing Z yields a finer dispersion of the POM in the PE. (a) Pure POM. The finer the dispersion, the more crystallization peaks arise. (Below left) PVDF/PA-6 = 75/25 vol-% blends after one and four extrusion cycles. (Below right) PVDF/PA-6 blends of different compositions after four extrusion cycles. The experimental curves are redrawn subtracting the baseline.

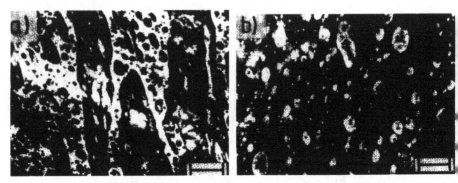

Fig. 12.2. PVDF/PA-6 = 75/25 vol-%; black phase is PVDF [3–5]. (Left) extruded one time; (right) extruded four times. Remark that the disperse phase PA-6 contains in its internal smaller droplets of the major phase PVDF. This finely dispersed PVDF gives rise to the second PVDF crystallization peak around 115 °C in Fig. 12.1 below left, $Z = 1$. Scale bar (left) 5 μm and (right) 2 μm.

The explanation is straightforward. Most polymers crystallize heterogeneously nucleated. The nucleating efficiency of a heterogeneous substrate may be characterized by the quantity $\Delta\sigma_s$,

$$\Delta\sigma_s = \sigma(\mathrm{m,c}) - \sigma(\mathrm{s,m}) + \sigma(\mathrm{s,c}) \tag{12.1}$$

(σ: interfacial energy, s: substrate, m,c: polymer melt and crystal, respectively). The critical work of nucleation F^* is

$$F^* \sim \Delta\sigma_s/(\Delta T)^2. \tag{12.2}$$

where

$$\Delta T = T_c^0 - T_c \tag{12.3}$$

is the undercooling (T_c: crystallization temperature). The number N of primary nuclei is moreover

$$N \sim \exp(-F^*/kT_c). \tag{12.4}$$

The crystallization starts effectively if N exceeds a critical value N_{cr}, $N \geq N_{cr}$. Then it is also $F^* = F_{cr}^*$ at $T = T_c$, and we have for the substrates A, B, C, ...:

$$\Delta\sigma_A : \Delta\sigma_B : \Delta\sigma_C : \ldots = T_c^{(A)}[\Delta T_c^{(A)}]^2 : T_c^{(B)}[\Delta T_c^{(B)}]^2 : T_c^{(C)}[\Delta T_c^{(C)}]^2 : \ldots \tag{12.5}$$

The highest possible undercooling is that of the own melt, ΔT_{hom}, i.e. homogeneous nucleation. Normalizing all values to those of homogeneous nucleation for which $\Delta\sigma_{\mathrm{hom}} = 2\sigma(\mathrm{m,c}) = 2\sigma_0$, Fig. 12.3 results.

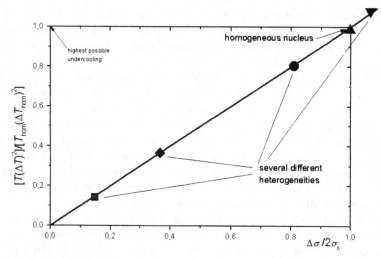

Fig. 12.3. Crystallization temperature in dependence on interfacial energy normalized to the homogeneous nucleation at T_{hom} with $\Delta\sigma_{\text{hom}} = 2\sigma_0$, cf. 12.5. All heterogeneities, which may act as heterogeneous nuclei, are located on a straight line as distinct points. In this figure, the heterogeneity associated to the squared dot has the lowest undercooling and is the effective one.

Only that heterogeneity (say: "A") with the lowest associated undercooling is effective if there are several ones present. If once nucleated, crystallization proceeds fast through the whole volume, and the other heterogeneities don't have a chance to act. But all heterogeneities are in their turn dispersed particles. The Poisson distribution

$$f_A(z) = \frac{(c_A v)^2}{z!}\exp(-c_A v) \tag{12.6}$$

describes the fraction of those droplets of the dispersed blend component with volume v, which contain z particles of the heterogeneity A (average concentration of A: c_A). The fraction $f_A(z > 0)$ of droplets, which contain at least one particle A, is then

$$f_A(z > 0) = 1 - \exp(-c_A v). \tag{12.7}$$

Nucleation by A is inhibited if

$$v \ll 1/c_A, \tag{12.8}$$

which may be achieved by sufficient dispersion. Then possibly other heterogeneities for which the same considerations hold may become effective at their typical undercooling. Finally, homogeneous nucleation may occur which concludes that fractionation of crystallization. The phenomenon allows consequently the determination of the number density of the heterogeneous nuclei of any kind and – for known σ_0 – that of the interfacial parameter $\Delta\sigma$. The number densities of the nucleating heterogeneities in the PVDF blended with PA-6 (cf. Fig. 12.1 and Fig. 12.2) as determined that way were, e.g., found to be $c_A < 30\mu m^{-3}$ and $c_B \approx 0, 2\mu m^3$, and the corresponding interfacial energy parameters are $\Delta\sigma_A = 3, 8\,mJ/m^2$ and $\Delta\sigma_B = 11mJ/m^2$.

The basic phenomenon may exhibit several complications. It is usually restricted to a crystallizable disperse phase, i.e. the minor phase. Sometimes, the continuous phase can however form droplets, too, if included in droplets of the other component. Then, the major component may also exhibit fractionated crystallization (cf. Fig. 12.1 below left, $Z = 1$). If the droplet surface acts as nucleating heterogeneity, $\Delta\sigma$ will depend on the radius of the droplet's curvature and, consequently, on the droplet's size, i.e., its associated point in Fig. 12.4 extends to a line, which reflects the size distribution. That point may jump to another position after crystallization of the second component.

The phenomenon has been observed in a number of blend systems at different conditions [1–3,5,6]. Addition of a compatibilizer may cause finer dispersion and such enhance or induce fractionated crystallization [7–11]. The efficiency of a compatibilizer may such be judged. Upon injection molding, the crystallization may be fractionated depending on the screw configuration [12]. The feature may be exploited for the erasable/rewritable storage of information in artificial papers [13].

12.3 Blends with Miscibility Gap: Interface Crossing Crystallization

In phase-separated blends of systems with miscibility gap, the two phases have different contents of the crystallizing component under consideration (Fig. 12.4a). Since the critical work of nucleation and, consequently, the number density of nuclei depend on the composition, the phase with the higher content of the crystallizing component will start to crystallize before the second one (Fig. 12.4b), and the crystallization may spread over the whole phase before the crystallization in the second phase starts (Fig. 12.4c). Upon reaching the interface, however, crystallization in the second phase is induced by secondary nucleation, i.e. by the own crystals, and the crystallization spreads now into the other phase. There, however, the supermolecular structures develop which are typical for the respective composition (Fig. 12.4d), and the two crystalline phases will look quite different. These differences will be discussed in more detail in paragraph 8. The second phase (here: the continuous phase) will later occasionally be bulk nucleated, too, and develop their own spherulites (Fig. 12.4e, upper right corner). Figure 12.4f shows finally the growth of a spherulites branch through a narrow connection between two adjacent dispersed droplets.

This interface crossing crystallization may consequently lead to a number of interesting and strange supermolecular structures. It causes moreover a good adhesion between the two phases and may such be of particular importance for the mechanical strength of the material.

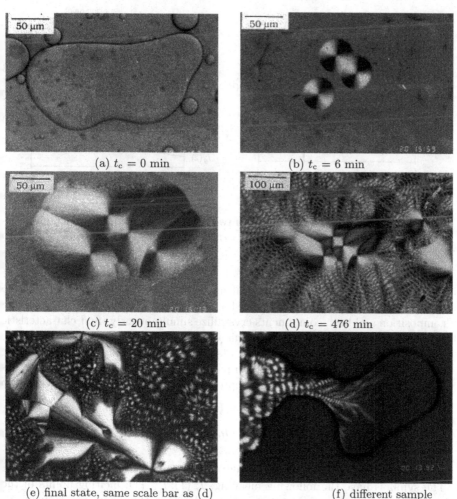

(a) $t_c = 0$ min

(b) $t_c = 6$ min

(c) $t_c = 20$ min

(d) $t_c = 476$ min

(e) final state, same scale bar as (d)

(f) different sample

Fig. 12.4. Interface crossing crystallization of PVDF in a blend with poly (ethyl acry-late), PEA [14]. $\phi_{PVDF} = 30$ wt-%, $T_c = 150\,°C$. The system has a lower critical solution temperature of $154\,°C$ at a composition of $\phi_{PVDF} = 65$ wt-% (Fig. 12.26) [15]. The initial melt temperature has been $T_m = 225\,°C$, and the two phases have compositions of $\phi_{PVDF} = 15$ wt-% (continuous phase) and $\phi_{PVDF} = 90$ wt-% (disperse phase). The last figure below right shows the crystallization in a different sample with a lower PVDF content in the disperse phase.

12.4   Miscible Crystallizing Blends: Pseudoeutectic Behaviour

The crystallization of blends of two crystallizable components within a homogeneous melt is of particular importance. For steric (i.e., basically, thermodynamic) reasons, mixed crystallization of polymers is a seldom exception. With low molec-

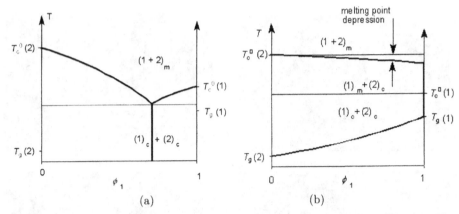

Fig. 12.5. (a) Eutectic phase diagram of two materials (1) and (2). Since the (equilibrium) melting point depression of polymers is usually rather small, their phase diagram degenerates (b).

ular weight materials, eutectic crystallization may result (Fig. 12.5a). Then, the melting point depression curves of the blend components meet at an intermediate composition and both components crystallize simultaneously (first characteristic feature of eutectic crystallization). This yields a finely dispersed structure of the crystals of the both blend components (second characteristic feature of eutectic crystallization). Such a crystallization path can however be realized in polymers only rarely since the melting point depression $\Delta T_c^0 \approx T_c^0 - T_{c,m}^0$ of alloys of two polymers (1) and (2) [16],

$$\frac{1}{T_{c,m}^0(1)} - \frac{1}{T_c^0(1)} = \frac{V_1}{V_2} \cdot \frac{\chi_{12}}{\Delta H^0(1)} \cdot R\phi_2^2 = \frac{\Delta T^0(1)}{[T_c^0(1)]^2}, \tag{12.9}$$

is very small due to the small mixing enthalpy as – in principal – expressed by χ. The symbols have the following meaning: $T_c^0(1), T_{c,m}^0(1)$: melting points of the crystallizing component (1) as pure material and in the blend, respectively, V: to molar volume, $\Delta H^0(1)$: crystallization enthalpy of the pure component (1), χ: Flory-Huggins-Staverman interaction parameter, R: gas constant]. The respective phase lines do then not meet at an intermediate composition unless the equilibrium melting temperatures are almost equal (Fig. 12.5b). Following equilibrium thermodynamics, the blend partners would then crystallize successively. In reality, however, the crystallization of the blended polymers is kinetically hindered distinctly with increasing content of the other component. This, in its turn causes distinctly inclined "kinetic melting point depression lines". In contrast to the equilibrium lines, these lines may intersect at an intermediate composition and simultaneous crystallization may such be kinetically enforced ("kinetically pseudoeutectic" crystallization, pseudoeutectic of type I). These kinetic phase diagrams are however not unique as those of the thermodynamic equilibrium; simultaneous crystallization may be brought about by different thermal paths. The resulting structure must then not necessarily consist of finely dispersed crystals of the components, e.g., mixed lamellar stacks of the components. That type of pseudoeutecticity fulfils therefore only the first feature of eutectic crystallization. Beside it, there is therefore a second one that gives the finely dispersed crystal structure of mixed lamellae or mixed lamellar stacks ("structural pseudoeutectic" crystallization, pseudoeutectic of type II). Then, however, the crystallization of the components may proceed successively.

There are only few polymer blend systems of two individually crystallizable components, which fulfill the presupposition of a homogeneously mixed melt in the whole composition range and at all accessible temperatures. Examples are the systems poly (vinylidene fluoride)/polyhydroxybutyrate (PVDF/PHB), polycarbonate of biphenyl A/poly-ε-caprolactone (PC/PCL), poly (ethylene terephthalate)/poly (butylene terephthalate) (PET/PBT) und PHB/poly (ethylene oxide) (PHB/PEO).

Figure 12.6 gives the equilibrium phase diagram of the blend PC/PCL [17]. Such phase diagrams can however not immediately be observed. They are extrapolated from isothermal crystallization experiments by suitable means [18]. The melting point depression of the PC is very small, and the PCL crystallizes from its own melt and shows therefore no depression. It should be noted that PC usually crystallizes only to a small extent due to the small gap between $T_c^0(\text{PC}) = 248\,^\circ\text{C}$ and $T_g(\text{PC}) = 146\,^\circ\text{C}$. Addition of PCL broadens this gap since T_g of PCL is very low, and PC can crystallize to degrees of up to 30% despite the dilution. This causes a hardening of the blend although PCL is usually used as a plastifier ("antiplastification").

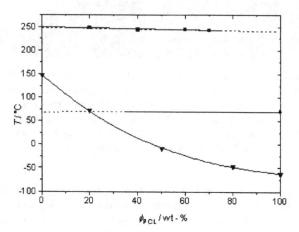

Fig. 12.6. Equilibrium phase diagram of a PC/PCL blend [17]. —▼--▼—▼— : T_g ; —■-■-■— : $T_c^0(\text{PC})$; —●—●—●— : $T_c^0(\text{PCL})$.

Figure 12.7 gives an example of a kinetic phase diagram. Simultaneous crystallization of the blend components is described in [19–22] for a blend of poly (butylene succinate) and poly (vinylidene chloride-co-vinyl chloride). That simultaneous crystallization yields the strange phenomenon of "interpenetrating

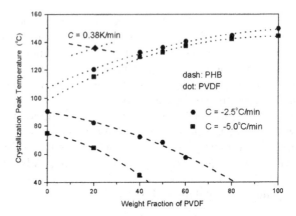

Fig. 12.7. Kinetic phase diagrams of a PHB/PVDF blend upon anisothermal crystallization by different cooling rates. A pseudoeutectic of type I may result for an extrapolated cooling rate of 0.4 K/min and a PVDF content of 20 wt-%.

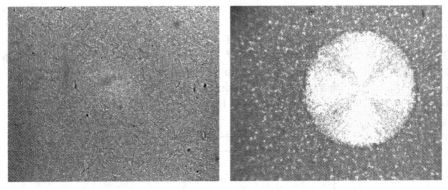

Fig. 12.8. Spherulitic structures in PVDF/PHB blends. PVDF/PHB = 20/80 wt-%; 2-stage isothermal crystallizations at 120 °C and 90 °C, both until completion. After the crystallization step at 120 °C, the volume is filled with small irregular PVDF spherulites (average diameter: about 40 μm). At 90 °C develop then large PHB spherulites (size here: about 200 μm), which at later stages fill also the whole volume but have incorporated all former small PVDF spherulites (pseudoeutectic of type II).

spherulites": whereas it is well known that spherulites of the same polymer stop to grow when they meet, spherulites of different polymers interpenetrate and grow into each other. The resulting structure is however dominated by the habit of the spherulites of one component. Such spherulite interpenetration can however result after a two-step crystallization, too. The spherulites of the two components grow then successively. Those of the first crystallizing component may even fill the whole volume but those of the second one incorporate them then nevertheless completely. This is demonstrated in Fig. 12.8 for the system PVDF/PHB and reported in [23] for a blend of PHB and cellulose ester. These structural features are detected by light and electron microscopy. Small angle

X-ray scattering, on the contrary, can unveil the fine structure on the lamellar level. Depending on the thermal history and on the mobility of the involved macromolecular chains, stacks of mixed lamellae of the two components as well as individual adjacent lamellar stacks of the blend partners can be detected. Corresponding observations were reported for PC/PCL blends [24], PHB/PEO blends [25], PVDF/PHB blends [26,27], and PVDF/poly (1,4-butylene adipate) blends [28].

12.5   Miscible Blends: Crystallization Induced Composition Inhomogeneities

The crystallization of one component in a homogeneous polymer blend causes, and bases on, far reaching diffusive displacements of the involved macromolecules. The non-crystallizing component "A" must be removed from the crystallization front and diffuse away into the remaining melt, whereas the molecules of the crystallizing component "C" must migrate to that location (Fig. 12.9). Clearly, it must be distinguished between the diffusion ability of the crystallizable units (which determines the local crystallization conditions) and that of the whole chains (which determines the overall crystallization kinetics). The appearance of a spherulite depends on both.

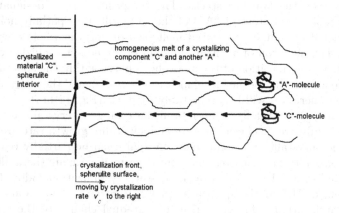

Fig. 12.9. Chain diffusion in the vicinity of a crystal growth front.

This particular condition results in a large variety of supermolecular structures. The actual morphology depends decidingly on the rates ratio between that diffusive displacement of chains (v_d) and growth of the crystalline entities (v_c). This ratio, in turn, can be controlled by suitable choice of the crystallization temperature T_c. The diffusive displacement rate v_d,

$$v_d = \frac{d}{dt}\left[\langle r^2\rangle(T_c,t)\right]^{1/2} = \frac{d}{dt}\left[6D(T_c)t\right]^{1/2} = \frac{1}{2}\sqrt{6D(T_c)}t^{-1/2} \ , \qquad (12.10)$$

with

$$D(T_c) \approx D^0 \exp\left[c_1(T_c - 50 - T_g)/(c_2 + T_c - 50 - T_g)\right]$$
$$= D^0 \exp\left[\frac{-\Delta G_d(T_g, T_c)}{RT_c}\right] \tag{12.11}$$

(D: whole chain diffusion constant, $\langle r^2 \rangle$: mean squared displacement of a chain) can sufficiently well be described by a Vogel-Fulcher-Tamman law (which became known as Williams-Landel-Ferry equation [30]). v_d increases consequently evenly with temperature [29]. The spherulite growth rate v_c, in contrast [31,32],

$$v_c(T_c) = v_c^0 \phi_C \exp\left[\frac{-k_c}{T_c(T_{c,m}^0 - T_c)f} + \frac{c_3 T_{c,m}^0 \ln\phi_C}{(T_{c,m}^0 - T_c)}\right] \exp\left[\frac{-\Delta G_d}{RT_c}\right], \tag{12.12}$$

(ϕ_C: volume fraction of the crystallizing component, k_c, f, c_i: constants) exhibits a maximum anywhere between the glass transition temperature T_g and the equilibrium melting temperature T_c^0, and is zero outside this temperature range (Fig. 12.10). Consequently, three different regimes can be defined [33]:

I: If $v_d \ll v_c$ which is usually observed at rather low T_c, the excess part of "A" which is released upon crystallization of "C" will completely be caught and included in the growing spherulites and trapped in their interlamellar regions or between the lamellar stacks. The composition ϕ (ϕ designates in this paragraph the content of "A") of the melt will stay spatially homogeneous and equal to that of the initial melt (ϕ^0) during the whole crystallization time. The inclusion of a surplus amount of the non-crystallizing component causes the composition of the amorphous phase within the spherulites to be different from that of the initial melt (increased ϕ), and the internal structure of the spherulites will be comparably irregular (Fig. 12.11a).

II: In that regime, usually at an intermediate range of crystallization temperatures, diffusive displacement rate and spherulite growth rate are balanced. The non-crystallizable component "A" will then be partially rejected from the spherulite growth front back into the melt without the ability to distribute evenly. This is equivalent to the crystallization induced impoverishment of "C" at the spherulite growth front and its necessary migration to that location. This gives rise to a gradual change of the composition $\phi(r = 0, t)$ at the spherulite surface (r: distance from the spherulite surface) and to the creation of composition profiles $\phi(r, t)$ around the spherulites with maximal "A"-content at their surfaces (Fig. 12.11b). These profiles may reach more or less wide into the melt. Again, it must be distinguished between very local compositional changes, which determine the fibrillar internal structure of a spherulite, and overall ones, which rule the growth of a spherulite as a whole. D/v_c yields an estimate for the width r_0 of the composition profile. It turns out that r_0 may amount up to 100 μm.

III: Here we have $v_d \gg v_c$ which is usually observed at high T_c, and all material of "A" which is released during the crystallization of "C" can diffuse far, and

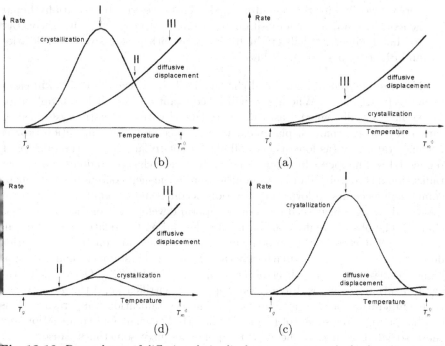

Fig. 12.10. Dependence of diffusive chain displacement rate and of spherulite growth rate on temperature (schematically). In a very general situation, three regimes I to III can be observed depending on the actual crystallization temperature T_c (a). In some cases, we have $v_d \gg v_c$ in the whole T_c range, and only regime III is observed (b) or it is $v_d \gg v_c$ for all T_c, and only regime I occurs (d). The crystallization time is a parameter in these drawings as is evident from (12.11) and (12.12): v_d depends explicitly on time, and v_c implicitly since ϕ_C may change with time. The ratio between v_d and v_c may consequently change with proceeding crystallization which may cause a shift between the several regimes.

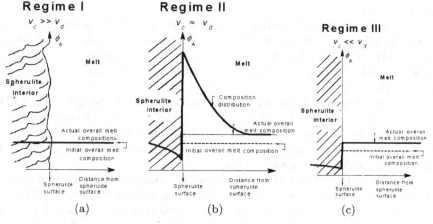

Fig. 12.11. Composition profiles at the various regimes.

spread evenly, into the remaining melt. This gives rise to a spatially homogeneous overall melt composition ϕ_∞ similarly to case "I" but, in contrast to that regime, ϕ^∞ will gradually increase with progressive crystallization and differ from ϕ^0 (Fig. 12.11c).

The described local and global changes in ϕ, by virtue of 12.12, will cause likewise changes in v_c. Whereas v_c will be constant during whole crystallization in regime I, it will change drastically in regime II with time until crystallization rate and diffusive chain displacement rate balance and the composition profiles become stationary (as long as crystallinity is not to large; then, the same as in regime III will happen). In regime III, finally, v_c will change gradually until crystallization has completed or v_c approaches zero if the melt content of crystallizing component becomes to low. The situation becomes more complicated since the described changes in v_c feed back the temporal development of the composition change. The observed effects will crucially depend on the relative contributions of the explicit composition dependence of v_c as such and on the composition dependencies of T_g (particularly whether T_g of the crystallizing component is higher or lower than that of the other component) and of $T_{c,m}^0$. Occasionally, the crystallization process will finally be stopped by vitrification if the glass transition temperature of the melt approaches the crystallization temperature. This can happen in later stages of regime II and III. The change of composition and the associated change of the $v_c(T_c)$ dependence may sometimes cause a shift from one regime to another.

The detection and quantitative characterization of these composition inhomogeneities is possible by a variety of means. The composition change at the spherulites surface may be determined by evaluation of the spherulite growth rate, and the composition profiles around the spherulites by evaluation of the number density of athermal nuclei or by IR microscopy. The latter allows also the measurement of the composition changes within the spherulites. If the blend system exhibits a miscibility gap, the composition inhomogeneities are unveiled by the spatial distribution of demixing zones if being sufficiently strong (cf. paragraph 7).

In Fig. 12.12, the composition changes at the surface of a growing spherulite as concluded from the temporal change of the spherulite growth rate is displayed for a PVDF/PEA blend with $\phi_{PEA} = 0.7$ wt-ct upon isothermal crystallization at $155\,^\circ\text{C}$ [34]. The increase of the composition to more than 0.95 wt-ct is astonishingly high.

The next Fig. 12.13 shows the composition change in and around a PVDF spherulite in a blend with PMMA as determined by IR microscopy [35]. The composition at the spherulite surface changes again remarkably from $\phi_{PMMA}^0 = 0.4$ wt-ct in the initial blend to $\phi_{PMMA} \approx 0.6$ wt-ct after 12 h of crystallization. The profile extends over about 50 μm.

A mathematical description of the described composition profiles by solving the diffusion equation must take into account the unusual boundary conditions: the boundary (the growing spherulite surface) moves by v_c and is simultaneously a source ε of the diffusing substance (the component "A" as released by crys-

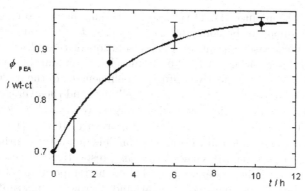

Fig. 12.12. Composition change at the surface of a growing spherulite [34]. Homogeneous PVDF/PEA blend, $\phi^0_{PEA} = 70$ wt-%, $T_c = 155\,°C$.

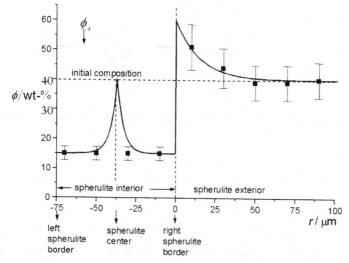

Fig. 12.13. Composition profile around and in a PVDF spherulite in a homogeneously mixed blend with poly(methyl methacrylate) (PMMA) as determined by IR microscopy [35]. The evaluation made use of the composition dependence of the absorption ratio between the IR bands at $\bar{v} = 880\text{cm}^{-1}$ (PVDF) and $\bar{v} = 765\text{cm}^{-1}$ (PMMA). Initial composition $\phi^0_{PMMA} = 0.4$ wt-ct, $T_c = 161\,°C$, $t_c = 12$ h.

tallization of "C"). From a physical point of view, the problem is the same as that of the migration of a phase transition front where latent heat of conversion is produced which has to be dissipated by heat conduction ("Stefan's problem" [36,37]). However, the present situation becomes more complicated since v_c and ε (i.e. essentially the degree of crystallization inside the spherulites) depend on each other and on the actual value ϕ_R of the composition at the boundary, which in turn changes with time. So far, the diffusion equation has not been solved under such complicated boundary conditions. D may be assumed to be independent on the composition since the D values of the components are roughly equal. If at first ε is assumed to be constant, too, and $v_c \sim t^{-1/2}$, a numerical modeling is possible which bases on the particular solution of the diffusion equation in spherical coordinates as given 1950 by Frank [38]. After suitable evaluation, this allows calculation of the effective diffusion coefficient D, of the internal composition of the spherulites, their degree of crystallinity, and the average composition of the remaining melt, which rises with progressive crystallization. The determination of D this way is of particular importance. It is rather difficult and laborious to get experimental information on this quantity by other means like inelastic neutron or forced Rayleigh scattering, respectively. The simulations allow moreover the prediction and evaluation of the temporal development of the thickness of the phase-separated layer around a growing spherulite in systems with miscibility gap.

In the approach of Frank [38], the diffusion equation

$$\frac{\partial \phi}{\partial t} = D \nabla^2 \phi \tag{12.13}$$

is solved by using the dimensionless variable s,

$$s = \frac{r'}{\sqrt{Dt}}, \tag{12.14}$$

where r' is the distance from the spherulite center and t is the time. The solution is

$$\phi = c \left\{ \frac{\exp(-s^2/4)}{s} - \frac{\sqrt{\pi}}{2} \left[1 - \mathrm{erf}(s/2) \right] \right\} + \phi_\infty. \tag{12.15}$$

$$\mathrm{erf}(x) = \frac{2}{\sqrt{\pi}} \int_{-\infty}^{x} \exp(-z^2) dz \tag{12.16}$$

is the error function, ϕ_∞ is the content of the noncrystallizable component far in the melt, and c is given by

$$c = \frac{qS^3}{2} \exp\left(\frac{S^2}{4}\right) \tag{12.17}$$

where q is that portion of "A" in units of ϕ_R (ϕ_R: the content of "A" immediately at the spherulite surface) which is expelled from the spherulite, and $R = S\sqrt{DT}$. Since a part $1 - \varepsilon$ of "A" is trapped within the spherulite, the expelled amount is given by $q = \varepsilon \phi_R$, and ϕ_R can be expressed by

$$\phi_R = \phi(s = S) = \frac{\phi_\infty}{1 - \frac{\varepsilon S^2}{2} \left\{ 1 - \frac{S\sqrt{\pi}}{2} \left[1 - \mathrm{erf}\left(\frac{S}{2}\right) \right] \right\}}. \tag{12.18}$$

The overall composition ϕ_i inside the spherulite is $\phi_i = (1 - \varepsilon)\phi_R$. ϕ_i must always be smaller than ϕ_∞ from mass conservation considerations. The composition $\phi_{i,A}$ of the amorphous phase within the spherulites, however, should exceed

ϕ_∞. ε is therefore restricted to

$$\varepsilon \le \frac{x_c(1 - \phi_\infty)}{1 - x_c\phi_\infty}, \tag{12.19}$$

where x_c is the degree of crystallinity within the spherulites.

In numerical treatments of this solution, the temporal change of v_c with temporal change of ϕ_R can formally be considered, this yielding a good adaptation to the real situation.

Figure 12.14 shows the result of such a calculation for a PVDF/PEA blend [39]. A best fit yields $D = 300 \ \mu m^2/h$. This value compares well with that of 120 $\mu m^2/h$ as determined in PVDF/PMMA sandwich arrays at 190 °C by evaluation of the interdiffusion profile [40].

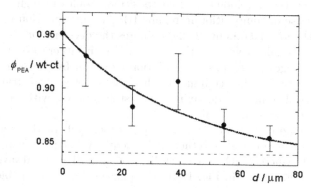

Fig. 12.14. Composition profile around a spherulite as determined by evaluation of the spatial number density of athermal primary nuclei after suitable quench from the crystallization temperature [34]. PVDF/PEA blend with $\phi_{PEA}^0 = 70$ wt-%; $T_c = 155$ °C, $t_c = 21$ h.

The full line is got by least square numerical modeling and the dotted line indicates the change of the overall composition to $\phi_{\infty,PEA} = 84$ wt-% with proceeding crystallization.

12.6   Miscible Blends: Competition Between Crystallization and Vitrification

As already pointed out, the crystallization induced changes in composition of the remaining melt cause by virtue of 12.12 a likewise change of the spherulite growth rate which in its turn results in a self-deceleration of crystallization; v_c decreases with time [39,41,42,46]. Figure 12.12 bases on the evaluation of this dependency. Such non-linear growth of spherulites has also been reported in some polymer/oligomer mixtures, and has there been explained in terms of diffusion-controlled growth of spherulites [47].

Several factors contribute to the composition and temperature dependence, respectively, of v_c:

$$v_c(T_c) = v_c^0 \phi_c \exp\left[\frac{-k_c}{T_c(T_{c,m} - T_c)f} + \frac{c_3 T_{c,m}^0 \ln\phi_C}{(T_{c,m}^0 - T_c)}\right] \exp\left[\frac{-\Delta G_d}{RT_c}\right]$$

$$= F_p(T, \phi) \cdot F_n(T, \phi) \cdot F_e(T, \phi) \cdot F_v(T, \phi)$$

$$\frac{-\Delta G_d(T_g, T_c)}{RT_c} = c_1(T_c - 50 - T_g)/(c_2 + T_c - 50 - T_g). \tag{12.20}$$

The factors F_p, F_n, F_e, F_v are a prefactor as well as the nucleation, entropic, i.e. dilution, and vitrification contributions to v_c. v_c depends not only explicitly but also implicitly on composition since both T_g and $T_{c,m}^0$ depend on composition, too. It may of particular interest whether that deceleration – if occurring at all – is caused by the development of composition profiles in regime II or by the gradual increase of composition in regime III. Another question to be answered is which of the four factors in 12.12 dominates the deceleration of the spherulite growth and particularly whether the deceleration – in either case – is immediately due to the changes in composition (dilution effect) or by the associated change in T_g (vitrification). That coupling of the kinetics of a phase transition phase transition – either mixing/demixing or crystallization – with vitrification has been basically discussed in [48].

A corresponding analysis has been performed [49] in the unconditionally miscible blend of polyhydroxyether of biphenyl A (PHE) and poly (ethylene oxide) (PEO) where PEO is a crystallizable and PHE an always-amorphous component (Fig. 12.15). In Fig. 12.16, the phase diagram of the blend is given. The development of crystallinity upon annealing causes a change of the average composition of the remaining melt, which in its turn causes a change of its T_g and its $T_{c,m}^0$ (Fig. 12.17). Figure 12.18 left shows that v_c behaves non-linear for a very narrow temperature range with width of less than 2K around a temperature $T_{c,n}$. This indicates regime II. The temperature $T_{c,n}$ depends on composition. $T_{c,n}$ shifts to higher temperature with increasing PEO content. This reflects both the increasing chain mobility with temperature and the increase of crystal growth rate with decreasing dilution. For still higher T_c, another slight deviation from the linear growth is found at longer times (Fig. 12.18 right). This indicates transition to regime III.

Since the non-linearity of Fig. 12.18 left is indicative for regime II, and the corresponding temperature range is rather small, Fig. 12.10a is applicable. In Fig. 12.19, all determined (initial) growth rates are collected, and the temperatures $T_{c,n}(\phi^0)$ are indicated. Using the T_g's of Fig. 12.16, the diffusive displacement rates can be adjusted by means of 12.11 with $c_1 = 20.4$ and $c_2 = 101.6$ K [30] to meet the schematic Fig. 12.10a. Introducing the T_g's and $T_{c,m}^0$'s of Fig. 12.17 into 12.12, it turns out that the several contributions relate approximately as

$$F_p(T_c) \cdot F_n(T_c) \cdot F_e(T_c) \cdot F_v(T_c) = 1 : 1 : 3 : 15$$

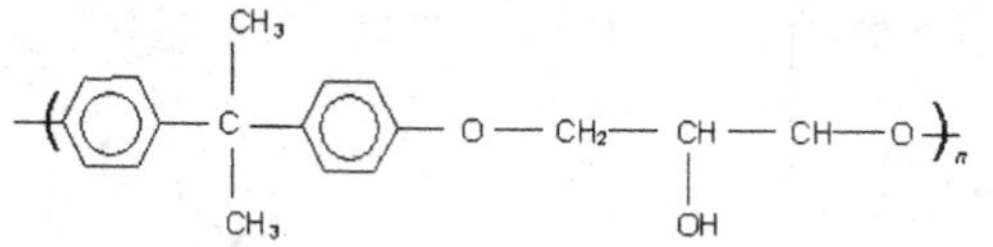

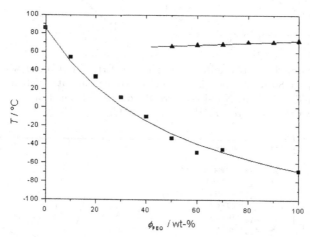

Fig. 12.15. Chemistry of PHE (left) and PEO (right).

Fig. 12.16. Phase diagram of the PHE/PEO blend [49].
——■—■—■—— : T_g; —▲—▲—▲— : $T_{c,m}^0$(PEO).

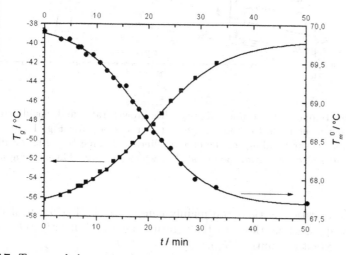

Fig. 12.17. Temporal change in glass transition temperature and equilibrium melting temperature with proceeding crystallization at $T_c = 48\,°C$ [49]. Initial composition $\phi_{PEO}^0 = 80$ wt-%.

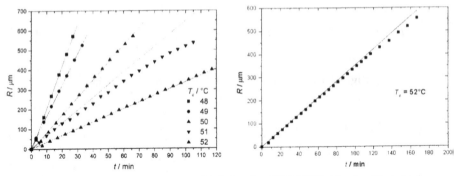

Fig. 12.18. Growth of spherulite radii R in the PHE/PEO blend with initial composition $\phi^0_{PEO} = 80$ wt-% upon isothermal crystallization at different crystallization temperatures T_c [49]. Left: Non-linear growth due to regime II at $T_c = 51\,^\circ$C. Right: Non-linear growth due to regime III at $T_c = 52\,^\circ$C.

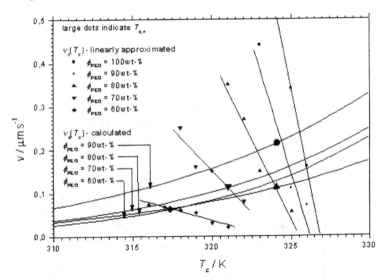

Fig. 12.19. Temporal Experimental spherulite growth rates and calculated diffusive chain displacement rates to model regime II in the schematic Fig. 12.10a [49]. The figure shows the surroundings of the crossing lines at regime II in Fig. 12.10a for the several blends. The measuring points at $T_{c,n}$ which indicate that crossing are drawn enlarged.

The deceleration is therefore mainly caused by slowing down of the chain motion by approaching the glass temperature, i.e. by vitrification (F_v) and only to a small extent to dilution (F_e).

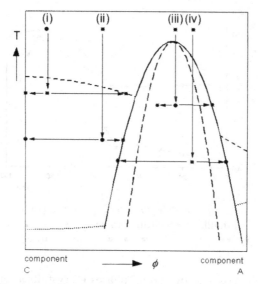

Fig. 12.20. Solidification paths in crystallizing polymer blends with miscibility gap. (i): Quench induced crystallization. (ii): Crystallization induced demixing. (iii): Demixing induced crystallization (iv): Immediate competition between crystallization and demixing. ———: binodal, — —: spinodal, - - - -: melting point depression curve, ...: glass transition temperature.

12.7 Blends with Miscibility Gap: Competition Between Crystallization and Mixing/Demixing

Interesting phenomena can be observed in systems with miscibility gap. Both crystallization and mixing or demixing are governed by their own kinetics, which may proceed on different time scales. The competition between these two kinetics may cause a number of particular features, which for the case of a blend system with upper critical solution temperature (UCST) are sketched schematically in Fig. 12.20 [33,41,42]. Of particular interest are the cases (ii) to (iv), i.e. crystallization induced demixing (ii), demixing induced crystallization (iii), and immediate competition between crystallization and demixing (iv). Case (i) relates in principal to a homogeneous blend and is discussed in detail in the foregoing section of this article. The following discussion as well as Fig. 12.20 relates to systems with UCST but corresponding observations may be made with blends with lower critical solution temperature. For the sake of simplicity, the consideration is restricted to systems with one crystallizable component; if both components are crystallizable, additional considerations are necessary which are treated in Sect. 12.4.

In path (ii), the homogeneous melt of suitable composition outside the miscibility gap will be crystallized by quench to a temperature where the system is in the homogeneous region. Due to the crystallization of the one component, the remaining melt will enrich in the other until the miscibility gap is reached

230   Bernd-J. Jungnickel

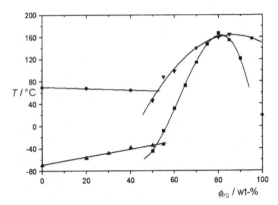

Fig. 12.21. Phase diagram of a PCL/OS blend [43]. M_w (PCL) = 15 000, M_w (PS) = 840. —•—•—•— : Equilibrium melting temperature, —▲—▲—▲— : glass transition temperature, —▼—▼—▼— : binodal, —■—■—■— : spinodal

and the melt decomposes in two liquid phases ("crystallization induced demixing"). Depending on the rates ratio between diffusive displacement of the chains and spherulites growth as discussed in more detail in Sect. 5, composition profiles may develop, and the mentioned enrichment of the melt may be restricted to the immediate vicinity of the spherulite surface (Fig. 12.22a). That demixing will consequently be observed first at that location. This is indeed the case (Fig. 12.23a) as illustrated here by corresponding observations in a blend of PCL and oligostyrene (OS; Fig. 12.21) [33,43,44]. This blend exhibits a phase diagram similar to that of Fig. 12.20. That demixing at the surface of the spherulites, moreover, starts in the gap between two nearly adjacent spherulites since the mentioned composition profiles superimpose there (Fig. 12.23b). This leads to a still more distinct composition change between the two spherulites (Fig. 12.22b). Finally, the volume between the several spherulites is filled with the second phase, which possibly crystallizes its own way dependent on its composition (Fig. 12.23c,d).

The mentioned composition profiles develop only in regime II, i.e. at intermediate crystallization temperatures. At lower T_c, crystallization will proceed according to regime I, and the surplus non-crystallizing component – here: the OS – will completely or at least partially be included in the developing spherulites (Fig. 12.24 left). If the binodal is passed inside the spherulites, demixed droplets my develop there. It has, however, been show that that demixing can occasionally be suppressed for kinetic reasons, and the amorphous phase has an nonequilibrium composition [45]. In any case, the spherulites will have an irregular shape and low crystallinity.

Figure 12.24 right shows finally an example for immediate kinetic competition between demixing and crystallization. The demixing proceeds faster that crystallization at the chosen T_c. The structure shows a disperse phase with low PCL content around which spherulites grow which finally engulf and include the dispersed droplets.

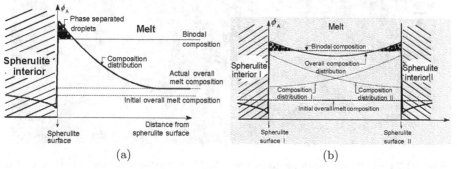

(a)   (b)

Fig. 12.22. Crystallization induced phase separation upon passing the binodal at the surface of spherulites after development of composition profiles around an isolated spherulite (a) and between adjacent spherulites (b).

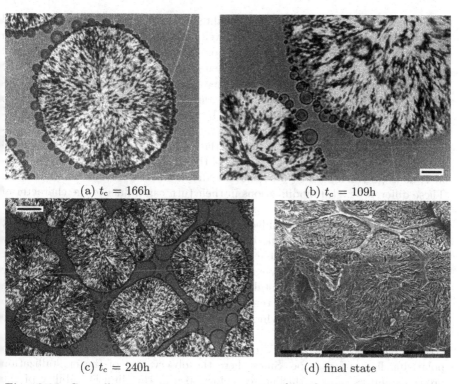

(a) $t_c = 166$h   (b) $t_c = 109$h

(c) $t_c = 240$h   (d) final state

Fig. 12.23. Crystallization induced phase separation [43,44]. PCL/OS blend, $\phi_{PCL} = 60$ wt-%, $T_c = 51\,^\circ$C. (a) to (c): light micrographs, scale bar: (a,b) 25 μm, (c) 125 μm. (d): electron micrograph, scale bar 100 μm.

12.8   Picture Gallery

It has been proved in the foregoing paragraphs by a number of examples (Fig. 12.5, Fig. 12.8, Fig. 12.23, Fig. 12.24) that the spherulitic morphology in polymer

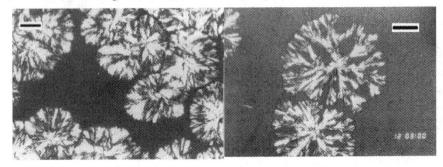

Fig. 12.24. (Left) same sample as in Fig. 12.23 but crystallization at 44 °C; scale bar: 50 μm. (Right) immediate competition between crystallization and demixing; PCL/OS = 40/60 wt-%; $T_c = 46$ °C, $t_c = 27$ h, scale bar: 100 μm.

blends can have a lot of different and interesting faces. It should be recalled that the habit of a spherulite, which develops in a blend at given composition and thermal conditions, depends mainly on the ratio D/v_c [50]. That ratio with respect to the crystallizing chain segments (D_s) determines the habit of the spherulites interior whereas that ratio as applied to the whole chain (D_c) is linked with the progress of the spherulite development as a whole. The habit in its turn may be characterized essentially by the branching probability of the radial lamellae, on their twist period, and on the surface curvature. Since v_c bases basically on the nucleation conditions, the named thermal and compositional conditions may also determine the type of the developing crystal lattice in polymorphous polymers, which are able to exhibit different crystal modifications. These different crystal modifications in their turn can influence the character of a spherulite, too.

This article shall be concluded with a selection of pictures by which that manifoldness of spherulite habits in polymer blends is underlined. They display PVDF spherulites as grown in blends with PEA or PMMA. The blend of PVDF and PEA exhibits a miscibility gap (Fig. 12.25) [15], which, however, in many cases may be ineffective for crystallization. The blend PVDF/PMMA is miscible in the whole composition and temperature range. PVDF can crystallize in up to five crystal modifications, which causes additional particular features in the developing spherulites.

The pictures of Fig. 12.26 are taken from crystallization experiments in a polarizing light microscope. Since there the observation of the crystallization process requires sample thicknesses below 50 μm, the diffusion ability of the macromolecules is restricted largely to two dimensions. This enhances the influence of the chain displacement ability on the developing structure. The actual habit in a given sample at given thermal conditions will moreover depend on the actual sample thickness. The pictures of Fig. 12.26 are therefore not necessarily representative for the given thermal and compositional conditions.

Pictures (a) to (d) yield examples for the sudden change of the crystal modification or the lamellar branching probability at a certain instant. At low con-

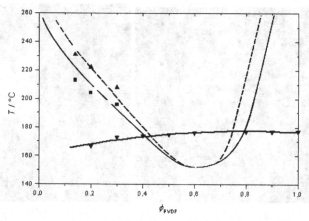

Fig. 12.25. Phase diagram of a PVDF/PEA blend [14,15]. —▼—▼—▼— : $T^0_{c,m}$ of PVDF, —■—■—■— : binodal, —▲—▲—▲— : spinodal.

centrations of the amorphous component (i.e. dilution of the PVDF), nucleation or growth (or both) of the crystalline α-modification of the PVDF are favored whereas the γ-modification grows predominantly at higher concentration. This sudden change of modification is then most probably due to passing a corresponding limiting composition at the spherulite growth front due to the development of composition profiles. The α-spherulites are characterized by a longer lamellar twist period and (in most instances) a more dendritic growth with respect to the γ-spherulites. The former is unveiled by the ringed structure of a picture, and the ring separation is equal to the twist period. The latter is linked with the lamellar branching probability, and a low branching probability results in an open and more dendritic habit. It is moreover seen that, in any case, the impoverishment of PVDF at the spherulites growth front and the corresponding increase of the migration paths of the macromolecules to or from that location brings about a more dendritic habit of the spherulites.

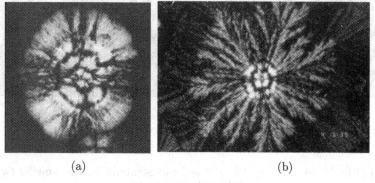

(a)   (b)

Fig. 12.26. (Cont.)

(c)

(d)

(e)

(f)

(g)

(h)

Fig. 12.26. (Cont.)

In Fig. 12.26e and Fig. 12.26f it is demonstrated that the lamellar twist perio in this particular blend system decreases with decreasing content of the crysta lizable component at the spherulite growth front: the rings become closer an

Fig. 12.26. Selection of spherulite habits, which are found in polymer blends. Here, blends of crystallizable PVDF and amorphous PEA or PMMA are shown.
(a) PVDF/PEA = 30/70 wt-%, T_c = 160 °C, t_c = 8 d. (b) PVDF/PEA = 20/80 wt-%, T_c = 153 C. (c) as (b) but other image contrast and other sample location. (d) PVDF/PMMA = 50/50 wt-%, T_c = 164 °C t_c = 6 d. (e) PVDF/PEA = 30/70 wt-%, T_c = 162 °C, t_c = 60 d. (f) as (e), but other image contrast. (g) PVDF/PEA = 30/70 wt-%, first demixed at 240 °C, T_c = 147 °C, picture taken from the continuous phase, cf. Fig. 12.5e. (h) PVDF/PEA = 30/70 wt-%, first demixed at 230 °C, T_c = 147 C, t_c = 2 d. (i) PVDF/PMMA = 50/50 wt-%, T_c = 160 °C, t_c = 118 h. (j) PVDF/PMMA = 50/50 wt-%, T_c = 160 °C, t_c = 118 h. (k) PVDF/PEA = 30/70 wt-%, T_c = 160 °C. (l) PVDF/PEA = 30/70 wt-%, T_c = 160 C, t_c = completion, sample thickness 5 μm. For more details cf. text. Figs. (d), (i), (j) cf. [52].

closer with the impoverishment of PVDF in the remaining melt (here: through regime III, cf. Fig. 12.11c). In a final state, they are unable to be resolved and the spherulite border becomes finally a frayed appearance. It can moreover seen that this border has a waved course. This may be explained by the enhancement of particular frequencies in the surface growth fluctuation spectrum [51].

Figure 12.26g and Fig. 12.26h give two interesting examples of interface crossing crystallization and are such a continuation of Fig. 12.5. The droplets from which the crystallization of the displayed continuous phase starts are at the upper rim of the picture (g) or in the upper left corner (h). It is obvious from the foregoing considerations that the droplet internal crystallizes in the γ-modification whereas the continuous phase exhibits α-crystals.

Figure 12.26i and Fig. 12.26j stem from PVDF/PMMA blends of different thickness. The dendritic spherulites of Fig. 12.26i did not grow circularly and are separated by uncrystallized material. This is due to the composition change, which is most pronounced between the spherulites (cf. Fig. 12.22b) and finally impedes further crystallization. It can seen in Fig. 12.26j that at suitable conditions α-spherulites and γ-spherulites may grow side by side. The growth of the γ-spherulites is however favored, and the α-spherulites turn occasionally to grow γ-like. The said depletion zone between adjacent spherulites is particularly pronounced in Fig. 12.26l where the final structure of a very thin PVDF/PEA sample is shown (cf. also Fig. 12.23d). To conclude, Fig. 12.26k yields another sample of a pronounced dendritic structure.

12.9   Concluding Remarks

It has been the purpose of the present article to give an impression of the diversity of supermolecular structures, which may develop in polymer blends. They are not only of aesthetical allurement. They are a convenient mean to get deeper information on, and insight into, the rules, which govern the kinetics and the thermodynamics of crystal growth not only in blends or mixtures but in one-component polymer systems, too. The particular features of that solidification may be of practical importance both for the processing of these materials and for their use. Polymer blends and mixtures can therefore serve as suitable model systems in theoretical research, and their treatment is moreover of considerable practical importance.

References

1. A. Ghijsels, N. Groesbeek, C.W. Yip: Polymer **23**, 1913 (1982)
2. N. Klemmer, B.-J. Jungnickel: Coll. Polym. Sci. **262**, 381 (1984)
3. H. Frensch, P. Harnischfeger, B.-J. Jungnickel: Fractionated Crystallization in Incompatible Polymer Blends. In: *Multiphase Polymer Materials: Blends, Ionomers, and Interpenetrating Networks* (Eds.: L.A. Utracki, R.A. Weiss). ACS Symposium Series Books 395, 101 (1989)
4. H. Frensch, B.-J. Jungnickel: Plast., Rubb.& Compos. Process. & Appl. 16, 5 (1991)
5. H. Frensch, B.-J. Jungnickel: Coll. Polym. Sci. **267**, 16 (1989)
6. M. Avella, E. Martuscelli, M. Raimo: Polymer **34**, 3234 (1993)
7. P. Jannasch, B. Wesslen: J. Appl. Polym. Sci. **58**, 753 (1995)
8. R.M. Holsti-Miettinen, M.T. Heino, J.V. Seppälä: J. Appl. Polym. Sci. **57**, 573 (1995)
9. X. Zhang, H. Chen, T. Tang, B. Huang: Macromol. Chem. Phys. **196**, 3585 (1995)
10. T.M. Ko, P. Ning: Polym. Eng. Sci. **40**, 1589 (2000)
11. N. Furgiuele, A.H. Lebovitz, K. Khait: Polym. Eng. Sci. **40**, 1447 (2000)
12. R.A. Morales, M.L. Arnal, A.J. Müller: Polym. Bull. **35**, 379 (1995)

13. T. Kowalewski, G. Ragosta, E. Martuscelli, A. Galeski: J. Appl. Polym. Sci. **66**, 2047 (1997)
14. Y. Li, L. Schneider, B.-J. Jungnickel: Polym. Networks & Blends **2**, 135 (1992)
15. B. Endres, R. Garbella, J.H. Wendorff: Coll. Polym. Sci. **263**, 361 (1985)
16. T. Nishi, T.T. Wang: Macromolecules **8**, 316 (1975)
17. K. Hatzius, Y. Li, M. Werner, B.-J. Jungnickel: Angew. Makromol. Chem. **243**, 177 (1996)
18. J.D. Hoffman, J.J. Weeks: J. Res. NBS 66A, 13 (1962)
19. J.-C. Lee, H. Tazawa, T. Ikehara, T. Nishi: Polymer Journal **30**, 780 (1998)
20. T. Ikehara, T. Nishi: Rep. Progr. Polym. Phys. Japan **43**, 253 (2000)
21. S. Hirano, Y. Terada, T. Ikehara, T. Nishi: Polym. J. **33**, 371 (2001)
22. T. Ikehara, T. Nishi: Polym. J. **32**, 683 (2000)
23. M. Pizzoli, M. Scandola, G. Ceccorulli: Macromolecules **27**, 4755 (1994)
24. Y.W. Cheung, R.S. Stein, B. Chu, G.Wu: Macromolecules **27**, 3589 (1994)
25. M. Avella, E. Martuscelli, P. Greco: Polymer **32**, 1647 (1991)
26. H. Marand, M. Collins: ACS Polym. Prepr. **31**, 552 (1990)
27. H.J. Chiu, H.L. Chen, J.S. Lin: Polymer **42**, 5749 (2001)
28. L.-Z. Liu, B. Chu, J.P. Penning, R.St.J. Manley: Macromolecules **30**, 4398 (1997)
29. J. S. Vrentas, J. L. Duda: J. Appl. Polym. Sci. **2**, 1715 (1977)
30. M. L. Williams, R. F. Landel, J. D. Ferry: J. Am. Chem. Soc. **77**, 3701 (1955)
31. J. I. Lauritzen, J. D. Hoffman: J. Appl. Phys. **44**, 4340 (1973)
32. J. Boon, J. M. Azcue: J. Polym. Sci. Part A-2 **6**, 885 (1968)
33. B.-J. Jungnickel: Research Trends: Current Trends in Polymer Science **2**, 157 (1997)
34. R. Wellscheid, J. Wüst, B.-J. Jungnickel: J. Polym. Sci. B: Polym. Phys. Ed. **34**, 267 (1996)
35. P. Kalivianakis, B.-J. Jungnickel: J. Polym. Sci. B: Polym. Phys. **36**, 2923 (1998)
36. J. Stefan: Ann. Phys. Chem. N.F. **42**, 269 (1891)
37. H.S. Carslaw, J.C. Jaeger: Conduction of heat in solids, Chapter XI: Change of state. p. 282ff. 2nd edn. (Oxford University Press, Oxford, New York, Toronto 1959)
38. F.C. Frank: Proc. Royal Soc. London Ser. A **201**, 586 (1950)
39. R. Wellscheid, J. Wüst, B.-J. Jungnickel:, J. Polym. Sci. B: Polym. Phys. **34**, 893 (1996)
40. S. Wu, H. Chuang, C.D. Han: J. Polym. Sci. Polym. Phys. Ed. **24**, 143 (1986)
41. H. Tanaka, T. Nishi: Phys. Rev. Lett. **55**, 1102 (1985)
42. H. Tanaka, T. Nishi: Phys. Rev. A **39**, 783 (1989)
43. Y. Li, M. Stein, B.-J. Jungnickel: Coll. Polym. Sci. **269**, 772 (1991)
44. H.M. Shabana, R.H. Olley, D.C. Bassett, B.-J. Jungnickel: Polymer **41**, 5513 (2000)
45. Y. Li, B.-J. Jungnickel: Polymer **34**, 9 (1993)
46. C. H. Lee, T. Okada, H. Saito, T. Inoue: Polymer **38**, 31 (1997)
47. T. Okada, H. Satio, T. Inoue: Macromolecules, **23**, 3865 (1990)
48. Y. D. Shibanov, Y. K. Godovsky: Makromol. Chem. Macromol. Symp. **44**, 61 (1991)
49. S. Zheng, B.-J. Jungnickel: J. Polym. Sci., Polym. Phys. Ed. **38**, 1250 (2000)
50. H.D. Keith, F.J. Padden: J. Polym. Sci. Polym. Phys. Ed. **25**, 229, 2265, 2371 (1987)
51. D. Braun, M. Jacobs, G.P. Hellmann: Polymer **35**, 706 (1994)
52. G.P. Hellmann: personal communication (1994)

13 Dendritic Growth of Polyethylene Oxide on Patterned Surfaces

Hans-Georg Braun, Evelyn Meyer, and Mingtai Wang

Max Bergmann Center for Biomaterials and Institute of Polymer Research Dresden, Germany Microstructure Group
D-01069 Dresden, Germany, braun@ipfdd.de

Abstract. Microstructured films are formed on micropatterned substrates. Film formation on hydrophilic micrometer-sized isles allows to prepare amorphous films of polyethylene oxide (PEO) which are metastable with respect to crystallisation. Heterogeneous nucleation can be initiated by contact with an AFM tip. The controlled nucleation in metastable micropatterned films enables us to study the propagation of crystallisation fronts within confined areas. The crystallisation process within the ultrathin micropatterned film segments favours branched lamellae structures typical for diffusion limited aggregation (DLA) processes. We observed a strong influence of confined geometries on the morphological features of branched PEO lamellae.

13.1 Introduction

During the last decades the crystallisation of thermoplastic polymers into lamellar crystals and their spherulitic superstructures has been of great interest in order to understand and to influence the bulk properties of engineering polymers such as fracture toughness, yield strength or optical transparency. While in most engineering applications of polymers structure-property relationships of the bulk state are in focus of attention new applications in micro- and nanotechnology focus on morphology formation in thin films.

Many previous studies of crystallisation in ultrathin films were related to epitaxial growth processes on interfaces. Within this context a number of TEM studies focused on the investigation of epitaxial growth processes of ultrathin polymer films in contact with oriented polymers films, organic molecular crystal surfaces or inorganic crystals. Typical examples are the epitaxial growth of polyethylene on benzoic acid [1,2] or hydrogen-bonding polymers as polymamides and polyesters in contact with various surfaces [3]. Polymer/polymer epitaxy has been observed for it-polypropylene crystallizing in contact with PTFE [4] and has also been extensively studied during polyethylene crystallisation on oriented polypropylene needle crystals [5] or for the crystallisation of syn-PP in contact with HDPE (100) lattice planes [6]. Epitaxial studies focus on the understanding of specific molecular interactions of an ordered substrate with polymer chains organising in polymer crystals. Epitaxy can influence the molecular packing and the orientation of polymer crystals at the interface.

With the rapidly growing interest in nanotechnology thin and ultrathin polymer films become more and more important as interfacial layers for the control of surface directed growth processes as for example in biomineralisation [7,8]. Consequently questions related to the structure formation of thin polymer films in contact with surfaces and its influence on physicochemical behaviour are of growing importance. The interest in polymer structure formation at interfaces includes the adsorption of polymers at surfaces, the wetting and dewetting behaviour and also the crystallisation behaviour. Recently it has been demonstrated that the crystallisation of flexible polymers such as polyethylenoxide (PEO) in ultrathin films results in highly branched lamellar crystals [9,10], which can be described theoretically by a model based on a diffusion limited aggregation process (DLA) [11].

The concept of nanotechnology as formulated by Drexler [12] introduces nanounits which are either single molecules or small aggregates of macromolecules, supramolecular structures of small molecules, single colloidal particles etc. which can assemble to larger nanosystems preferably with special functionality. Examples of such nanosystems are protein assemblies which may form rigid tubular structures of microtubuli, small containment with nanosized reaction chambers for the synthesis of inorganic particles biological systems [13] or 2d-crystalline arrays of biofunctional layers (S-layers). The 2-dimensional ordered S-layers [14,15] as well as the one dimensional microtubullis [16] can be used as templates for metal cluster synthesis. An essential prerequisite for the design of new artificial nanosystems is the detailed knowledge of the assembly processes in order to create defined 2- or 3-dimensional nanosystems.

The scope of this contribution is to demonstrate major influences on the diffusion controlled growth of crystalline lamellae of several nm height within well defined lateral patterns (Fig. 13.1). The surface design that is realized in order to obtain this type of micropatterned PEO films is schemed in Fig. 13.2. With regard to basic ideas of nanotechnology the micropatterned lamellae should be addressed as nano-building blocks. The strategy that is applied to realise the lamellae nano-building blocks includes the following steps which will be discussed in detail:

- creation of defined micrometersized patterns by chemical surface heterogenisation
- preparation of ultrathin micropatterned amorphous polyethylene oxide (PEO) layers by controlled dewetting on heterogeneous surfaces
- heterogeneous nucleation of the metastable PEO layers and diffusion controlled crystallisation within lateral defined areas

13.2   Creation of Defined Micrometersized Patterns by Chemical Surface Heterogenisation

The generation of microstructured polymer films by dip-coating from solution requires the heterogenisation of surfaces into geometrical defined areas of different wettability [17]. Variations in surface wettability are generally related to

240   Hans-Georg Braun, Evelyn Meyer, and Mingtai Wang

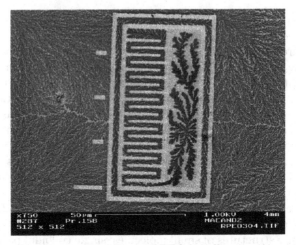

Fig. 13.1. Characteristic growth patterns of PEO lamellae on micropatterned surfac areas

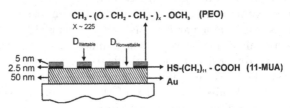

Fig. 13.2. Basic structure of the micropatterned substrate including the crystallisabl polymer film

differences in the surface chemical composition [18]. Surface chemical propertie can be designed by self-assembled monolayers of appropriate chemical structures In case of gold 11-mercaptoundecanoic acid (11-MUA) is a suitable compoun to create a hydrophilic surface that favours wetting by PEO (Fig. 13.2). Loca defined micrometersized areas of hydrophilic 11-MUA domains surrounded b hydrophobic areas can be prepared either by micro-contact printing [19] of 11 MUA or by electron beam lithography [20] of 11-MUA layers. By micro-contac printing 11-MUA molecules are transferred from a flexible polydimethylsiloxan (PDMS) stamp to a substrate surface (Au). Electron beam irradiation change the hydrophilic property of a self-assembled 11-MUA layer into hydrophobic.

13.3   Preparation of Ultrathin Micropatterned Amorphous Polyethylene Oxide (PEO) Layers by Controlled Dewetting on Heterogeneous Surface

Ultrathin polymer films are commonly prepared by spin- or dip-coating. In ou experiments we prepared PEO films by dip-coating from chloroformeous PE(solutions concentrations of 0.15 %(weight) PEO. The molecular weight M_w of th

PEO is about 10000 which is identical to a degree of polymerisation P_n of 225. With a length of 0.2783 nm of a single extended PEO unit in the crystalline state and a lamellae height of 15 nm the number of stems of a single molecule that fold into a lamellae is equals to 4. Keeping all other parameters [lift-off velocity of the substrate (ca. 20 mm/min.), temperature] of the dip-coating process constant the thickness of the initial amorphous film is about 4 nm.

During the dip-coating process the heterogeneous surface of the patterned substrate is lifted from the polymer solution with constant velocity. Initially the complete surface is covered by a thin liquid layer of polymer (PEO) solution. This layer is continuously thinning by solvent evaporation and it finally ruptures on a poor wettable surface. Dewetting of thin films on homogeneous substrates typically results either in a network of holes growing in time or in a bicontineous structure similar to that observed in spinodal decomposition processes [21–23]. On patterned surfaces thin film ruptures are initiated in relation to the geometrical features defined by wettable and nonwettable domains. In case of periodic stripes of wettable and nonwettable motifs a theoretical approach describes the conditions which allow that the surface wettability pattern is replicated into the film [24]. Following this theoretical approach the geometry of our motifs ensure the replication of the lateral surface heterogenities into the corresponding dewetting film structures. The surface topography of the solution respectively the local thickness of the polymer film formed on the wettable areas is strongly dependent on following influences:

- Ratio of wettable and non-wettable area within a motif
- Shape of the wettable area

Figure 13.3 demonstrates the different distribution of material within a motif in dependence of the ratio α of wettable and nonwettable areas. While a low ratio α of wettable to nonwettable areas results in a high amount of solution on the wettable areas due to the dewetting process and consequently in a high film thickness after drying (Area A), ultrathin films are formed at high α (Area B). As we will discuss later the difference in film thickness strongly controls the lamellar growth patterns. The influence of the shape of motifs on the 3d-topography of the liquid layer on the wettable structures is clearly seen in the light microscope Fig. 13.4. Within the edges of U-shaped channels or square like structures the hight of the liquid layer inside the U-shaped turn is larger than along the parallel channel structures. Consequently the thickness of the polymer film that are formed during the drying process reflects the topography of the preceeding liquid phase of the polymer solution. The topography of liquid phases on patterned surfaces and their stability is currently of great interest from the theoretical point of view [25,26].

The PEO films prepared by dip-coating are **initially amorphous** both on micrometer sized and large sized areas. On a large scale of several mm the probability to find a nucleus to initiate heterogeneous crystallisation in the film is high. Because a single nucleus is sufficient to start the crystallisation process, the amorphous films crystallise a few minutes after film preparation. The propagation of the crystallisation front can be followed by light microscopy. In contrast

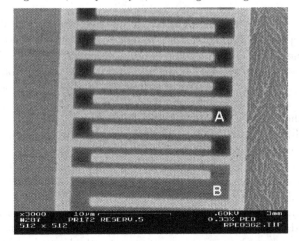

Fig. 13.3. SEM micrograph: Local thickness variation of a dewetted polymer film resulting from different ratio of wettable and nonwettable areas within a motif (scalebar eq. 2μm). The SEM micrograph demonstrates qualitatively the different height of the polymer film on the patterned surfaces (darker areas represent higher film thickness)

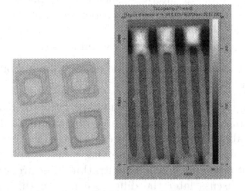

Fig. 13.4. Surface topography of a solution on micropatterned surface and of the crystallised polymer lamellae

to large continuous areas film segments within isolated micrometer sized isles are stable with respect to crystallisation. Within these isolated areas which are separated from the macroscopic PEO film by hydrophobic μm-sized barriers the heterogeneous nucleation probability is relatively low. The amorphous film segments can be kept in their metastable state for several days or even weeks and they can be nucleated within pre-selected areas by contact with an AFM tip upon request. This experimental conditions allow us to initiate diffusion controlled crystallisation processes and to follow crystalline growth patterns which are formed in lateral confined motifs. In contrast to other systems in which the wetting process and structure formation are simultaneous processes [27] dewet-

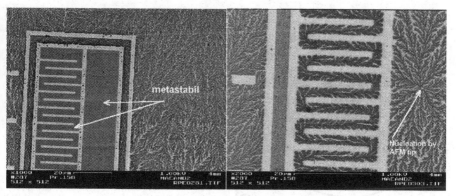

Fig. 13.5. SEM micrograph: a)Amorphous ultrathin PEO segments isolated through hydrophobic barriers b)Crystallized PEO layers. Heterogeneous nucleation was done by contact with an AFM tip inside the metastable amorphous isles

ting and crystallisation are clearly separated processes in the PEO ultrathin films.

13.4   Characteristic Growth Patterns of PEO Lamellae on Micropatterned Surface Areas

13.4.1   Basic Ideas on Diffusion Controlled Crystallisation

Highly branched morphologies similar to those described for the crystallisation of PEO in ultrathin layers correspond to a general type of pattern which is observed in a number of very different systems as for example

- growth patterns of bacterial colonies [28]
- neuron pattern of the nerve system [29]
- aggregation patterns of colloidal particles [30]
- electrocrystallisation [31]
- atomic aggregates on surfaces [32]

The characteristic length scales observed for structural elements may differ from the atomic scale to the macroscopic so from nm to mm. The highly branched structures that are observed in all previous examples result from diffusion controlled aggregation processes and the theoretical approach to simulate these features is know as the diffusion limited aggregation (DLA) theory originally described by Witten and Sander. Within the DLA model patterns are generated by the following algorithm (Fig. 13.6): One particle (nucleus) is fixed in the center of a coordinate system. Another diffusing particle is launched as a random walker from a circular line of diameter D around the nucleus. If the random walker contacts the nucleus it sticks to it and the cluster growth while the next random walker is launched from the circular line. This simple algorithm generates the highly branched patterns typical for DLA processes mentioned

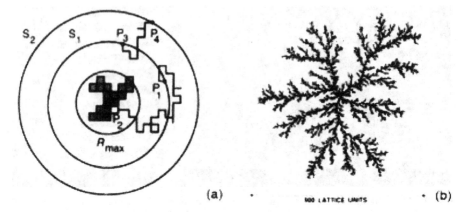

Fig. 13.6. Diffusion limited aggregation algorithm

above. With respect to crystallisation processes the model does only include the diffusional process to the solid (crystalline) interface but it does not consider thermodynamic stability criteria relevant to the the solidification process. The irreversible attachement of particles to a growth front results in growth tips with thermodynamic unfavorable small radii of curvature (high local interfacial energy). The interfacial energy can be reduced by local rearrangement processes increasing the tip radii and consequently decreasing the interfacial energy. In a modified DLA algorithm [33] the influence of tip curvature onto the growth process is taken into account by introducing a sticking probability less than 1 for the attachement of a random walker to an existing cluster. The sticking probability becomes dependent on the local curvature of the attachement site. A modified DLA model which describes the characteristic morphological features which can be observed in the crystallisation of polyethyleneoxide in ultrathin layers has been described by Sommer and Reiter [1]. In order to compare branching structures developed under different growth conditions the following terms have been introduced [34](Fig. 13.7):

- average distance between central stems (correlation width)
- average distance between side branches along a central stem as given by a correlation length
- tip radius of branches

With respect to the growth process of the branched structures the following correlations should be kept in mind:

- High undercooling causes highly branched structures with small tip radii and large correlation lengths. The number of molecular attachements per time to the solidification front is large.
- Low undercooling causes less branched and more compact structures with larger tip radii and small correlation lengths at low attachement rate.

---

[1] see Chaps. 8 and 9.

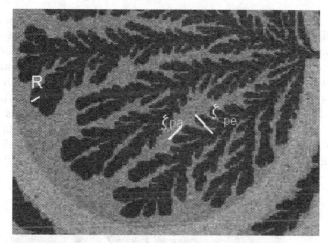

Fig. 13.7. Morphological features for DLA

13.4.2   Branching Structures of PEO Crystallised in Ultrathin Micro-patterned Films

Crystallisation features in ultrathin micropatterend films were studied for geometries schemed in Fig. 13.8. The size of nonwettable structural elements (black) varied between 1 μm and 3 μm , the size of the wettable stripes in between ranged from 1 um to 10 um.

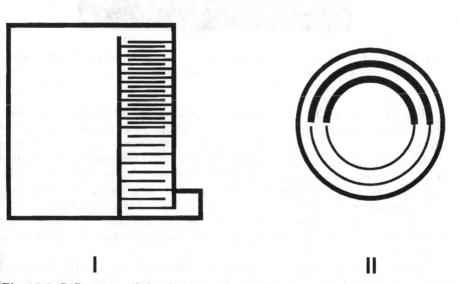

Fig. 13.8. Different motifs for the study of crystallisation patterns under geometrical constraint as obtained by selective dewetting of hydrophilic PEO by hydrophobized surface areas

The typical lamellar growth patterns which appear on homogeneous non patterned films areas (A) Fig. 13.9 are similar to those investigated by Reiter and Sommer [9,11] in PEO films prepared by spin-coating and crystallized at ambient temperature. The side branches within DLA structures are always directed towards the growth front. Therefore the history of the growth process even within complicated artificial structures can be easily followed by the analysis of the branching direction. As shown in Fig. 13.9 crystallisation was nucleated as indicated by B. Initially the crystallisation front propagates radial around the nucleation site similar to the growth of spherulites. As the crystallisation front reaches the entrance of the channel system C further propagation is directed by the channel geometry.

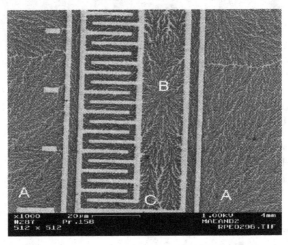

Fig. 13.9. SEM micrograph of crystalline growth morphologies in ultrathin PEO layers. A non-patterned film, B nucleation center, C channel entrance

During the dewetting process rims can appear at the boundary between wettable and non-wettable surface domains depending e.g. on geometry or concentration of polymer. As demonstrated in Fig. 13.10 crystallisation propagates preferentially along the rim in which the film is thicker than in surrounding areas. As soon as the first branch of the spherical dendrite contacts the rim crystallisation proceeds along it. Nuclei initiated along the rim grow into adjacent non-crystalline PEO areas. The dependence of the crystallinity on the film thickness in homogeneous films has been demonstrated by Frank [35,36]. In agreement with his results we also found the preferential crystallisation in thicker film layers (rims).

Peculiar growth features with respect to the diffusional growth in confined areas are observed for crystallisation within patterned films of motif II (13.8) are shown in Fig. 13.11a-c. The degree of undercooling decreases from Fig. 13.11a to c. Figure 13.11a shows fine branched tree-like lamellae structures typical for non-equilibrium crystallisation at high undercooling. The thickness of the

Fig. 13.10. SEM micrograph of crystalline grwoth morphologies in ultrathin PEO layers. Preferential growth along the rim

crystalline lamellae grown inside the circular structure is almost constant and equals 8 ± 1 nm. Morphological features typical for DLA structures (correlation length, tip radius) remain almost constant during crystallisation within the confinement at high undercooling. The motif is homogeneously covered by the branched lamellae. With decreasing undercooling the branches appear coarser (Fig. 13.11b,c). During the crystallisation process the growing lamellae are much thicker than the initial amorphous layer and consequently extended depletion zones are formed around the crystal growth front. Tip-radius, correlation length, correlation width and lamellae thickness are increased as expected from theory. Additionally these parameters are also influenced by the change of the diffusion gradient during the solidification process due to the limited material reservoir in closed areas connected with a further extension of the depletion zones.

During the diffusion controlled growth process polymer molecules have to pass the depletion zone before they reach the growth front. For a given diffusion rate the number of molecules/time which attach to the crystal growth front is related to the width of the depletion zone. An increase in the extension of the depletion zone slows down the growth velocity of the solidification front which finally influences the morphology of the branched structures. A lower velocity of the solidification front causes a larger tip radius of the growing branches. While for unlimited material reservoir attaching the growth front the diffusion gradient is constant this parameter decreases contineously during growth within a limited reservoir due to an increasing depletion zone resulting in a change of morphological features during the growth process as shown in Fig. 13.12.

As the tip diameter approaches to the channel width the solidification front completely fills the channels and growth proceeds without branching.

The growth process within confined channel systems is strongly dependent on the width of the channel in relation to the correlation width of the growing

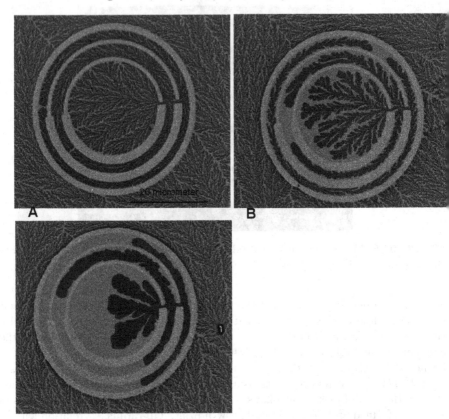

Fig. 13.11. SEM micrographs of PEO films crystallised in confined geometries at d
creasing undercooling (a-c)

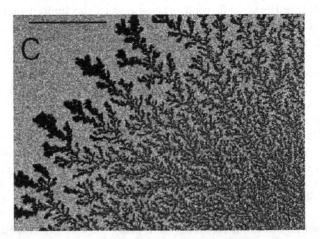

Fig. 13.12. SEM micrograph of morphological changes in the branching structure
a consequence of variation in the concentration gradient at limited reservoir

branches. Two situations resulting in a different growth process with respect to the kinetics of the growth front and the resulting morphologies should be distinguished:

- The correlation width of the growing branches is much smaller than the width of the channel. Under these conditions dendritic growth within a closed channel proceeds homogeneous without a morphological change during the growth process.

- The correlation width of branches is larger than the channels width. An extended and increasing depletion zone in front of the solidifying branch decreases the growth process.

13.4.3 Crystallisation of PEO in Microdroplets Obtained by Dewetting on Heterogeneous Surfaces

The polymer is not only metastable with respect to crystallisation within ultrathin films but also within the confinement of small droplets which form on hydrophilic ares on a micropatterend surface. As can be seen from (Fig. 13.13 a,b) the surface profile of an initially formed μm-sized PEO-dot can be fitted by a spherical cap function. The crystallisation inside the small droplet results in multistacked lamellae. Sharp steps originating from each lamellae are formed inside the droplet. Depending on the initial amount of polymer amorphous droplets are transformed either into a branched lamellae (A Fig. 13.13b) or into a compact basal lamellae (B Fig. 13.13b) which may be decorated on top by additional branched lamellae structures (C Fig. 13.13b). The crystallisation within the limited volume of the μ-droplet appears as a step by step process. After the formation of stacked lamellae the residual polymer on the upper lamellae crystallises.

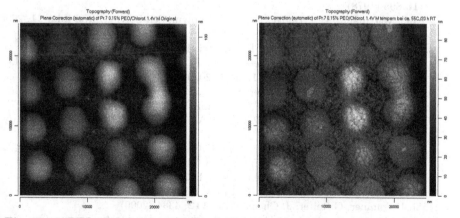

Fig. 13.13. PEO microdroplets a) in the initial state (the profile contour is defined by a spherical cap function which typically fits for the initial dewetting morphology) b) after crystallisation initiated by contact with an AFM tip;monolamellae (A), multilamellae (B) and branched lamellae (C) are observed.

The appearance of branched lamellae growth and the structural features of the branches are strongly dependent on

- the initial thickness of the crystallising polymer layer
- lateral patterns of the amorphous precursor film and
- confinement of the film

13.5   Summary

The basic approach of our study aim at to realize experimental conditions in which dewetting, nucleation and non-equilibrium crystal growth processes are separable processes initiable in individual steps on request. These conditions allow to initiate diffusion controlled crystallisation and to follow in-situ crystalline growth processes on lateral confined motifs. Complex morphological features can be generated by appropriate surface design and control of process parameters. It is shown that the diffusional gradient in confined areas is changed during the growth process due to the limited material reservoir causing a variation in morphological features.

Acknowledgement

We wish to acknowledge the support of E. Meyer within the DFG priority programm "Wetting and structure formation at surfaces"

References

1. S.E. Rickert, E. Baer, J.C. Wittmann, A.J. Kovacs:J.Polym.Sci.Polym.Phys.Ed. **16**,895 (1978)
2. J.C. Wittmann, A.M. Hodge, B. Lotz: J. Polym. Sci. Polym. Phys. Edn. **21**, 2495 (1983)
3. J.C. Wittmann, B. Lotz: Prog. Polym. Sci. **15**, 909 (1990)
4. S. Yan, F. Katzenberg, J. Petermann, D. Yang, Y. Shen, C. Straupe, J.C. Wittmann, B. Lotz: Polymer **41**, 2613 (2000)
5. J.Petermann, R. Gohil: J. Mater. Sci. **14** , 2260 (1979)
6. S. Yan, D. Yang, J.Petermann: J. Polym. Sci., B, Polym. Phys. **35**, 1415 (1997)
7. M. Breulmann, H. Coelfen, H. P. Hentze, M. Antonietti, D. Walsh, S. Mann: Adv. Mater. **10**, 237 (1998)
8. S. Mann, S.A. Davis, S.R. Hall, M. Li, K.H. Rhodes, W. Shenton, S. Vaucher, B. Zhang: J. Chem. Soc. Falton Trans. 3753 (2000)
9. G. Reiter, J.-U.Sommer: J. Chem. Phys. **112**, 4376 (2000)
10. G. Reiter, J.-U.Sommer: Phys. Rev. Lett. **80**, 3771 (1998)
11. J.-U.Sommer, G.Reiter: J. Chem. Phys. **112**, 4384 (2000)
12. K.E. Drexler:*Nanosystems-Molecular Machinery, Manufacturing and Computation* (John Wiley Sons Inc., 1992)
13. T. Douglas, M. Young: Nature **393**, 152 (1988)

14. R. Wahl, M. Mertig, J. Raff, S. Selenska-Pobell, W. Pompe: Adv. Mat. **13**, 736 (2001)
15. W. Shenton, D. Pum, U.B. Sleytr, S. Mann: Nature **389**, 585 (1997)
16. M. Mertig, R. Kirsch, W. Pompe: Appl. Phys. A **66**, S723 (1998)
17. E.Meyer, H.-G. Braun: Macromol.Mater. Eng. Appl. Phys. A **276/277**, 44 (2000)
18. A.Ulman: Thin Solid Films **273**, 45 (1996)
19. Y.Xia, G.M. Whitesides: Angew. Chemie **110**, 568 (1998)
20. C. K. Harnett, K. M. Satyalakshmi, H. G. Craighead, Appl. Phys. Lett. **76**, 2466 (2000)
21. G. Reiter: Phys. Rev. Lett. **68**, 75 (1992)
22. U. Thiele, M. Mertig, W. Pompe:Phys. Rev. Lett. **80**, 2869 (1998)
23. R. Seemann, S. Herminghaus, K. Jacobs:Phys. Rev. Lett **86**, 5534 (2001)
24. K. Kargupta, A. Sharma: Phys. Rev. Lett. **86**, 4536 (2001)
25. R. Lipowsky, P.Lenz, P.S. Swain: Colloids and Surf. A Physicochem. Eng. Aspects **161**, 3 (2000)
26. R. Lipowsky: Current Opinion in Colloid and Interface Science **6**, 40 (2001)
27. D.W. Zheng, W. Wen, K.N. Tu: Phys. Rev. E **57**, R3719 (1998)
28. I. Golding, Y. Kozlovsky, I. Cohen, E. Ben-Jacob: Physica A **260**, 510 (1998)
29. H.-J. Wolf: Discrete Models for the Morphogenisis of Neurites Regulated by Attraction-Diffusion-Depletion, Ph.D. Thesis University Heidelberg 2001
30. T.A. Whitten, L.M. Sander: Phys. Rev. Lett. **47**, 1400 (1981)
31. M. Castro, R. Cuerno, A. Sanchez, F. Dominguez-Adame: Phys. Rev. E **62**, 161 (2000)
32. Z. Zhang, G. Lagally: Science **276**, 377 (1999)
33. T. Vicsek: Phys. Rev. A **32**, 3084 (1985)
34. P. Meakin: *Fractals scaling and growth far from equilibrium* (Cambridge University Press 1998)
35. M.M. Despotopoulou, C.W. Frank, R.D. Miller, J.F. Rabolt: Macromolecules **29**, 5797 (1996)
36. C.W. Frank, V. Rao, M.M. Despotoupolou, R.F.W. Pease, W.D. Hinsberg, R.D. Miller, J.F. Rabolt:Science **273**, 4536 (1996)

14 Vitrification and Devitrification of the Rigid Amorphous Fraction in Semicrystalline Polymers Revealed from Frequency Dependent Heat Capacity

Christoph Schick, Andreas Wurm, and Alaa Mohammed

University of Rostock, Department of Physics, 18051 Rostock, Germany
christoph.schick@physik.uni-rostock.de

Abstract. For semicrystalline polymers the observed relaxation strength at glass transition is often significantly smaller than expected from the non-crystalline fraction. This observation leads to the introduction of a rigid amorphous fraction (RAF) which does not contribute to the heat of fusion or X-ray crystallinity nor to the relaxation strength at glass transition. The RAF is non-crystalline and in a glassy state at temperatures above the common glass transition. Complex heat capacity in the high frequency limit allows for the measurement of base-line heat capacity also at temperatures above the glass transition. From that the temperature and time dependence of the RAF can be obtained. For PC, PHB and sPP it is possible to study the creation and disappearance of the RAF in situ during isothermal crystallization and on stepwise melting. If crystallization is not limited by the stability (melting point) of the crystals to be formed the total RAF is created during the isothermal crystallization. Simultaneously with the melting of the smallest crystals the RAF disappears. Vitrification and devitrification of the non-crystalline material detected as the RAF at glass transition is structural (conformational) and not temperature induced for these polymers. The formation of the last growing crystals, which melt first, are responsible for the vitrification of the amorphous material around them and, consequently, by that they limit their own growth.

14.1 Introduction

The morphology of semicrystalline polymers is often described as a lamellae stack of crystalline and non-crystalline layers [1]. This so called "two-phase model" is successfully applied for the interpretation of X-ray diffraction as well as heat of fusion or density measurements [2]. On the other hand it is well known that several mechanical properties as well as the relaxation strength at glass transition can not be described by such a two-phase approach as recently discussed by Gupta [3]. From standard DSC measurements [4], dielectric spectroscopy [5–8], shear spectroscopy [8], NMR [9] and other techniques probing molecular dynamics at glass transition (α-relaxation) the measured relaxation strength is always smaller than expected from the fraction of the non-crystalline phase. The difference in mobility is caused by different conformations of the chains as detected by IR and Raman spectroscopy [10–12] or due to spatial confinement because of the neighboring lamellae. As an example the heat capacity at glass transition

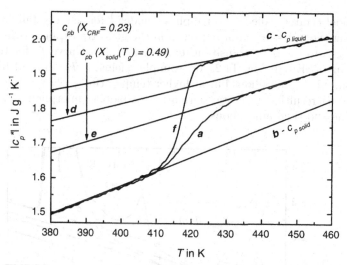

Fig. 14.1. TMDSC scan measurement of semicrystalline PC at underlying heating rate 0.5 K min^{-1}, temperature amplitude 0.5 K and period 100 s, curve a. Curve b and c correspond to heat capacities from ATHAS data bank for crystalline and liquid PC, respectively. Curve d was estimated from a two-phase model, (14.1) and curve e from a three-phase model, (14.2) using $\chi_{solid}(T_g)$. Curve f shows the measured heat capacity for the amorphous PC.

of amorphous and semicrystalline Bisphenol-A Polycarbonate (PC) is shown in Fig. 14.1.

According to the crystallinity of 0.23 one expects a reduction of the relaxation strength at glass transition (step height of c_p in case of calorimetric measurements) for only 23 % compared to that of the totally amorphous sample. Line d represents the expected base-line heat capacity for the semicrystalline sample according to such a two-phase model and crystallinity $\chi_{CRF} = 0.23$

$$c_{p\,b}(T,t) = \chi_{CRF}(T,t)c_{p\,crystal}(T) + (1 - \chi_{CRF}(T,t))c_{p\,liquid}(T) \qquad (14.1)$$

Obviously, the reduction is much larger. It is close to 50 %. To explain the disagreement between the expected values of relaxation strength as well as base-line heat capacity and the measured values Takayanagi et al. [5] and Wunderlich et al. [4] discussed not only crystalline and non-crystalline phases in semicrystalline polymers. The non-crystalline phase has to be subdivided in one part contributing and a second one not contributing to the relaxation strength at glass transition. Furthermore Wunderlich et al. distinguished between a mobile and a rigid fraction of the polymer. The rigid fraction consists of the crystalline phase and that fraction of the non-crystalline phase not contributing to the glass transition. We end up with a model distinguishing between the crystalline (CRF), the rigid amorphous (RAF) and the mobile amorphous (MAF) fractions.

This model is often called a "three-phase model"[1] for semicrystalline polymers. Obtaining the necessary information from the glass transition, obviously, limits the analysis to the glass transition temperature (T_g). Curve f for the initially amorphous PEEK in Fig. 14.2, for example, yields at T_g no rigid fraction. At T_g the sample is amorphous but at higher temperatures cold crystallization can be observed resulting in a transformation of mobile amorphous material into crystalline and rigid amorphous.

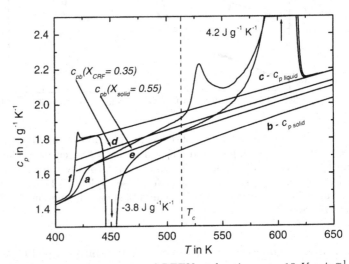

Fig. 14.2. DSC scan measurements of PEEK at heating rate 10 K min^{-1}. Curve a after isothermal crystallization for 30 min at 513 K. Curve b and c correspond to heat capacities from ATHAS data bank for crystalline and liquid PEEK, respectively. Curve d was estimated from a two-phase model, (14.1) and curve e from a three-phase model, (14.2) using $\chi_{solid}(T_g)$. Curve f shows the measured heat capacity for the initially amorphous PEEK.

For the semicrystalline PEEK, line a in Fig. 14.2, a rigid fraction of about 0.55 is estimated at T_g. This is approximately twice the crystalline fraction. With this value and (14.2) curve e was calculated. While for the semicrystalline PC in Fig. 14.1 heat capacity above the glass transition is well described by (14.2) the heat capacity for PEEK starts to deviate from curve e just above the glass transition. The deviation may be caused by latent heats due to melting or by a broad devitrification of the RAF or both processes may appear simultaneously. Unfortunately, it is not possible to distinguish these explanations from curves like in Fig. 14.2. It is still a matter of debate when the RAF disappears or when

---

[1] The two non-crystalline fractions (MAF, RAF) can not be considered as different phases in thermodynamics because there is no phase transition between them. Both belong to the non-crystalline phase and can be distinguished because of difference in molecular mobility (presence or absence of the degrees of freedom typical for a liquid compared to a glass or crystal).

it is formed because it is not possible from such curves to obtain information about the RAF except at T_g. There may be a broad glass transition of the RAF at temperatures higher than the glass transition of the semicrystalline polymer [6,13]. Then vitrification of the amorphous material detected as the RAF at T_g occurs during cooling from the crystallization temperature down to T_g and de-vitrification occurs on heating in the same broad temperature interval. Cebe et al. [14] have performed annealing experiments showing that the disappearance of the RAF is somehow connected with the crystals responsible for the lowest melting endotherm in PPS. Then vitrification and devitrification of the RAF is directly coupled to crystallization and melting, respectively. In that case vit-rification of the RAF occurs during crystallization and one expects significant interactions between vitrification and the crystallization process itself. The aim of this paper is to answer these questions by following the development of the RAF as a function of time during isothermal crystallization and as a function of temperature on melting. Heat capacity spectroscopy will be used to reach this goal.

14.2   Determination of the Rigid Amorphous Fraction

Differences in the molecular mobility are used to distinguish the mobile and the rigid amorphous fraction of a semicrystalline polymer. According to Wunderlich's definition [4] only two states are assumed. Namely, that fraction of the non-crystalline phase which contributes to the glass transition observed as a step change of heat capacity at temperatures slightly higher compared to the fully amorphous polymer. And a second fraction which does not contribute to the step change in heat capacity at the glass transition. This way a possibly very complex situation in respect to molecular mobility is described by only two parts. All gradients or gradual changes in molecular mobility between the crystal and the melt are neglected. Also the question if the polymer chains in the RAF have a conformation close to that of the crystal or close to that of the melt is neglected. Of course, these conformations are the reason for the differences in molecular mobility and consequently for their assignment to the RAF or to the MAF. Actually there is no generally accepted theory of glass transition and therefore, at the moment, it is not possible to make a close relationship between conformation or changes in conformation with the contribution of particular parts of the polymer to the glass transition. Assuming the molecular processes responsible for the increase in heat capacity of a liquid compared to that of a glass are cooperative with a correlation length of about 2 nm [15] we have not to discuss the behavior of single chains but we can distinguish between regions of nanometer size which do or do not contribute to the heat capacity step at glass transition. A very schematic sketch of such a situation is shown in Fig. 14.3.

Accepting such a simplified picture, heat capacity of the semicrystalline poly-mer can be described as a superposition of the heat capacity of the mobile frac-tion (χ_{MAF}) contributing and the solid fraction (χ_{solid}) not contributing to the step change of heat capacity at glass transition. For polymers heat capacity of the

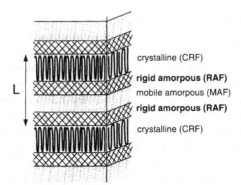

Fig. 14.3. Schematic sketch of the arrangement of crystalline, rigid amorphous and mobile amorphous fractions (three-phase model) in a lamellae stack.

glassy material often equals that of the crystalline and is considered as $c_{p\ solid}$ [16].

$$c_{p\ b}(T,t) = \chi_{solid}\ (\text{T,t})\ c_{p\ solid}(T) + \chi_{MAF}\ (T,t)\ c_{p\ liquid}(T) \qquad (14.2)$$

where $\chi_{solid} = 1\text{-}\chi_{MAF}$ and $c_{p\ b}(T,t)$ is base-line heat capacity. Base-line heat capacity corresponds to the heat necessary to increase the temperature of the sample without changing crystallinity. In other words, it is the heat capacity without any contribution from latent heats. The measured heat capacity (c_p) in general equals

$$c_p(T,t) = c_{p\ b}(T,t) + c_{p\ excess}(T,t). \qquad (14.3)$$

How to obtain $c_{p\ b}(T,t)$ will be discussed below. At glass transition $c_{p\ excess}$ is often very small and can be neglected. Base-Line heat capacity is then measured directly by DSC. The step change of heat capacity at glass transition for a fully amorphous sample equals

$$\Delta c_{p\ a} = c_{p\ liquid}(T_g) - c_{p\ solid}(T_g) \qquad (14.4)$$

and for a semicrystalline sample

$$\Delta c_{p\ sc} = c_{p\ b\ sc}(T_g) - c_{p\ solid}(T_g)$$
$$= \chi_{solid}(T_g)c_{p\ solid}(T_g) + \chi_{MAF}(T_g)c_{p\ liquid}(T_g) - c_{p\ solid}(T_g) \qquad (14.5)$$

with $\chi_{solid} = 1\text{-}\chi_{MAF}$ one can write

$$\Delta c_{p\ sc} = \chi_{MAF}(T_g) \cdot (c_{p\ liquid}(T_g) - c_{p\ solid}(T_g)) = \chi_{MAF}(T_g) \cdot \Delta c_{p\ a} \qquad (14.6)$$

and for the mobile fraction at glass transition

$$\chi_{MAF}(T_g) = \frac{\Delta c_{p\ sc}}{\Delta c_{p\ a}} \qquad (14.7)$$

If crystallinity $(\chi_{CRF}(T_g))$ is known from independent measurements like X-ray diffraction, density or heat of fusion the two amorphous fractions of a semicrystalline polymer at T_g can be quantified.

$$\chi_{RAF}(T_g) = \chi_{solid}(T_g) - \chi_{CRF}(T_g) = 1 - \chi_{MAF}(T_g) - \chi_{CRF}(T_g) \qquad (14.8)$$

The analysis described above is limited to the glass transition temperature because at other temperatures base-line heat capacity is not available from standard DSC measurements. As shown in Fig. 14.2, curve a, large excess heat capacities may occur above glass transition because of cold crystallization, reorganization and melting. On the other hand information about the mobile fraction is also available from relaxation strength of other susceptibilities like dielectric function, shear modulus or NMR responses. Due to the wide frequency range available with these techniques it is possible to determine the mobile fraction in a broader temperature range by shifting the relaxation temperature by varying frequency [6–8]. Unfortunately, it is not easy to obtain quantitative results. For dielectric measurements, as an example, the temperature dependence of the relaxation strength of the mobile fraction has to be known. By varying frequency the size of the probed volume will decrease with increasing frequency and other parts of the polymer may be detected as mobile. These problems are avoided by measuring heat capacity at constant frequency (heating rate). Consequently, it is an attractive task to measure base-line heat capacity in a broader temperature interval including cold-crystallization, reorganization, and the different steps of melting to study the formation and disappearance of the RAF.

Let's assume to be able to measure base-line heat capacity as function of temperature or time [17]. Base-line heat capacity is the superposition of the heat capacities of the different fractions as discussed above. For the three fractions, MAF, RAF, CRF, of a semicrystalline polymer we obtain

$$c_{p\,b}(T,t) = \chi_{MAF}(T,t)c_{p\,liquid}(T)$$
$$+ \chi_{RAF}(T,t)c_{p\,rigid\,amorph}(T) + \chi_{CRF}(T,t)c_{p\,crystal}(T) \qquad (14.9)$$

and with

$$1 = \chi_{MAF} + \chi_{RAF} + \chi_{CRF} \qquad (14.10)$$

and

$$c_{p\,solid} = c_{p\,rigid\,amorph} = c_{p\,glass} \qquad (14.11)$$

$$\chi_{RAF}(T,t) = \frac{c_{p\,b}(T,t) + \chi_{CRF}(T,t)\,[c_{p\,liquid}(T) - c_{p\,solid}(T)] - c_{p\,solid}(T)}{c_{p\,solid}(T) - c_{p\,liquid}(T)}.$$
$$(14.12)$$

For most polymers $c_{p\,solid}(T)$ and $c_{p\,liquid}(T)$ are available from the ATHAS data bank [18] or can be measured. The temperature or time dependent crystallinity $\chi_{CRF}(T,t)$ can be obtained from the enthalpy changes during calorimetric measurements

$$\chi_{CRF}(T,t) = \frac{h_{liquid}(T) - h(T,t)}{h_{liquid}(T) - h_{crystal}(T)} \qquad (14.13)$$

where $h_{liquid}(T)$ and $h_{crystal}(T)$ are the specific enthalpies of the liquid and crystalline phase, respectively, which are also available from ATHAS data bank $h(T, t)$ is obtained from the measured enthalpy change due to crystallization or melting, for details see [19]. The remaining task for calculation of $\chi_{RAF}(T, t)$ according to (14.12) is the experimental determination of base-line heat capacity $c_{p\ b}(T, t)$. How to obtain $c_{p\ b}(T, t)$ for semicrystalline polymers by means of heat capacity spectroscopy will be shown next.

14.3   Experimental

Different techniques can be used to measure heat capacity as a function of temperature, time, or frequency. In the present study we used temperature modulated DSC (TMDSC) and an AC-calorimeter. For details of the AC-calorimeter used, see [20]. TMDSC, a technique introduced in 1971 by Gobrecht et al. [21] and the necessary data treatments are described elsewhere [21–26]. If one wants to perform measurements in a broad frequency range the results from high sensitive apparatuses with different time constants like AC-calorimeter, PerkinElmer Pyris 1 DSC and Setaram DSC 121 must be combined, for details see [27]. For measurements at a fixed frequency of 0.01 Hz a TA Instruments DSC 2920 was used. For the comparison of various experimental data sets, a careful temperature calibration of all instruments is necessary. The DSC's are calibrated at zero heating rate according to the GEFTA recommendation [28]. The calibration was checked in TMDSC mode with the smectic A to nematic transition of 8OCB [29,30].

The Polycaprolactone (PCL) is a commercial sample from Aldrich with a molecular weight average $M_W = 55700$ g mol^{-1}. More details about the sample are reported in [31]. The Bisphenol-A Polycarbonate (PC) was obtained from General Electric (trade name LEXAN^{TM}) and was purified by dissolution in chloroform, filtering and precipitation in methanol [32,33]. The weight average molar mass and polydispersity index for the Polycarbonate were obtained by Gel Permeation Chromatography in chloroform ($M_w = 28,400$ g/mol and $M_w/M_n = 2.04$). The Poly(3hydroxybutyrate) (PHB) was received from the University of Cairo, Prof. A. Mansour. The syndiotactic Polypropylene (sPP) ($M_W = 150,000$ g/mol) is a commercial product from Atofina and the PEEK Victrex 381G is from ICI. The heat capacity data for these polymers in the liquid and the crystalline state, except for PHB, are available from ATHAS data bank [18]. The heat capacities for PHB where measured outside the transition regions by TMDSC and interpolated for the liquid and extrapolated for the solid state [17].

The measured heat capacity is the superposition of base-line and excess heat capacities. For measurements through phase transitions it is not possible to distinguish them. In some cases base-line heat capacity can be obtained from model calculations. For low molecular mass compounds where base-line heat capacity can be measured outside the transition region the change in the transition region can be estimated from the progress of the phase transition (sigmoidal base-line for peak integration). But this is only possible for two-phase systems. As soon as a third fraction is present the information available from a single heat capacity curve is not enough to distinguish three fractions straight forward. Assuming a certain coupling between the RAF and the CRF allows to solve the problem in an iterative way [34]. But the validity of the assumption must be independently checked and the experimental data must be of high accuracy. Therefore a direct measurement of base-line heat capacity is favorable.

To avoid latent heat contributions it would be nice to measure at constant temperature. But for the measurement of heat capacity changes in temperature are a prerequisite. On the other hand it is known that high molecular mass polymers need a super-cooling in the order of 10 K for crystallization. Therefore quasi-isothermal experiments with small temperature amplitudes should allow to measure base-line heat capacity if the isothermal period is long enough to finish all crystallization, reorganization or recrystallization in the crystallization or melting region of a polymer.

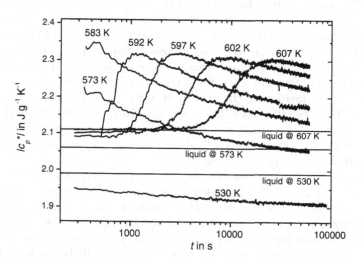

Fig. 14.4. Modulus of complex heat capacity at quasi-isothermal crystallization of PEEK at different temperatures. Temperature amplitude 1 K and period 200 s.

It was shown by Wunderlich et al. [35] that also under quasi-isothermal conditions contributions due to latent heats appear in the measured heat capacity. The heat capacity values above the liquid heat capacity in Fig. 14.4 can not at all be explained without contributions from latent heat. But also at long times heat capacity is larger than base-line heat capacity from (14.1). Consequently, these quasi-isothermal measurements do not allow to measure base-line heat capacity during the crystallization of PEEK.

A more detailed study of the excess heat capacity under quasi-isothermal conditions yields a frequency dependence of the excess heat capacity [27, 36-38]. The origin of the excess heat capacity during quasi-isothermal measurements and its frequency dependence is not yet understood. Probably the molecular processes involved are related to the surface of the polymer crystallites and often the term reversing melting [35] is used. For polymers showing a sliding diffusion in the crystallites (α-relaxation in case of Polyethylene or Polyethylene oxide) large contributions to reversing melting are due to surface melting [38]. For other semicrystalline polymers processes at the lateral surfaces may be responsible for the process of reversing melting and the corresponding excess heat capacity [39,40].

From glass transition, it is well known and generally accepted to describe heat capacity by complex numbers. The typical frequency dependence as known from other dynamic measurements is observed – a sigmoid step in real and a peak in imaginary part of heat capacity [21,41,42]. A similar frequency dependence of heat capacity of semicrystalline polymers was observed outside the glass transition range [27,36,38]. These observations are related to the occurrence of the excess heat capacity which is present in a rather wide temperature range between glass transition and melting temperature. In order to obtain information about the characteristic time scale of the molecular process related to excess heat capacity we have studied the frequency dependence of complex heat capacity during quasi-isothermal crystallization for PCL, sPP, PHB and PC, see Fig. 14.5. To extend the frequency range available with TMDSC (10^{-5} Hz 10^{-1} Hz) AC calorimetric measurements were performed at frequency 1 Hz [20]. For Polycaprolactone (PCL), a mean relaxation time in the order of a few seconds can be estimated for the process at 328 K. For sPP at 363 K it is in the order of 500 s. For PC and PHB we did not observe any frequency dependence after isothermal crystallization at 457 K and 296 K, respectively. For these polymers a frequency dependent excess heat capacity was only observed at higher temperatures in the melting range.

The frequency range available is still not broad enough for a detailed discussion of the curve shape, see Fig. 14.5. But from the curves one expects to measure base-line heat capacity without contributions due to reversing melting at high frequencies. For PCL high means higher than about 100 Hz which is outside the accessible frequency range for our experimental techniques. But for sPP PHB, and PC measurements at frequencies of about 0.01 Hz, which is inside the frequency range of TMDSC, allow for the measurement of the high frequency asymptotic value.

In general, the frequency dependence of the excess heat capacity as shown in Fig. 14.5 always allows for the measurement of base-line heat capacity at sufficient high frequencies. Whether or not for a particular polymer the high frequency limit can be reached by the calorimeters available and whether or not the curve is shifted along the frequency axis with temperature has to be checked for each single experiment. It is still an open questions if a time-temperature superposition is possible for the frequency dependent excess heat capacity.

For polymers like sPP, PC, and PHB base-line heat capacity is experimentally accessible in the temperature range between conventional glass transition and melting. For these semicrystalline polymers it is therefore possible to study

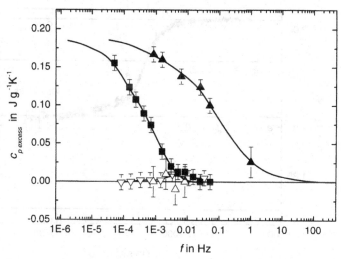

Fig. 14.5. Excess heat capacity during isothermal crystallization of PCL (▲) after 120,000 s at 328 K, of sPP (■) after 10,000 s at 363 K, of PC (▽) after 900,000 s at 457 K, and of PHB (△) after 100,000 s at 296 K as a function of modulation frequency.

the formation and the disappearance of the rigid amorphous fraction (RAF) according to (14.12).

14.4   Results

In Fig. 14.6 the time evolution of heat capacity during quasi–isothermal crystallization of PC at 457 K is shown. To check whether or not base-line heat capacity is measured the frequency dependence was studied at the end of crystallization, see Fig. 14.5. No frequency dependence of measured heat capacity can be seen indicating the absence of reversing melting and that base-line heat capacity was obtained.

To answer the question if the RAF was formed isothermally during the crystallization process or on cooling from the crystallization temperature to the glass transition the measured base-line heat capacity at the end of the measurement was first compared with the expected values according to a two-phase model, (14.1), curve *d*. Base-line heat capacity is assumed to be the superposition of the heat capacities of the crystalline and the non-crystalline fraction. If the non-crystalline material detected as the RAF at T_g vitrifies on cooling and not during isothermal crystallization one expects an agreement between (14.1) and the measured base-line heat capacity during and at the end of the isothermal crystallization. As can be seen in Fig. 14.6 measured heat capacity becomes significantly smaller than curve *d*, – indicating the occurrence of a significant RAF at the end of the isothermal crystallization process. Next, we compare the measured base-line heat capacity with the expected value according to a three-phase model, (14.2), curve *e*. Let us now assume vitrification of the non-crystalline

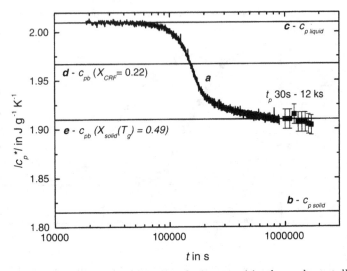

Fig. 14.6. Time evolution of heat capacity during quasi-isothermal crystallization of PC at 457 K, temperature amplitude 0.5 K and period 100 s, curve **a**. Curve **b** and **c** correspond to crystalline and liquid heat capacities from ATHAS data bank, respectively. Curve **d** was estimated from a two-phase model, (14.1) and curve **e** from a three-phase model, (14.2), using $\chi_{solid}(T_g)$ from (14.7)). The squares represent measurements at modulation periods ranging from 30 s to 12,000 s.

material detected as the RAF at T_g occurs during the isothermal crystallization and nothing happens on cooling from the crystallization temperature to the glass transition. Then we expect an agreement between the measured heat capacity and the heat capacity from (14.2) using the MAF determined at T_g from the step height of heat capacity at the glass transition, (14.7). Line **e** in Fig. 14.6 was accordingly estimated. The agreement is perfect within the accuracy of the measurement. For PC we conclude that the total RAF was established (vitrified) during the isothermal crystallization. No additional vitrification occurs on cooling from the crystallization temperature (457 K) down to the glass transition at 420 K.

For polymers crystallizing faster than PC it is difficult to follow isothermal crystallization by TMDSC at the temperature of maximum crystallization rate as we did for PC. We have either to choose temperatures closer to the melting or closer to the glass transition temperature to reduce crystallization rate to a reliable value. To choose crystallization temperatures close to the melting temperature is often not possible because of large excess heat capacities at typical TMDSC frequencies, see Fig. 14.4 for PEEK. Furthermore a comparison between data obtained at such high temperatures with that obtained at T_g is problematic because of possible changes in morphology on cooling. Choosing crystallization temperatures close to T_g, on the other hand, requires polymers which can be quenched without crystallization on cooling from the melt. PHB and sPP are polymers where this can be easily done. The result from quasi-isothermal crys-

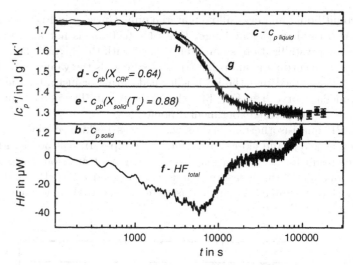

Fig. 14.7. Time evolution of heat capacity during quasi-isothermal crystallization of PHB at 296 K, temperature amplitude 0.4 K and period 100 s, curve *a*. Curve *b* and *c* correspond to solid and liquid heat capacities, respectively. Curve *d* was estimated from a two-phase model, (14.1) and curve *e* from a three-phase model, (14.2), using $\chi_{solid}(T_g)$ from (14.3). The squares represent measurements at modulation periods ranging from 240 s to 1,200 s. Curve *f* shows the exothermal effect in the total heat flow and curves *g* and *h* the expected values from model calculations, see text.

tallization of PHB at 296 K, what is close to the glass transition, is shown in Fig. 14.7.

Base-line heat capacity was measured as function of time and compared with the predictions of (14.1) and (14.2). As for PC the RAF in PHB was established during the quasi-isothermal crystallization as can be seen from the agreement of line *e* with the measured heat capacity at the end of the crystallization process. Because of the faster crystallization rate compared to PC we were able to measure the exothermic effect due to the crystallization process. The Pyris 1 DSC allows for a quantitative measurement over 17 hours also the maximum of the heat flow rate was less than 40 μW. From the integral we obtain the enthalpy change, $h(t)$, and with (14.13) the CRF as a function of time. The time dependence of base-line heat capacity can be determined from

$$c_{p\,b}(t) = c_{p\,liquid} - \frac{\chi_{CRF}(t)}{\chi_{CRF}(\infty)}\left(c_{p\,liquid} - c_{p\,b}(\infty)\right). \tag{14.14}$$

The calculation can be performed for two cases. (i) the RAF is formed during the whole crystallization process or (ii) first the crystalline morphology is build up during main crystallization and in a second step, e.g. during secondary crystallization at longer times, the RAF is formed. Then during main crystallization no or only a little RAF should be present. This situation (ii) should be described by (14.14) where $c_{p\,b}(\infty)$ equals the value from (14.1), line *d*, taking into account liquid and crystalline material only. Curve g in Fig. 14.7 shows the result. Also the behavior at longer times ($>$10,000 s) is not known the result during main crystallization is not in agreement with the measured curve. To calculate $c_{pb}(t)$ according assumption (i) $c_{p\,b}(\infty)$ equals the value from (14.2), line e. Here it is assumed that the RAF is formed during or just after the formation of the lamella. Curve h in Fig. 14.7 shows the result. The agreement is perfect within the scatter of the experimental points.

For sPP quasi-isothermal crystallization was performed close to the glass transition at 280 K and closer to the melting temperature at 363 K. Figure 14.8 shows the result for crystallization close to the glass transition. Despite the fact that the quality of the measurement is not as good as for PC and PHB the behavior is very similar. All or at least most of the RAF is formed during the isothermal crystallization.

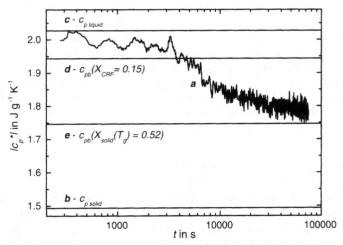

Fig. 14.8. Time evolution of heat capacity during quasi-isothermal crystallization of sPP at 280 K, just above T_g. Temperature amplitude 0.5 K and period 120 s, curve a. Curve b and c correspond to solid and liquid heat capacities, respectively. Curve d was estimated from a two-phase model, (14.1) and curve e from a three-phase model (14.2), using $\chi_{solid}(T_g)$ from (14.3).

For crystallization at 363 K the situation changes. First of all it is not possible to follow crystallization with time because it is to fast. In Fig. 14.9 the frequency dependence of the heat capacity after isothermal crystallization at 363 K for 3 hours is shown. For frequencies above 3 mHz no frequency dependence is observed. At these frequencies base-line heat capacity is measured. The value is again lower than expected from a two-phase model, (14.1) but it is significantly above the value for the three-phase model, (14.2), taking into account the RAF detected at T_g.

There must be further vitrification of amorphous material on cooling from 363 K to the glass transition at 270 K. To find out what is the reason for

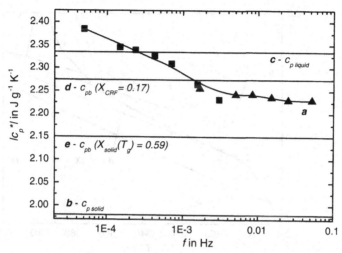

Fig. 14.9. Modulus of specific complex heat capacity of sPP after crystallization at 363 K for 3 hours as a function of frequency. Quasi-isothermal rectangular multi frequency temperature-time profile [43] with period 600 s, triangles, and 20,000 s, squares. Temperature amplitude 1 K.

vitrification we performed TMDSC scan measurements on cooling from 363 K to 230 K and successive heating to the melt at 420 K. The heat capacity data obtained on cooling and heating in the temperature range 363 K – 230 K – 363 K are the same within the thickness of the line in Fig. 14.10.

In the temperature range between the crystallization temperature and 320 K significant deviations from line *e* are observed. Near the crystallization temperature heat capacity is close to the value expected from a two-phase model, line *d*. Devitrification and vitrification occur in a broad temperature range of about 40 K. If this is due to a broad glass transition one would not expect latent heats in this temperature range. But also below 363 K the total heat capacity, dashed line *f* in Fig. 14.10, is significantly higher than the modulus of complex heat capacity, line *a* in Fig. 14.10. This indicates latent heats because of crystallization on cooling and melting on heating. For sPP part of the RAF devitrifies on heating below the crystallization temperature (363 K). Here devitrification and melting are superimposed over a broad temperature range. Possible explanations will be discussed below.

For PC and PHB, where all RAF vitrifies at the crystallization temperature, the question arises at what temperature the RAF devitrifies on heating? Is devitrification smeared over a broad temperature interval? Does it occur before the crystals melt? Or is devitrification of the RAF part of the main melting? To answer these questions heat capacity was measured on stepwise heating and compared with expected base-line heat capacities, see Figs. 14.11 and 14.12 for PC and PHB, respectively. To avoid contributions from irreversible melting to the measured complex heat capacity quasi-isothermal measurements at stepwise increasing temperatures were performed, see [44] for details.

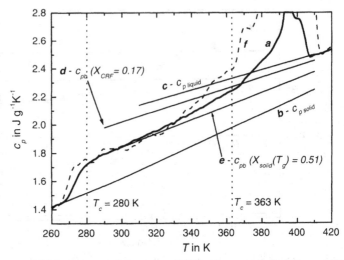

Fig. 14.10. TMDSC scan measurement of sPP after crystallization at 363 K for 3 hours at underlying heating rate 1 K min^{-1}, temperature amplitude 0.4 K and period 60 s, curve a (thick line). Curve b and c correspond to heat capacities for solid and liquid sPP and curves d and e to expected heat capacities from a two and a three-phase model, respectively. Curve f, dashed line, shows the total heat capacity. The vertical dotted line shows the crystallization temperature.

The heat capacity measured at stepwise heating in the melting region of PC Fig. 14.11 curve a, shows deviations from base-line heat capacity, line e, in the temperature range above 460 K. At the lowest endotherm, between 460 K and 485 K, a pronounced increase in heat capacity is observed. The heat capacity starts to deviate from the base-line heat capacity obtained from a three-phase model including RAF, (14.2) curve e, and around 480 K it is close to the base-line heat capacity obtained from a two-phase model, (14.1) curve d and g. In (14.1) only crystalline and liquid material is taken into account. The increase in heat capacity cannot be explained by the decrease of crystallinity due to the lowest melting endotherm between 465 K and 485 K. The expected increase due to the change in crystallinity at the lowest endotherm corresponds to the difference between curve g, which was calculated, from (14.1) assuming temperature dependent crystallinity, and curve d assuming constant crystallinity. At 480 K the difference is 0.01 J g^{-1}K^{-1} only. The observed step in heat capacity is about 5 times larger. At about 490 K heat capacity becomes larger than liquid heat capacity because of excess heat capacity, see discussion and Fig. 14.13.

Basically the same behavior as for the PC is observed for PHB on heating Again, at the lowest endotherm around 320 K a second step in heat capacity towards the expected value from the two-phase model is observed. Because PHB crystallizes fast, the isothermal crystallization was performed close to the glass transition to be able to follow the crystallization process. Consequently, pre melting is not well separated from glass transition in temperature.

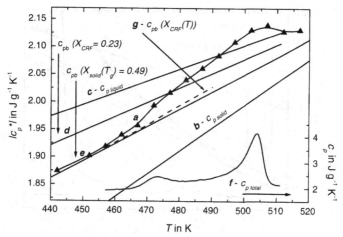

Fig. 14.11. Melting region of PC. Quasi-isothermal TMDSC measurements on step-wise increasing temperatures taken after 15 minutes (▲) at $t_p = 100$ s. Curve b and c correspond to heat capacities for solid and liquid PC, respectively. Curve d was estimated from a two-phase model, (14.1) and curves e and g from a three-phase model, (14.2), using data from T_g or temperature dependent crystallinity, respectively. Curve f shows the total heat capacity at underlying heating rate 0.5 K min^{-1}.

14.5   Discussion

For sPP, PC, and PHB a significant rigid amorphous fraction can be determined from the step of heat capacity at the glass transition. Taking into account the crystalline, the rigid amorphous and the mobile amorphous material information about the fractions of different molecular mobility can be obtained. For PC after 11 days crystallization at 457 K crystallinity was 0.23, rigid amorphous fraction 0.26 and mobile amorphous fraction 0.51. For PHB after crystallization at 296 K for 28 h, the values were 0.64, 0.22, 0.12 and for sPP after crystallization at 280 K for 20 h 0.15, 0.37, 0.48, respectively. From the TMDSC scan measurements no reversible melting can be detected for these semicrystalline polymers at temperatures above glass transition and below the lowest endotherm. While PC crystallizes extremely slow, PHB crystallizes reasonable fast like PET, PEEK, PEN, and high degrees of crystallinity are reached. We do not know why these polymers do not show excess heat capacities under the given experimental conditions, indicating the absence of reversing melting, while other polymers like PCL, PET, PTT, and PEEK show large excess heat capacities [44,35,45].

The absence of excess heat capacities in a temperature range suitable for crystallization experiments allows us to study base-line heat capacity as a function of time and to compare measured with expected values, see Figs. 14.6–14.8. For PC, PHB and sPP at low crystallization temperatures the measured heat capacity becomes significantly smaller than base-line heat capacity expected from a two-phase model, (14.1). For these polymers a significant portion of the RAF is formed during the isothermal crystallization process. Furthermore, a perfect

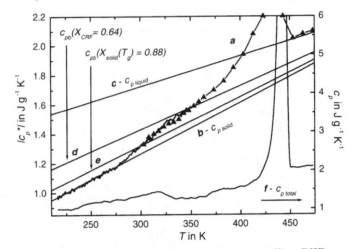

Fig. 14.12. TMDSC scan measurement (lines) of semicrystalline PHB at underlying heating rate 1 K min^{-1}, temperature amplitude 0.4 K and period 60 s, curve *a*. Curve *b* and *c* correspond to heat capacities for solid and liquid PHB, respectively. Curve *d* was estimated from a two-phase model, (14.1) and curve *e* from a three-phase model (14.2) using data from T_g. The triangles show the heat capacities from quasi-isothermal TMDSC measurements on step-wise increasing temperatures. The data were taken after 30 minutes. Curve *f* shows the total heat capacity.

match between the measured heat capacity at the end of crystallization and the expected base-line heat capacity from a "three-phase model", (14.2), can be seen. Because in Figs. 14.6–14.8 line *e* was obtained with the RAF determined from the heat capacity increment at T_g, there are no indications for changes in the amount of the RAF on cooling from the crystallization temperature to the glass transition. In other words: the whole RAF, detected at T_g, was established during the quasi-isothermal crystallization. From these observations we conclude that, at least for PC, there is no broad glass transition of the RAF somewhere in between crystallization temperature and T_g. Consequently, vitrification of the RAF results from the crystallization process itself and prevent further crystallization. For PC and PHB, at the crystallization temperatures studied, the low degree of crystallinity seems to be caused by the formation of the RAF during crystallization. Vitrification of the RAF is the result of morphological change and not due to cooling below a second glass transition temperature. This is in agreement with the observations of Lu and Cebe [14] for PPS but does not support the view of Song and Hourston [13] and Huo and Cebe [6] considering broad glass transition of the RAF for PET and PEEK, respectively.

For sPP at higher crystallization temperatures the situation is not as simple. As can be seen from Fig. 14.10 the low frequency asymptotic value of the specific heat capacity after isothermal crystallization at 363 K is only a little below the expected value from a two-phase model, line *d*, 14.1, and significantly larger than the expected value obtained from the heat capacity increment at glass transition line *e* 14.2 and 14.8. For sPP a significant part of the RAF detected at the glass transition is still mobile at the end of the isothermal crystallization at 363 K and vitrifies on cooling. The same may be true for PEEK, see Fig. 14.2, and other polymers. There are significant changes in heat capacity between crystallization temperature and the conventional glass transition. But the latent heat observed in this temperature range indicates simultaneous changes in morphology. The sequence length distribution of the crystallizable sequences may be responsible for this behavior. For polymers where the crystallization is controlled by tacticity or distribution of non-crystallizable co-units along the chain the size of possible crystals may be defined by the length of the crystallizable sequences. Therefore, for crystallization temperatures above the melting temperature of the smallest crystals (shortest sequences) these crystals can not be formed. On cooling, as soon as temperature is below the corresponding crystallization temperature, the small crystals are formed. The formation of such small crystals results in the reduction of molecular mobility of the surrounding melt and consequently in the vitrification of the amorphous material which is detected as the RAF at T_g. For sPP, contrary to PC and PHB, the RAF mainly vitrifies on cooling from the crystallization temperature (363 K) to the glass transition within ca. 40 K. Consequently, crystallization can not be limited by the vitrification of the melt surrounding the growing crystals. For sPP other reasons must be responsible for the very low degree of crystallinity finally reached. It may be the length distribution of crystallizable sequences of the necessary stereo regularity which limits crystallization of sPP.

Because there is no generally accepted theory for polymer crystallization, we will discuss these findings in the light of Marand's description of PC crystallization [33] and Strobl's view on polymer crystallization [46].[2]

Marrand and coworkers describe the crystallization of PC as a two-step process: First lamellae are formed which build up lamella stacks and finally spherulites. After completion of this step, small crystals are formed in between the existing lamellae. The fold surface of the existing lamellae provides the necessary large number of nuclei. This second step, the formation of small crystals with a large specific surface, seems to be the crucial step in respect to the formation of the RAF. A prerequisite for the formation of these small crystals is some mobility of the melt surrounding the growing crystals. If the RAF is assumed to be a consequence of the formation of the primary lamellae, there would be no chance to form new crystals because the material is vitrified. Consequently, the RAF results from the vitrification of the amorphous material in the neighborhood of the small secondary crystals. The reason for the vitrification is primarily the increase in constrains for the amorphous material on a molecular level and/or a decrease in the volume of the remaining amorphous domains below the necessary volume to establish a cooperatively rearranging region (confinement effects) [15]. The vitrification of the RAF further stops crystallization because vitrification of the amorphous material around the growing crystals prevent all large scale

---

[2] See also the contribution by M. Al-Hussein and G. Strobl on page 48 in this book

molecular motions which are necessary for the attachment of chain segments to the growth face.

On heating, the constraints should disappear as soon as the small crystals melt. According to Marand this happens at the lowest endotherm [32, 33, 47, 48]. From Figs. 14.11 and 14.12 it can be seen that just in the temperature range of the lowest endotherm heat capacity shows a step-like increase and becomes very close to the value expected from a two-phase model. Consequently, also on heating there is no temperature induced separate glass transition of the RAF. The devitrification of the RAF is, as vitrification, a direct consequence of morphological and not temperature changes. For PC, as soon as the RAF is devitrified at the lowest endotherm and molecular mobility is high again, re-organization and re-crystallization is possible as it was observed by Marand [33]. Because small crystals are formed again the just devitrified material will vitrify and heat capacity will decrease. Such behavior was indeed found for PC after annealing for 10 hours, down triangles in Fig. 14.13. The heat capacity comes back to the expected value according to a three-phase model, line *e*.

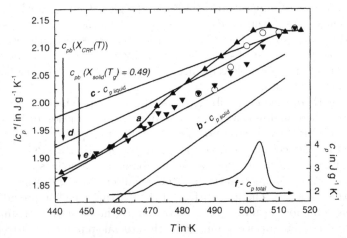

Fig. 14.13. Melting region of PC, curves as in Fig. 14.11. Quasi-isothermal TMDSC measurements on step-wise increasing temperatures. The data from independent measurements were taken after 15 minutes (▲) at $t_p = 100$ s and after 10 hours at $t_p = 100$ s (▼) and at $t_p = 1000$ s (o).

Up to 490 K no frequency dependence can be observed in the frequency range 0.01 Hz to 0.001 Hz. Above 490 K, in the main melting peak, also for the quasi-isothermal measurements at a time scale of 100 s an excess heat capacity (reversing melting) as well as a frequency dependence can be observed. Therefore no further conclusions can be drawn for the main melting process.

Strobl's model for the formation of crystal lamellae assumes a three-step processes [46]. At the first step, a layer with mesomorphic structure arises. This phase consists out of incompletely stretched and refolded chains and corresponds in its thermodynamic properties rather to the isotropic melt as to the crystal. These layers should have a minimal thickness to be stable in the melt. The attachment of further partly stretched segments of appropriate length is assumed to result in the lateral growth. Due to the still high flexibility of the chains, cooperative rearrangements yield a thickening and perfection of the layers. The gaining of a critical thickness results in the rigidity of the layer and stops the thickening. In the second step of the model, granular crystalline layers (blocks) arise. This transition should appear spontaneous with no need for nucleation and is thought to be of second order. The third step characterizes the merging of the blocks to homogeneous crystal lamellae with the thickness of the original blocks. For the first two steps the growing structures are assumed to be in dynamic equilibrium with the surrounding melt and fluctuations of the structure may appear. Only the last step, the merging of the blocks, yield a more stable structure that melts at higher temperatures [49]. We relate our observations to the second and third step. In case the surrounding of the granular lamellae is vitrified, the vitrified material will prevent the blocks to merge because of a lack of mobility. Finally, the lack of mobility preserves the granular structures and they will melt at temperatures close to the crystallization temperature. As in Marand's model the melting of these structures will release the constrains for the amorphous material and it devitrifies. It is assumed that this process occurs at the lowest endotherm. Because temperature is now approximately 20 K above the previous crystallization temperature re-organization sets in and some of the newly formed blocks may be able to merge and to increase the fraction of stabilized lamellae. Finally, a structure with a smaller number of non merged blocks appears which again vitrifies the amorphous material what prevent further block merging and stops stabilization again.

In both models devitrification of the RAF occurs in conjunction with the melting of a small part of the crystalline fraction. The increase in the MAF increases the possibility of large scale molecular motions which are necessary for the formation of new crystals or for the further merging of blocks. The growth of this newly formed crystals will again vitrify the surrounding melt and finally prevent further crystallization. The quasi-isothermal measurements for 10 hours, Fig. 14.13, support this view. Only after devitrification of the RAF, seen at 470 K in the step in heat capacity in the measurements after 15 minutes, significant reorganization (decreasing heat capacity) can be observed. After 10 hours annealing the heat capacity reaches values which are close to that expected from (14.2) taking in to account the RAF. Only near the main melting peak the annealing time is not long enough to reach steady state. A significant excess heat capacity, showing frequency dependence, can be observed.

14.6   Conclusion

For PC, sPP and PHB the asymptotic value of heat capacity at high frequencies can be measured by TMDSC during crystallization. This allows to measure base-line heat capacity and to study the formation of the rigid amorphous fraction (RAF). For PC, PHB and sPP the RAF is established during isothermal crystallization. Devitrification of the RAF occurs at the lowest endotherm. The immobilization of the amorphous material around less perfect crystals, which are formed during isothermal crystallization, results in the vitrification of the RAF during crystallization and in its devitrification during melting. For sPP, crystallized at 363 K, only a small fraction of the RAF detected at the glass transition is vitrified during isothermal crystallization. These differences regarding the vitrification of the RAF indicate differences in the crystallization process. While for PC, PHB and sPP at 280 K crystallization is limited by the vitrification of the melt surrounding the growing crystals for sPP at 363 K other mechanisms must be responsible for the low degree of crystallinity reached. The length distribution of the crystallizable sequences must be considered. In any case vitrification of the RAF results from the crystallization process itself. Vitrification of the RAF is the result of morphological changes and not due to cooling below a sometimes assumed second glass transition temperature in the semicrystalline polymers.

Our results can be explained by Marand's [33] and Strobl's [46] models of polymer crystallization. From our data it is not possible to favor one of these models. But, our results accentuates the interplay between molecular mobility of the melt and polymer crystallization. Hopefully, a more complex view, taking into account the structure and properties of the melt surrounding the growing crystals, will help to solve some of the still open questions of polymer crystallization [50–52].

Frequency dependent heat capacity, in the high frequency limit, yield quantitative information about fractions of different mobility during the crystallization process and how mobility of the melt is influenced by the crystallization process itself. From the step in heat capacity versus frequency information about the characteristic time scale of attachment and detachment processes of polymer segments at the crystal surface can be obtained. In order to perform such measurements the frequency range of heat capacity measurements has to be enlarged

Acknowledgement

We are thankful to Prof. H. Marand, Blacksburg, VA, for supplying the PC sample and to Prof. A. Mansour, Cairo, for supplying the PHB sample and to both for stimulating discussions. This research was supported by the European Commission (grant IC15CT96-0821), the German Science Foundation (grant DFG Schi-331/5) (AW) and the Government of Egypt (AM).

References

1. G. Strobl: The Physics of Polymers, Springer, Berlin 1996.
2. V. Sharma, P. Desai, A.S. Abhiraman: J. Appl. Polym. Sci. **65**, 2603 (1997)
3. V.B. Gupta: J. Appl. Polym. Sci. **83**, 586 (2002)
4. H. Suzuki, J. Grebowicz, B. Wunderlich: Makromol. Chem. **186**, 1109 (1985)

5. Y. Ishida, K. Yamafuji, H. Ito, M. Takayanagi: Kolloid-Z. & Z. Polym. **184**, 97 (1962)
6. P. Huo, P. Cebe: Macromol. **25**, 902 (1992)
7. K. Nogales, T.A. Ezquerra, F, Batallan, B. Frick, E. Lopez-Cabarcos, F.J. Balta-Calleja: Macromol. **32**, 2301 (1999)
8. C. Schick, J. Dobbertin, M. Potter, H. Dehne, A. Hensel, A. Wurm, A.M. Ghoneim, S. Weyer: J. Thermal. Anal. **49**, 499 (1997)
9. W. Gabrielse, H.A. Gaur, F.C. Feyen, W.S. Veeman: Macromol. **27**, 5811 (1994)
10. K.C. Cole, A. Ajji, E. Pellerin: Macromol. **35**, 770 (2002).
11. G. Strobl, W. Hagedorn: J. Polym. Sci. B-Phys. **16**, 1181 (1978)
12. M. Glotin, L. Mandelkern: Coll. Polym. Sci. **260**, 182 (1982)
13. M. Song, D.J. Hourston: J. Thermal. Anal. **54**, 651 (1998)
14. S.X. Lu, P. Cebe: Polymer **37**, 4857 (1996)
15. E. Donth: The Glass Transition, Springer, Berlin 2001.
16. C. Schick, A. Wurm, A. Mohammed: Thermochim. Acta, in press.
17. C. Schick, A. Wurm, A. Mohammed: Coll. Polym. Sci. **279**, 800 (2001)
18. B. Wunderlich: Pure & Appl Chem **67**, 1019 (1995); see on WWW URL: http://web.utk.edu/~athas/databank/intro.html
19. V.B.F. Mathot: Calorimetry and Thermal Analysis, Hanser Munich: (1994), Chap 5.2.
20. A.A. Minakov, Yu. Bugoslavsky, C. Schick: Thermochim. Acta **317**, 117 (1998)
21. H. Gobrecht, K. Hamann, G. Willers: J. Physics E: Scientific Instruments **4**, 21 (1971)
22. B. Wunderlich, Y.M. Jin, A. Boller, Thermochim. Acta **238**, 277 (1994)
23. M. Reading: Trends Polym. Sci., **8**, 248 (1993)
24. J.E.K. Schawe: Thermochim. Acta **260**, 1 (1995)
25. M. Merzlyakov, C. Schick, Thermochim. Acta **330**, 55 and 65 (1999)
26. S. Weyer, A. Hensel, C. Schick: Thermochim. Acta **304/305**, 267 (1997)
27. M. Merzlyakov, A. Wurm, M. Zorzut, C. Schick: J. Macromol. Sci.- Phys. **38**, 1045 (1999)
28. S.M. Sarge, W. Hemminger, E. Gmelin, G.W.H Höhne, H.K. Cammenga, W. Eysel: J. Therm. Anal. **49**, 1125 (1997)
29. A. Hensel, C. Schick: Thermochim. Acta **304/305**, 229 (1997)
30. C. Schick, U. Jonsson, T. Vassiliev, A.A. Minakov, J.E.K. Schawe, R. Scherrenberg:, D. Lörinczy: Thermochim. Acta **347**, 53 (2000)
31. P. Skoglund, A. Fransson: J. Appl. Polym. Sci. **61**, 2455 (1996)
32. S. Sohn, A. Alizadeh, H. Marand: Polymer **41**, 8879 (2000)
33. A. Alizadeh, S. Sohn, J. Quinn, H. Marand, L. Shank, H.D. Iler: Macromol. **34**, 4066 (2001)
34. M. Alsleben, C. Schick: Thermochim. Acta **238**, 203 (1994)
35. I. Okazaki, B. Wunderlich: Macromol. **30**, 1758 (1997)
36. Y. Saruyama: Thermochim. Acta **305**, 171 (1997)
37. A. Toda, C. Tomita, M. Hikosaka, Y. Saruyama: Thermochim. Acta **324**, 95 (1998)
38. T. Albrecht, S. Armbruster, S. Keller, G. Strobl: Macromol. **34**, 8456 (2001)
39. C. Schick, M. Merzlyakov, A.A. Minakov, A. Wurm: J. Therm. Anal. Cal. **59**, 279 (2000)
40. R. Androsch, B. Wunderlich: Macromol. **34**, 5950 (2001)
41. N.O. Birge, S.R. Nagel: Phys. Rev. Lett. **54**, 2674 (1985)
42. S. Weyer, A. Hensel, C. Schick: Thermochim. Acta **304/305**, 251 (1997)
43. M. Merzlyakov, C. Schick, Thermochim. Acta, **377**, 193 (2001).

274 Christoph Schick, Andreas Wurm, and Alaa Mohammed

44. M. Pyda, B. Wunderlich: J. Polym. Sci. B-Phys. **38**, 622 (2000)
45. A. Wurm, M. Merzlyakov, C. Schick: J. Macromol. Sci.-Phys. B **38**, 693 (1999)
46. G. Strobl: Eur. Phys. J. E **3**, 165 (2000)
47. S. Sohn: Crystallization Behavior of Bisphenol A Polycarbonate: Effect of Time, Temperature and Molar Mass, PhD Thesis, Virginia Polytechnic and State University, April 2000
48. H. Marand, A. Alizadeh, R. Farmer, R. Desai, V. Velikov: Macromol. **33**, 3392 (2000)
49. A. Wurm. C. Schick, e-polymer, **024** (2002).
50. P.H. Geil: Polymer **41**, 8983 (2000)
51. S. Z. D. Cheng, C. Li, L. Zhu: Eur. Phys. J. E **3**, 195 (2000)
52. M. Muthukumar: Eur. Phys. J. E **3**, 199 (2000)

15 Probing Crystallization Studying Amorphous Phase Evolution

Tiberio A. Ezquerra[1] and Aurora Nogales[2]

[1] Instituto de Estructura de la Materia, C.S.I.C. Serrano 119, Madrid 28006, Spain.
[2] J.J. Thomson Physics Laboratory, University of Reading, Witheknights, Reading RG6 6AF, U.K. imte155@iem.cfmac.csic.es

Abstract. The isothermal crystallization process of polymers from the glass can be studied in real time by dielectric spectroscopy and X-ray scattering experiments. The combination of these two techniques aims to reveal a complete picture of the crystallization processes as far as both, the amorphous and the crystalline phases are concerned. In this contribution we will show that by using this experimental approach, information can be obtained about three key aspects of the isothermal polymer crystallization process: i) polymer chain ordering, through Wide Angle X-ray Scattering; ii) lamellar crystals arrangement,through Small Angle X-ray Scattering; iii) amorphous phase evolution through dielectric spectroscopy.

15.1 Introduction

It is well known that polymer systems may develop a characteristic folded chain crystalline lamellar morphology at the nanometer level upon thermal treatment within the temperature range between the glass transition temperature, T_g, and the equilibrium melting temperature [1-3]. This typical lamellar arrangement, schematically shown in Fig. 15.1a, consists of stacks of laminar crystals and amorphous regions intercalated between them. In particular, this semicrystalline state can be obtained when a glassy polymer is heated above its glass transition temperature (T_g) under the so called "cold crystallization" conditions. Above T_g, segmental mobility can enhance a large number of conformations in polymer chain segments. In some cases this conformation accessibility may promote the formation of three dimensional ordered crystallites thermodynamically more stable than the amorphous state.

X-ray scattering techniques may provide information about the structure of the ordered regions in polymers at different length scales [4,5]. Wide angle X-ray scattering (WAXS) offers the possibility to obtain information about molecular ordering in the scale of tenths of nanometers as illustrated in Fig. 15.1b. Thus, it is useful to estimate the fraction of material which possesses three dimensional ordering currently referred to as crystallinity X_c. Small angle X-ray scattering (SAXS), on the other hand, allows one to analyse the structure developed over the length scale of tens of nanometers. Hence, SAXS is useful to characterize the lamellar stack arrangement as shown in Fig. 15.1c. Synchrotron radiation further offers the possibility to perform real time SAXS and WAXS experiments simultaneously during crystallization [6] enhancing the understanding of the correlation

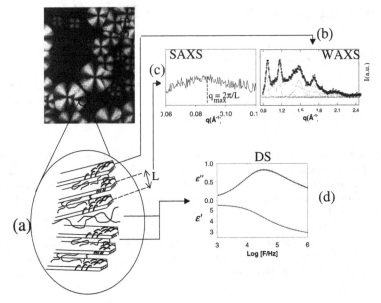

Fig. 15.1. Scheme showing the experimental approach to get access to both, cry talline and amorphous phase in semicrystalline polymers. Micrograph showing a typ cal spherulitic microstructure of a polymeric system. (a)Scheme of the idealized lame lar stacks (according to [3].). (b)Wide angle X-ray scattering (WAXS) provide info mation about molecular ordering in the crystalline lamellae. (c) Small angle X-r scattering (SAXS) may provide information about lamellar stack arrangement.(d) I electric spectroscopy (DS) can offer information about the dynamics of the amorpho phase. Experimental results and micrograph correspond to a Poly(3-hydroxybutyrat co-Poly(3-hydroxyvalerate) (22/78)HV/HB, [13]

between the nanostructure and crystal development [7]. By using simultaneous both techniques, the ordering of the macromolecules trough a very broad leng scale can be examined. X-ray scattering techniques can also be used to extra structural information in amorphous materials [8]. However, in semicrystalli systems mainly information about the ordered regions is obtained due to t. fact that the ordered regions provoke strong diffraction phenomena superi posed over a relatively weak contribution of the amorphous phase.

As far as the amorphous phase is concerned, dielectric spectroscopy expe ments (DS) have shown that, upon crystallization, amorphous phase dynamics strongly affected by the progressive development of the crystalline phase [9–1. Hence, if one would monitor the microstructure development, by X-ray sca tering methods, and the dynamic changes occurring in the amorphous pha by dielectric relaxation methods, a more complete picture of the crystallizati process can be obtained.

The objective of this contribution is to show that an improved understan ing of polymer crystallization can be obtained by monitoring changes of t. dynamics in the amorphous phase during the development of lamellar structu

A scheme of this experimental approach is shown in Fig. 15.1 and involve the combination of real time X-ray scattering and dielectric relaxation techniques experiments.

15.2 Brief Description of Dielectric Spectroscopy

When an electrical field is applied to a dielectric material, a displacement of the electric charge occurs which is characterized by the electric displacement vector D. Due to the fact that the response of the material to an external excitation is not instantaneous, there appears a phase shift between D and the exciting field E which can be accounted for introducing a complex constant of proportionality in the form $D = \epsilon^* \epsilon_{vac} E$, where $\epsilon^* = \epsilon' - i\epsilon''$ is the complex dielectric permittivity of the material and ϵ_{vac} is the permittivity of the vacuum. The real part of ϵ^*, ϵ', corresponds to the dielectric constant and is associated to the energy stored in the material through polarization. The imaginary part, ϵ'', is related to the energy dissipated in the medium and therefore is frequently referred to as dielectric loss [14–16].

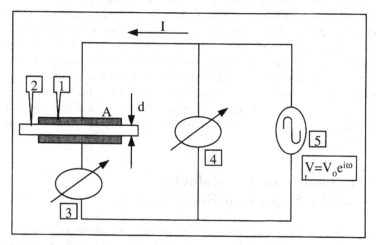

Fig. 15.2. Schematic view of a typical dielectric spectroscopy experiment. (1) Electrodes of area A, (2) Sample film of thickness d, (3) Current Analyzer, (4) Voltage Analyzer, (5) Alternating Voltage generator.

Dielectric spectroscopy is a technique developed to measure ϵ^* as a function of both, frequency of the exciting field and temperature [5,14–16]. A very schematic view about the principle of measurement is given in Fig. 15.2. When an alternating electric field $V = V_0 e^{i\omega t}$ of frequency $F = \omega/(2\pi)$ is applied to the system, a current $I = C\frac{dV}{d\omega}$ is produced where $C = \epsilon^* C_0$ is the complex capacity characterizing the sample, $C_0 = \epsilon_{vac}\frac{A}{d}$ is the capacity of the empty capacitor, d is the thickens of the sample and A is its surface area . By measuring the

278   Tiberio A. Ezquerra and Aurora Nogales

impedance of the sample , $Z = V/I = \frac{-i}{\omega C}$, it is possible to calculate ϵ^* as $\epsilon^* = \frac{C}{C_0}$ [14,15]. When dielectric spectroscopy is used to study molecular motion in polymeric materials then the frequency range of interest typically covers from 10^{-2} Hz to 10^9 Hz [17]. Unfortunately, this broad frequency range can not be covered by a single experimental set-up. Within the 10^{-2} <F/Hz< 10^6 range ϵ^* measurements can be performed by using impedance or frequency response analyzers (Solartron Schlumberger, Hewlet Packard 4192 A, Novocontrol, etc.) and lock-in amplifiers (Standford, EG&G, etc). In this case thin films with circular gold metallic electrodes in both free surfaces(typically 3 to 4 cm in diameter) are prepared and placed between two metallic electrodes building up a capacitor. For frequencies between 10^6Hz to 10^9Hz the above described "sandwich geometry" is not valid and reflectometer techniques are required. Here, ϵ^* can be obtained by reflection coefficient measurements [17,18]. To obtain the temperature dependence of ϵ^* the dielectric cell including sample and electrodes is introduced in a cryostat operating under controlled temperature conditions [17]

A typical dielectric spectroscopy example is shown in Fig. 15.3a which depicts the variation of ϵ'' with temperature and frequency for amorphous poly(ethylene naphthalene-2,6 dicarboxylate)(PEN). PEN is semiflexible polyester with a $T_g = 117°C$ which has found different applications for engineering purposes [19]. PEN exhibits three main dielectric relaxations, β, β^* and α in order of increasing temperature [20,22]. The relaxation processes appear as maxima in ϵ'' moving towards higher temperatures when the frequency is increased. The α process is the most prominent one while both the β and the β^* are much less intense. The molecular origin of every relaxation, corroborated by a great variety of techniques [20–23], has been schematically described at the bottom of Fig. 15.3 and corresponds to the local motion of ester groups (β), a local motion of the naphthalene ring (β^*) and the α-relaxation caused by segmental motions appearing at at T>T_g.

15.3   Influence of Crystallinity on the Segmental Relaxation

The molecular motions occurring in the amorphous phase of a semicrystalline polymer present characteristic aspects which depend, in a first approximation on the degree of crystallinity [10]. In semicrystalline polymers at temperature higher than the glass transition temperature, T_g, the amorphous polymer chain are confined to move between the crystalline regions [9,10,24]. This restriction modifies the dielectric α relaxation which is detected in polymeric systems a temperatures above T_g. The existence of crystallinity in a polymer is reflected in the dynamics of the α relaxation, when compared with that of the pure amorphous polymer, by three effects: a) a decrease of the intensity of the relaxation b) a decrease of the frequency of maximum loss and c) a concurrent change in it shape. [9–11,25]. All these affect are illustrated in Fig. 15.3b which present a dielectric spectroscopy spectrum for a PEN sample with a crystallinity of 0.2 % a estimated by WAXS measurements [26]. In the semicrystalline sample the inten

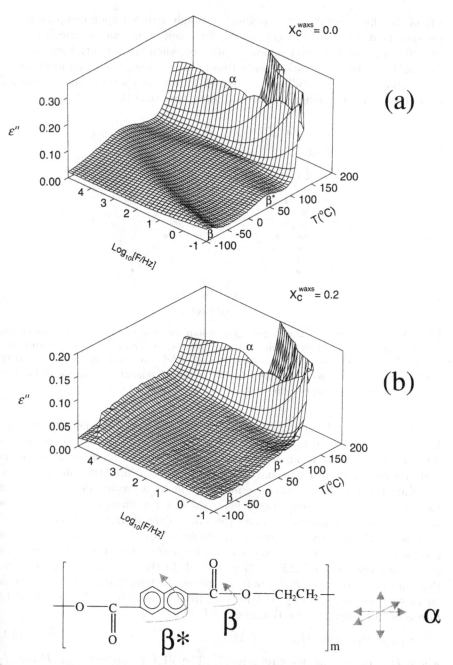

Fig. 15.3. ϵ''-values for two selected PEN samples with different crystallinity as a function of frequency and temperature. Scheme at the bottom shows an scheme of the monomer, and the origin of the observed dielectric relaxations.

sity of the three dielectric relaxations is strongly reduced upon comparing with the spectrum of the amorphous sample. The position in temperature of both, β and β^* relaxations remains almost unaffected when crystallinity increases. On the contrary, the α-relaxation shifts toward higher temperatures and its intensity decreases when crystallinity increases. This indicates that the α relaxation is more sensitive to changes in the microstructure than the β and the β^* ones.

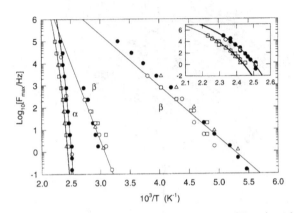

Fig. 15.4. Dependence of the frequency of maximum loss (F_{max}) with the reciprocal temperature for the relaxations in PEN samples with different degree of crystallinity: (○) $X_c^{WAXS} = 0.00$, (●) $X_c^{WAXS} = 0.13$, (□) $X_c^{WAXS} = 0.2$ and (△) $X_c^{WAXS} = 0.27$ Continuous lines represent Arrhenius fits for β and β^* relaxations and Vogel-Fulcher-Tamann fits for the α relaxation.

The position of the maxima in Fig. 15.3, referred to as frequency of maximum loss (F_{max}), have been represented in Fig. 15.4 as a function of the reciprocal temperature for both the amorphous and several semicrystalline samples [26]. In such a representation, the β and β^* relaxations follow an Arrhenius behaviour revealing them as local non-cooperative thermally activated processes. In the case of both, β and β^* relaxations, F_{max} values for samples with different crystallinity lie on a similar range. From the slope of the straight line obtained it is possible to get the activation energy of both processes. In the case of β relaxation a value of $\Delta E = 13$ Kcal/mol is obtained. For the β^* relaxation, the activation energy is larger, around $\Delta E = 33$ Kcal/mol. In the case of the α relaxation the dependence of $Log_{10}[F_{max}]$ with the reciprocal temperature shows a certain curvature (see inset in Fig. 15.4 for magnification)which can be described by means of the Vogel-Fulcher-Tamann (VFT) equation:

$$F_{max} = F_0 \exp\{-DT_0/(T - T_0)\} \qquad (15.1$$

where D is called fragility strength [27]. The VFT parameters, F_0, D and T_0 are presented in Table 15.1. The obtained values for the Vogel temperature T_0 increases with crystallinity and the pre-exponential factor F_0 shows different values for the semicrystalline samples but slightly similar D values are obtained for both samples.

Table 15.1. Vogel Fulcher Tamann parameters for PEN samples with different crystallinity

X_c^{WAXS}	F_0 (Hz)	A (K)	T_0 (K)	D
0	$9.8 \cdot 10^{13}$	1898.6	340.2	5.6
0.13	$9.8 \cdot 10^{13}$	1839.5	342.2	5.4
0.2	$1.0 \cdot 10^{13}$	1844.2	347.6	5.3
0.27	$1.0 \cdot 10^{13}$	1828.4	349.1	5.2

As mentioned above, crystallinity also influences dramatically the shape of the α relaxation inducing a strong broadening. Fig. 15.5 shows ϵ''-values as a function of frequency for three selected PEN samples of varying crystallinity, X_c^{WAXS}, in the temperature region of the α-relaxation. The dotted lines correspond to the separation of the different contributions to the measured relaxation by using the Havriliak-Negami equation as described elsewhere [26]. This analysis will be discussed in detail in a following paragraph . The α relaxation tends to shift to higher temperatures as compared with that of the amorphous sample. As crystallinity increases there is a reduction in intensity. This fact can be quantified by calculating the area under the α-relaxation peak, $\Delta\epsilon$, given by:

$$\Delta\epsilon = 2/\pi \int_{-\infty}^{\infty} \epsilon''(\omega) dl_n\omega \qquad (15.2)$$

From this magnitude it is possible to estimate the amount of mobile fraction by using the ratio between the relaxation intensity of the semicrystalline sample to that of the amorphous sample [28]. The dependence of the normalized $\Delta\epsilon$ with crystallinity is shown in Fig. 15.6 for PEN [26] and for several polymers including: PET [29], PEEK [30] and Poly(hydroxybutyrate)(HB)-co-poly(hydroxyvalerate)(HV) with different HB/HV ratios [12]. The data seem to follow a linear trend, which extrapolates to $\Delta\epsilon = 0$ for a crystallinity value different from 100%. It is evident that the presence of crystallinity reduces the amount of relaxing material because the material incorporated within the crystals does not contribute to the α relaxation process. In addition, it seems that for semicrystalline polymers, the crystalline phase is not the only fraction of material which remains rigid at temperatures above T_g. Results of Fig. 15.6 point towards the existence of another portion of the amorphous phase, without three dimensional order but with restricted mobility above T_g. This phase has been referred to as "Rigid Amorphous Phase" (RAP) and indicates the heterogeneity of the dynamic characteristics of the amorphous fraction in semicrystalline polymers [28,31,32][1].

---

[1] For a detailed dicussion of the RAP, see Chap. 14 on page 252.

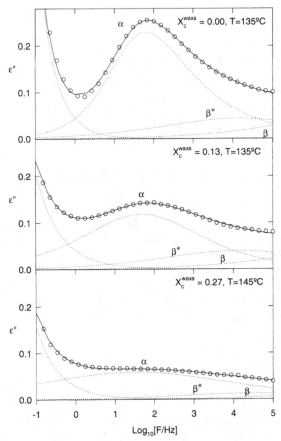

Fig. 15.5. ϵ''-values for three selected PEN samples as a function of frequency (F) for different temperatures. Continuous lines represent best fits according to Havriliak-Negami equation. Dashed lines shows the separated contribution of the relaxation and the conduction process.

15.4 Influence of Crystallinity on the Segmental Cooperativity

As shown in the previous paragraph the α relaxation as a function of temperature does not follow an Arrhenius behaviour but is better described by a VFT behaviour for both the amorphous and the semicrystalline state. In polymers, the degree of departure from the Arrhenius behaviour exhibited by the relaxation time of the α relaxation process has been associated with the strength of the correlation between non bonded species and has been defined as cooperativity [27] that can be quantified by the parameter D appearing in the VFT equation. In principle, lower D-values indicate a higher degree of segmental cooperativity [27]. By scaling F_{max}-values as a function of T^*/T, where T^* is a reference tempera-

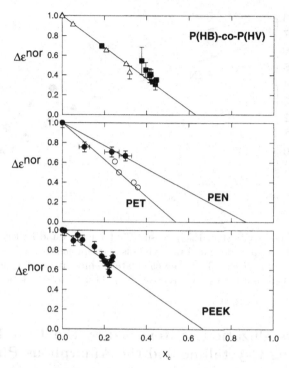

Fig. 15.6. Amorphous mobile fraction of the studied samples determined Dielectric Spectroscopy for different polymers as a function of crystallinity. Upper panel: Poly(hydroxybutyrate)-co-poly(hydroxyvalerate), 74:26 △ and 94:6 □. Intermediate panel: PEN at $T = 145°C$ (○, PET from [29]). Lower panel: PEEK $T = 155°C$ (●)

ture, the relaxation time temperature dependence of polymers with different T_g values can be compared [27,33]. Relaxation experiments comparing polymers in, both, the amorphous and the semicrystalline state have shown that, in spite of the observed shift in T_g, a rather similar dependence of the relaxation time with T^*/T (cooperativity plot) is observed [32,34].

To illustrate this effect, we have plotted in Fig. 15.7 the F_{max} data for PEN [26] and poly(aryl-ether-ether-ketone)(PEEK) [30] as a cooperativity plot (F_{max} versus T^*/T) where T^* is a reference temperature calculated from the VFT equation, as the absolute temperature at which $\tau=10s$ ($\tau = 1/(2\pi F_{max})$). Although the samples with different crystallinity exhibit different T^* values, upon scaling by T^*/T, F_{max}-values tend to collapse at low frequencies into a single curve [34]. This has been interpreted assuming that only those chain segments in the proximity of the crystallites decrease their characteristic frequency. However, in average the degree of intermolecular coupling for most chain segments remains almost unaffected [34].

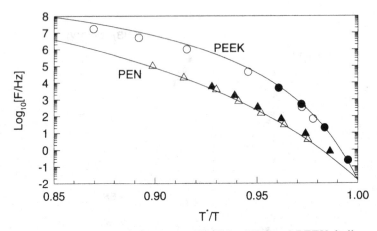

Fig. 15.7. Cooperativity plot: F_{max} versus T^*/T for PEN and PEEK, hollow symbols are the amorphous specimens and full symbols corresponds to 27 % of crystallinity for PEN and 0.34% for PEEK. T^* is the calculated temperature at which $\tau = 1/(2 \cdot \pi \cdot F_{max}) = 10s$. The continuous lines represent the best fit to the data according to the VFT equation.

15.5   Crystallization as Revealed by the Time Evolution of the Crystalline and the Amorphous Phases

15.5.1   Lamellar Structure Formation

The main objective of this paragraph will be to exemplify how the changes of the dynamics in the amorphous phase during crystallization can be followed by combining real time X-ray scattering and dielectric relaxation experiments. This experimental approach has been employed in a great variety of polymers including: poly(ethylene terephthalate)(PET) [9,35,36], polystyrene [37] poly(aryl-ether-ketone) [11,30], copolyesters of hydroxybutyrate (HB) and hydroxyvalerate (HV) [12,38], poly(vinylidene fluoride)(PVDF) [39] among others [40–43].

Experiments will be illustrated for poly(ether ether ketone) (PEEK). When amorphous PEEK is placed above its glass transition temperature (Tg=145 °C) but below its melting temperature (Tm=360°C) crystallization takes place through the formation of spherulites [44,45]. Figs. 15.8a and 15.8b shows the real time evolution of the WAXS and SAXS patterns during the crystallization process of an initially amorphous PEEK sample heated at T=160°C. For crystallization times t_c longer than 6 min, in the WAXS pattern, the development of three Bragg maxima it is clearly seen. These maxima correspond to the (110), (111) and (200) reflections of the orthorhombic unit cell of PEEK [46] respectively. From these experiments it is possible to extract the crystallinity values by estimating the ratio between the area under the crystalline peaks in relation to the total diffracted area [46]. The scattered intensity at small angles displays a maximum around q=0.06Å^{-1} for $t_c > 6$ min indicating the formation of a stacking lamellar structure. From the Lorentz corrected SAXS profile, the position o

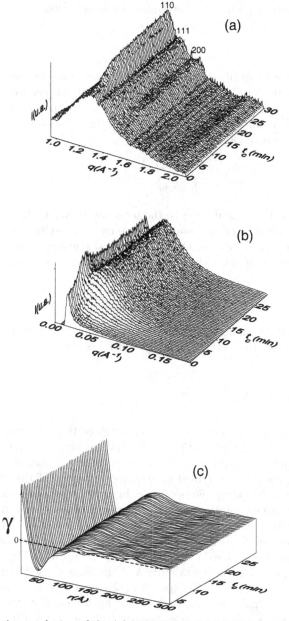

Fig. 15.8. Real time evolution of the (a) WAXS and (b) SAXS intensity as a function of q and crystallization time t_c during a crystallization process at T=160°C of initially amorphous PEEK. (c) Real time variation of the correlation function obtained from the SAXS patterns during this same crystallisation process.

the maximum can be used to determine the long period:

$$L = \frac{2\pi}{q_{max}} \tag{15.3}$$

where $q = (4\pi/\lambda)\sin(\theta)$ is the scattering vector, and 2θ the scattering angle. L represents the sum of the average thickness of the crystal lamellae, l_c, and of the interlamellar amorphous regions, l_a. Fig. 1c. shows the one-dimensional correlation function profiles obtained from the data of Fig. 1b calculated as explained elsewhere [30,47,48] from:

$$\frac{\gamma_{1,r}}{Q} = \frac{\int_0^\infty (I - I_b)q^2 \cos(qr)exp(\sigma^2 q^2)dq}{\int_0^\infty (I - I_b)q^2 dq} \tag{15.4}$$

Here, I_b is that contribution to the total scattering arising from density fluctuations (liquid scattering), and σ is a term, related to the thickness of the crystal/amorphous interface. Q is the so-called scattering invariant that can be determined by integrating the SAXS profile over all scattering angles:

$$Q = \int_0^\infty (I - I_b)q^2 dq \tag{15.5}$$

One of the advantages of using the correlation function is that, in addition to the long period L, estimates for l_a and l_c can be obtained can be obtained according to:

$$\frac{B}{L_c^M} = X_1(1 - X_1) \tag{15.6}$$

where B is the position of the first intercept of the correlation function with the r-axis and L_c^M is the position of the first maximum of the correlation function. From the two solutions of the above quadratic equation, the one with the higher value should be ascribed to the larger fraction of the two phases found within the lamellar stacks. For example, in highly crystalline samples, X_1 would correspond to the crystal fraction within the lamellar stacks (denoted as X_{cL},) and $1-X_{cL}$ would, then, represent the amorphous fraction within the stack. Once the assignment of X_1 is made for each particular case, one may calculate the l_a and l_c considering that $lc=X_{cL}L_c^M$ and $l_a=(1-X_{cL})L_c^M$. Fig. 15.9a illustrates the evolution of both the crystallinity and the invariant with the crystallization time for two temperatures. As expected, the degree of crystallinity, X_c, develops according to a typical sigmoidal shape for both temperatures. After an initial induction time, X_c, first increases rapidly (primary crystallization) and finally tend to level off for longer times (secondary crystallization). Preceding studies suggest that the induction time for crystallization from the glass is governed by the segmental mobility of the supercooled melt [49]. The crystallization process is faster at the higher crystallization temperature. The final value of crystallinity reached is about the same within the error of experiment, for the two crystallization temperatures.

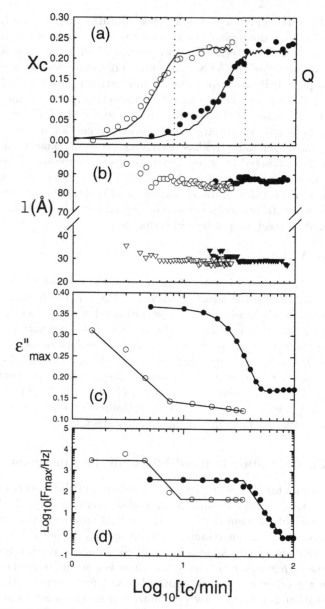

Fig. 15.9. PEEK.(a) Plot of the crystallinity X_c and invariant(continuous lines), (b)thickness of the two phases present on the crystalline stacks. (l_1)(upper) and (l_2)(lower),(c) maximum dielectric loss values ϵ''_{max}, and (d) frequency of maximum loss F_{max}, with the crystallisation time as a function of crystallization time t_c for T=155°C (solid symbols) and T=160°C (hollow symbols). Continuous lines are eye-guides.

After the initial induction period, whose duration depends on the temperature, crystallization takes place, as indicated by the appearance of the reflection peaks in the WAXS patterns. The occurrence of these peaks is accompanied by the development of a SAXS maximum. This indicates the formation of a larger scale periodicity corresponding to alternating crystalline and amorphous layers giving rise to lamellar stacks. The parameters l_1 and l_2 (thickness of the two constituent phases of the lamellar stack) can be obtained by a correlation function analysis [30]. Based on several considerations [30] and in recent experiments [44], the larger value (l_1) has been assigned to the lamellar thickness (l_c), and the smaller value (l_2) has been designated as the thickness of the amorphous layer (l_a). Both l_c and l_a values are illustrated in Fig. 15.9b for the two presented temperatures. Both values initially decrease and remain almost constant during the rest of the crystallization process. The overall crystallinity in the sample can be considered as composed of several terms as:

$$X_c = X_s X_L X_{cL} \tag{15.7}$$

where X_s is the fraction of material included in the spherulites, X_L is the fraction of the material in the lamellar stacks, and X_{cL} is the fraction of the crystalline material in the stacks. X_{cL} can be estimated as the ratio $l_1/(l_1+l_2)$. From Fig. 15.9b, one observes that, during the latest stages of primary crystallization (the end of the rapid increase in X_c) the linear fractions of each phase remain nearly constant. This indicates that the composition of the stacks is virtually the same during late stages of the crystallization process in agrement with other studies [45]. This would imply, according to (15.7) , that the increase observed in crystallinity during the late stages of crystallization ought to be assigned to an increase in the total amount of the lamellar stacks.

15.5.2   Crystallization Induced Effects in the Amorphous Phase

Fig. 15.10 shows for PEEK the real time evolution of the α relaxation followed by dielectric spectroscopy during the crystallization process at T=155°C . The imaginary part of the complex dielectric permittivity, measured for the amorphous PEEK, manifests a maximum centered at $F_{max} = 500$Hz. The α relaxation of the initially amorphous sample is found to endure some changes as crystallization time increases. On the one hand, there is a strong reduction of the peak height. On the other hand, a shift towards lower frequencies in the position of the α relaxation is observed, for crystallization times longer than 40min. The intensity of the maximum loss ϵ''_{max}, and the frequency of the maximum loss F_{max}, have been illustrated in Fig. 15.9c and d respectively as a function of crystallization time for two different temperatures. The variation rate of both magnitudes is dependent on the crystallization temperature. At T=160°C, ϵ''_{max} starts to decrease from t_c=0 min and it levels off at t_c >8min. At T=155°C ϵ''_{max} initially remains constant, and decreases at t_c >20min and finally reaches a plateau at t_c >60min. A very different trend is observed for F_{max}. At both

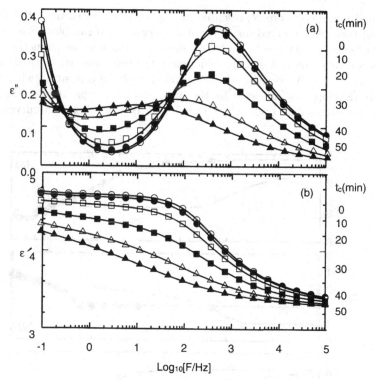

Fig. 15.10. Real time evolution of the dielectric loss (a) and the dielectric constant (b) during isothermal crystallization of PEEK ($T_c=155°C$) as a function of frequency at selected crystallization times .

temperatures, F_{max} remains constant when ϵ''_{max} exhibits a higher rate of variation. Only when ϵ''_{max} tends to reach its lowest plateau value, F_{max} decreases abruptly.

A deeper analysis of the relaxation curves can be accomplished performed on the basis of the phenomenological description of Havriliak Negami [50] given by:

$$\epsilon^* = \frac{\epsilon_0 - \epsilon_\infty}{[1 + (i\omega\tau_{HN})^b]^c} + \epsilon_\infty - i\frac{\sigma}{\epsilon_{vac}\omega^s} \tag{15.8}$$

where $\omega = 2\pi F$, ϵ_0 and ϵ_∞ are the relaxed($\omega=0$) and unrelaxed($\omega=\infty$) dielectric constant values, τ_{HN} is the central relaxation time of the relaxation time distribution function and b and c ($0<b,c<1$) are shape parameters which describe the symmetric and asymmetric broadening of the relaxation time distribution function, respectively [50]. The second term of (15.8) correspond to the conductivity contribution. Here, σ is related to the direct current electrical conductivity, ϵ_{vac} is the vacuum dielectric constant and s depends on the nature of the conduction mechanism. A value of $s<1$ is associated to a non-ohmic transport [51]. The continuous lines in Fig. 15.10 represent the best fits according to (15.8). The obtained parameters are represented in Fig. 15.11 as a function of the crystallization time. As observed for isothermally crystallized samples with controlled levels of crystallinity (Sect. 15.3), the main observations as a function of crystallization time are: i) A strong reduction of the relaxation strength $\Delta\epsilon$ for both temperatures; ii) A decrease of the asymmetry of the relaxation when crystallization develops, as indicated by the increase of the c parameter towards the higher possible value, c=1; iii) A broadening of the relaxation curves, as can

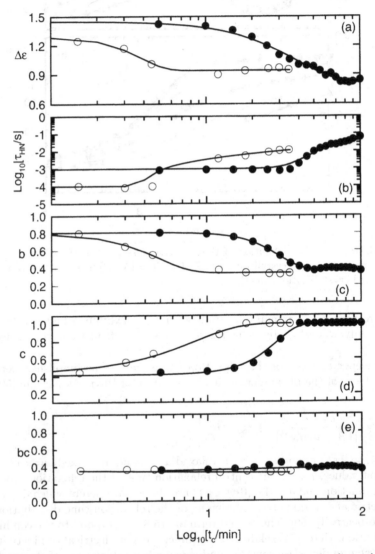

Fig. 15.11. Real time evolution of $\Delta\epsilon, \tau_{HN}$, b, c parameters during isothermal crystallization of PEEK at T=155°C (●) and T=160°C (○). Continuous lines are eye-guides

be seen by the decrease of the b parameter. Independently of the crystallization temperature, the final shape of the relaxation curves is almost the same, as indicated by the similar final shape parameters. The central relaxation time τ_{HN}, shows some different changes during crystallization. During the period in which the other HN-parameters exhibit main variations due to crystallization, the central relaxation time remains almost constant. Only when all the other parameters tend to reach their final values, τ_{HN} begins a sudden change.

As mentioned previously, the dielectric strength $\Delta\epsilon$ is proportional to the amount of dipoles involved in the relaxation process [14]. The decrease of $\Delta\epsilon$ with crystalization time observed in Fig. 15.11 indicates that, as a consequence of the crystallization process, the amount of relaxing dipoles decreases. This is evident, as the molecules incorporated to the crystals cannot relax. As shown in Fig. 15.6, $\Delta\epsilon$ follows a linear dependence with crystallinity. Should the only immobile material be the only one incorporated within the crystals, then the slope of the $\Delta\epsilon$-X_c linear dependence should be about -1. However, as one may see in Fig. 15.6, the dependence is stronger (-1.5) for all the represented polymers. This trend indicates that, the growing crystals do not only immobilize the material incorporated within them, but also constrain some additional amorphous material. Then, it is possible to conclude that not all the non-crystalline fraction behaves as a relaxing material. There is, indeed, an amount of material that neither contribute to the WAXS patterns nor relaxes in agreement with very early studies [9].

An interesting factor to consider concerns the characteristics of the relaxation process of the remaining mobile material. As observed in Fig. 15.9, the position of the relaxation function does not change during the primary crystallization, where the major increase in the crystallinity occurs. Considering the magnitude $(2\pi F_{max})^{-1}$ as an average relaxation time, the observed decrease in F_{max} for longer crystallization times, indicates a slowing down tendency of the chain mobility as crystals develop further. However, this decrease does not significantly occurs until the crystallinity tends to reach their final plateau values. Thus, the dynamics of the amorphous material during the primary crystallization remains, in average, at the same rate, as indicated by the similar F_{max}. Is during the secondary crystallization regime when the remaining amorphous phase slows down its relaxation, and consequently F_{max} diminishes.

15.5.3   Combined Picture About Crystallization from the Structure-Dynamics Relationships

Let us emphasize that, for the kind of polymers mentioned here and under cold crystallization conditions, the structural results , mainly based on SAXS [52,53] but also in recent microscopic observations [44,45], can be interpreted according to a structural model in which, in the semicrystalline polymer is filled up with spherulites. However, the lamellar stacks are separated from each other by larger amorphous gaps.

Based on this scenario, one may assume in a first approach the existence of two distinct amorphous regions: a) the interlamellar amorphous phase, that can be visualized as being highly constrained and b) the amorphous regions between the lamellar stacks that are less constrained. In accordance, one may expect two main segmental relaxation processes: one occurring in interlamellar regions whose intensity should decrease with crystallization time and a second relaxation taking place in the interstack amorphous regions which should increase with crystallization time at expenses of the first one. Some authors have proposed this model on the basis of dynamical mechanical [45,54], dielectric [36,29] and calorimetric [29] measurements [2]. However, the absence of an isobestic point characterized by a frequency at which no change in ϵ'' should be expected with crystallization time indicates that the relaxation model is not as simple as above described. Additionally, calculation done on the basis of the above model do not precisely describe the experimental dielectric measurements [55]. The process of crystallization can be envisioned according to the view proposed by Schultz et al [56] in which "loose" spherulites are firstly formed which become "dense" spherulites in later stages of crystalization. Such a model is also supported by investigations involving simultaneous measurements of wide- and small-angle X-ray scattering during isothermal crystallization of PET [57] and PEEK [53].

The dielectric experiments illustrated here, for PEN and PEEK, are more in favor of a single relaxation process that shifts to lower frequencies as crystallization develops. In principle, one may consider that, inside the stacks, the non crystalline chains are constrained to a level where the segmental mobility is strongly arrested. If this is the case, $\Delta\epsilon$ would represents the fraction of material outside the stacks, i.e. $\Delta\epsilon = 1-X_L$. Introducing this relationship in (7), it follows.

$$X_c = X_s X_{cL}(1 - \Delta\epsilon) \qquad (15.9)$$

When the material is fully covered by spherulites, Xs=1. Considering for PEEK the extrapolation performed in Fig. 15.6 ($\Delta\epsilon = 0$), the above equation leads to $X_c(\Delta\epsilon = 0) = X_{cL} = \approx 70\%$, which is consistent with the value estimated by assuming $l_1 = l_c$, ($X_{cL} = 75\%$). It is convenient to remember that, the linear crystallinity obtained from Fig. 15.6 is a mass fraction, since the X_c values used (from WAXS) are a mass fraction. However, the value obtained from SAXS is a volume fraction. But, given the crystalline and amorphous densities of our PEEK, the difference between volume and mass crystallinity is below 2% [30].

Fig. 15.12 depicts schematically the picture which emerge from the combined interpretation of the structural and dynamical experiments. Here, it is assumed that the noncrystalline regions in the primary lamellar stacks do not exhibit a relaxation process. Only the non crystalline regions in the gaps between the stacks show a heterogeneous relaxation behaviour, which is strongly dependent of the secondary crystallization process, as we discuss hereafter. As crystallinity starts to increase, the frequency of maximum loss, F_{max}, or its counterpart the relaxation time, τ_{HN} remain unaffected. Only when the crystallinity achieves a certain critical value, at the end of the primary crystallization, the relaxation time begins to increase, indicating a slowing down of the mobility of the un restricted amorphous fraction. This effect can be understood considering the

---

[2] See Chap. 14.

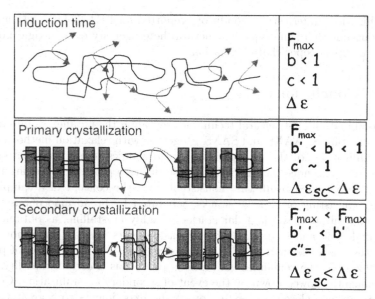

Induction time	F_{max} $b < 1$ $c < 1$ $\Delta \varepsilon$
Primary crystallization	F_{max} $b' < b < 1$ $c' \sim 1$ $\Delta \varepsilon_{sc} < \Delta \varepsilon$
Secondary crystallization	$F'_{max} < F_{max}$ $b'' < b'$ $c'' = 1$ $\Delta \varepsilon_{sc} < \Delta \varepsilon$

Fig. 15.12. Schematic model summarizing the obtained results from structural and dynamical experiments. Before crystallization, the polymer chains present segmental motion (schematically depicted by the curved arrows) leading to the dielectric α relaxation of shape parameters b and c and characterized by a certain $\Delta\epsilon$ value. During primary crystallization, the restrictions imposed by the crystalline lamella (represented the grey domains) mainly inhibit segmental motions within the intralamellar amorphous phase but not in the interlamellar amorphous phase. Here the α relaxation modifies its shape (b and c parameters) but essentially not F_{max} and the amount of relaxing material diminishes ($\Delta\epsilon_{sc} < \Delta\epsilon$). During secondary crystallization, the amorphous material located outside the stacks, is capable to undergo segmental movement, but its dynamics is effectively slowed down with respect to the purely amorphous phase.

following: During primary crystallization, the formation of lamellar stacks produces an immobilization, not only of the material incorporated into the crystals, but also of the amorphous material remaining between lamella crystals. This fact influences strongly $\Delta\epsilon$ since it is related to the amount of the dipoles involved in the relaxation process. It also affects the shape of the relaxation process which becomes broader and more symmetric, due to the increasing number of different environments created by the crystals. However, this situation does not affect the average position of the relaxation peak. The average relaxation time of the remaining relaxing areas, (broad amorphous areas located between the lamellar stacks) is the same. Once the space tends to be filled-in with spherulites, the crystallization can proceeds in the broad gaps between the stacks. The remaining amorphous phase located in these areas has similar or slightly restricted mobility as compared to the initial amorphous sample. Secondary crystallization has the effect of restricting the mobility on these amorphous gaps. As a consequence, the average relaxation time is now effectively increased. It is worth mention that even the appearance of two relaxation α processes during primary crystallization under certain conditions reported for PET [36] can be considered as a consequence of the typical structural heterogeneity of semi rigid polymers which requires deeper characterization.

15.6 Conclusions

The combination of structural techniques revealing the parallel evolution of the crystalline phase (WAXS and SAXS) together with relaxation methods detecting modifications of the amorphous phase dynamics (dielectric spectroscopy) show a more extended picture of cold crystallization processes in semirigid polymers. The obtained results are consistent with the concept of a sample morphology consisting of lamellar stacks, separated by broad amorphous regions. The composition of the lamellar stacks is highly crystalline, as obtained from both, dielectric and structural X-ray measurements and the amorphous phase inside the stacks seems to be highly constrained. After completion of primary crystallization, the remaining amorphous phase has a certain mobility, but it is significantly slowed down by the event of secondary crystallization. Certainly the topic is not closed and extension of the experiments to a greater variety of polymers, to a broader dynamic range and to shorter time resolution for the structural characterization will be experimental challenges for the incoming work.

Acknowledgements

We would like to express our gratitude to our colleagues I. Sics, Z. Denchev and F.J. Baltá-Calleja whose support contributed to many of the results presented here. Also we are indebted to II-00-015 EC for financial support of measurements at HASYLAB(DESY, Hamburg, Germany) and to MCYT(Spain) for the grant FPA2001-2139(Spain).

References

1. D.C. Bassett, *Principles of polymer morphology*. (Cambridge University Press Cambridge, 1981).
2. A. Keller: Philos. Mag. 2, 1171 (1957).
3. J.M. Schultz:'Structure development in Polyesters', In *Solid State behaviour of linear polyesters and polyamides* ed. by J.M. Schultz and S. Fakirov (Prentice Hall Inc London, 1990) pp.75-130.
4. F.J. Baltá-Calleja and C.G. Vonk , *X-ray Scattering of Synthetic Polymers* (Elsevier, Amsterdam, 1989).
5. G. Strobl, *The Physics of Polymers* (Springer Berlin, 1996).
6. C. Wutz, M. Bark, J. Cronauer, R. Döhrmann, and H. G. Zachmann: Rev. Sci Instrum. 66, 1303 (1995).
7. M. Bark and H. G. Zachmann: Acta Polymerica 44, 18 (1993).

8. E. Eckstein , J. Qian , T. Thurn-Albrecht, W. Steffen and E.W. Fischer: J. of Chem. Phys. **113**, 4751 (2000).
9. G. Williams:, G. Adv. Polym. Sci. 33, 59 (1979).
10. J.C. Coburn, R.H. Boyd:. Macromolecules 19, 2238 (1986).
11. T.A. Ezquerra, J. Majszczyk, F.J. Baltá-Calleja, E. López-Cabarcos, K.H. Gardner, B.S. Hsiao, Phys. Rev. B 50, 6023 (1994).
12. A. Nogales, T.A. Ezquerra, J.M. García, F.J. Baltá-Calleja: J. Polym. Sci. B: Polym Phys 37, 37 (1999).
13. I. Sics, T.A. Ezquerra, A. Nogales, F.J. Balta-Calleja, M. Kalni, V. Tupureina: Biomacromolecules 2, 581 (2001).
14. P. Hedvig: *Dielectric Spectroscopy of Polymers* (Bristol: Adam Hilger Ltd, 1997).
15. A.R. Blythe: *Electrical Properties of Polymers* (Cambridge: Cambridge University Press, 1979).
16. N.G. McCrum, B.E. Read, G. Williams: *Anelastic and Dielectric Effects in Polymeric Solids* (Dover, New York, 1991, Original Issue John Wiley, London 1967).
17. F. Kremer, D. Boese, G. Meier, E.W. Fischer:. Prog. in Colloid&Polym. Sci. 80,129 (1989).
18. T.A. Ezquerra, F. Kremer, G. Wegner:'AC electrical properties of insulator-conductor composites'. In: *Progress in Electromagnetics Research (PIERS VI)*ed. by J.I. Kong Ed. (Elsevier,Amsterdam 1992) pp. 273-301.
19. Z. Denchev, I. Šics, A. Nogales, T.A. Ezquerra: ' Microstructural characterization of poly(ethylene naphthalene 2,6-dicarboxylate) as revealed by the properties of both amorphous and crystalline phases'. In *Handbook of Thermoplastic Polyesters PET, PBT, PEN (Homopolymers, copolymers, blends and composites)* edited by S.Fakirov, Wiley-VCH (2001).
20. D. Chen, H.G. Zachmann: Polymer 32, 1612 (1991).
21. H. Dörlitz, H. G. Zachmann: J. Macromol. Sci.-Phys. B36, 205 (1997).
22. T.A. Ezquerra, F.J. Baltá-Calleja, H.G. Zachmann: Acta Polymerica 44, 18 (1993).
23. J.C. Cañadas, J.A. Diego, M. Mudarra, J. Belana, R. Díaz-Calleja, M.J. Sanchís, C. Jaimés: Polymer 41, 2899 (2000).
24. J.R. Havens, D.L. VanderHart: Macromolecules 18, 1663 (1985).
25. E. Schlosser, A. Schönhals: Colloid & Polymer Sci. 267, 963 (1989).
26. A. Nogales, Z. Denchev, I. Šics, T. A. Ezquerra: Macromolecules 33, 9367 (2000)
27. R. Böehmer, K.L. Ngai, C.A. Angell, D.J.J. Plazek: J. Chem. Phys. 99,4201 (1993)
28. P. Huo, P Cebe: Macromolecules: 25, 902 (1992).
29. J. Dobbertin , A. Hensel, C.J. Schick: Thermal Anal. 47, 1027 (1996).
30. A. Nogales , T.A. Ezquerra, Z. Denchev , I. Sics F. J. Baltá Calleja.B.S. Hsiao: J. of Chem. Phys.115, 3804 (2001).
31. D.S. Kalika, R.K. Krishnaswamy: Macromolecules 26, 4252 (1993).
32. A. Nogales, T.A. Ezquerra, F.J. Batallán, B. Frick, E. López-Cabarcos,F.J. Baltá-Calleja: Macromolecules 32, 2301 (1999).
33. W.T. Laughlin, D.R. Uhlmann: J. Phys. Chem. 76, 2317 (1972).
34. K.L. Ngai, C.M. Roland: Macromolecules 26,2688 (1993).
35. T.A. Ezquerra, F.J. Baltá-Calleja, H.G. Zachmann: Polymer 35, 2601 (1994).
36. K. Fukao and Y. Miyamoto: Phys. Rev. Lett. 79,4613 (1997).
37. K. Fukao, Y. Miyamoto:Polymer 34, 238 (1993).
38. A.A. Mansour, G.R. Saad, A.H. Hamed: Polymer 40,5377 (1999).
39. J. Mijovic, J.W. Sy: Macromolecules 30, 3042 (1997).
40. J. Dobbertin, J. Hannemann, C. Schick, M. Potter, H. Dehne :J. Chem. Phys. 108, 9062 (1998).

41. S. Andjelic S, B.D. Fitz:J. Polym. Sci.: Pol. Phys. 38 , 2436 (2000).

42. M. Mierzwa, G. Floudas, P. Stepanek, G. Wegner: Phys. Rev.B 62, 14012 (2000).

43. Y.L. Cui, J. Wu, A. Leyderman, G.P. Sinha, F.M. Aliev: J. Phys.D: Appl. Phys. 33, 2092 (2000).

44. Z. Xia, H. Sue, Z. Wang, C.A. Avila-Orta, B.S. Hsiao: J. Macromol. Sci.-Physics B40, 625 (2001).

45. D. A. Ivanov, R. Legras, and A. M. Jonas: Macromolecules 32, 1582 (1999).

46. D. J. Blundell and B. N. Osborn: Polymer 24, 953 (1983).

47. C.G. Vonk and G. Kortleve: Kolloid-Z 220, 19 (1967).

48. G. R. Strobl and M. Schneider: J Polym Sci. 18, 1343 (1980)

49. A. Nogales, T.A. Ezquerra, Z. Denchev, F.J. Balta-Calleja: Polymer 42, 5711 (2001).

50. S. Havriliak and S. Negami: Polymer 8, 161 (1967).

51. K.U. Kirst, F. Kremer, V. Litvinov: Macromolecules 26, 975 (1993).

52. C. Santa Cruz, N. Stribeck, H. G. Zachmann, and F. J. Baltá-Calleja: Macromolecules 24, 5980 (1991).

53. R.K. Verma, H. Marand, and B.S. Hsiao: Macromolecules 29, 7767 (1996).

54. G. Vigier, J. Tatibouet, A. Benatmane, R. Vassoille: Colloid&Polymer Sci. 270 1182 (1992).

55. T.A. Ezquerra, F.J. Baltá-Calleja, H.G. Zachmann: Acta Polymerica 44, 18 (1993)

56. J.M. Schultz: Makromol. Chem. Makromol. Symp. 15, 339 (1988).

57. H.G. Zachmann and C. Wutz:'Studies of the mechanism of crystallization by means of WAXS and SAXS empoying synchrotron radiation'. In *Crystallization of Poly mers*403, Edited by M. Dosiëre (Kluwer Academic Publichsers 1993)pp.403-.

16   Modeling Polymer Crystallization: DSC Approach

Joan Josep Suñol

Grup de Recerca en Materials i Termodinàmica. Universitat de Girona, E-17071 Girona, Spain
joanjosep.sunyol@udg.es

Abstract. The kinetics of the crystallization process of several polymers, polymer blends and copolymers have been analyzed by differential scanning calorimetry. A modified isoconversional method is applied to evaluate the parameters governing the crystallization of solidification from the assumed isotropic melt. Once the parameters are known, the modeling of the polymer crystallization can be performed. The temperature - cooling rate - transformation diagrams were constructed for a wide range of conditions and they showed a good agreement between the experimental data and curves obtained by calculation. This verifies the reliability of the method utilized and the validity of the rate constant model description. Furthermore, the specific surface energies were calculated from non-isothermal data of PP-PE based copolymers with different additives. The σ values obtained using this approach are similar to those reported by other techniques.

16.1   Introduction

The knowledge of the thermal behavior of materials is important to control their structure and properties. For instance, processes as solidification at different rates can promote changes in materials properties. Several techniques have been utilized to investigate the kinetics of crystallization processes. Experimentally, the crystalline morphology and the crystallization kinetics of polymeric materials can be assessed by numerous conventional techniques as small-angle x-ray scattering (SAXS) [1], transmission electron microscopy (TEM) [2] and scanning probe microscopy (SPM) [3]. Moreover, time-resolved scattering experiments are able to probe polymer crystallization kinetics at very early times [4]. The recent literature [5-8] demonstrates that this field presents a great scientific interest. In particular, the thermoanalytical techniques (TA), such as differential thermal analysis (DTA), thermogravimetry (TG) and differential scanning calorimetry (DSC) are ones of the most often applied [9-12]. Moreover, the recently developed temperature-modulated differential scanning calorimetry (TM-DSC) permits the measure of various thermal responses and the understanding of their kinetics and thermodynamics [13] of glass transition, melting and crystallization of polymers [14,15]. In the Chap. 14 of this book, TMDSC and other techniques are applied to the study of the vitrification and devitrification of the rigid amorphous fraction in semicrystalline polymers revealed from frequency dependent heat capacity.

From the DSC data, several models [16-21] have been developed to reproduce experimental data under isothermal or continuous heating rate conditions. Quite surprisingly, the number of publications devoted to the kinetic analysis of solidification of materials from the melt is relatively low. Moreover, it is known that the thermal history in the melt can affect the crystallization behavior of polymers [22-23]. This "memory" effect can increase the crystallization rate for insufficient melt time [24]. Thus, unmolten crystalline seeds or local chain organization could still remain at temperatures above the melting point [25]. The presence of such primary nuclei increases the crystallization rate. Such a phenomenon is also defined as self nucleation. Another analysis is related to the entanglement density. In the melting process, the entanglement density should increase with time, which can be regarded as a relaxation process [26]. To reduce it, an optimal thermal treatment, temperature-time, of the melting must be chosen.

Transformation diagrams are one of the most reliable way to predict the thermal behavior of the material. Once a model has been determined, one can construct the transformation diagrams: Time - temperature (T-T-T) and temperature - heating/cooling rate (T-HR/CR-T) transformation diagrams [22,27,28] In this chapter, the crystallization process of several polymers was analyzed and a modified isoconversional method is applied to perform kinetic analysis of non-isothermal processes. As expected, it was stated that the Avrami model provides a good description of the solidification process. Moreover, temperature-cooling rate-transformation diagrams were constructed and there was a good agreement between experimental data and the calculated T-CR-T curves. Furthermore, determining the specific surface energy is one of the most characteristic methods to describe the crystallization of previously molten materials [29] or other surface phenomena [30]. Such energy was determined using only DSC measurements.

The methods were applied to the study of the solidification process of several polymers: polyethylene glycol (PEG) with different mean molecular weight blends of PEG with pharmaceutical drugs (PEG-PD), copolymers of polypropylene-polyethylene with different colorant additives (PP-PE) and blends with high density polyethylene (HDPE).

16.2   Kinetic Analysis

The macroscopic study of the crystallization process is based in the evolution of the crystalline fraction of the material, α, as a function of the time in isothermal regime or of the temperature under dynamic constant rates. Most of the methods for non-isothermal regime analysis are based on an extension of the Avrami equation [16], which was first introduced to describe the transformation kinetic in the isothermal regime and has been theoretically treated by Avrami as the primary crystallization and growth of spherulites.

$$\alpha(t) = 1 - \exp\{-(kt)^n\} \tag{16.1}$$

where $\alpha(t)$ stands for the crystalline fraction for time t, k is the crystallization rate constant, t the time elapsed from the beginning of the process and n a kinetic exponent related to the mechanisms governing the transformation. Usually k and n are considered as constants with respect to time. Moreover, the exponent n can be considered as a function ($n = c + d$) which accounts for the time dependence of the nucleation process, c, and for the number of dimensions in which growth takes place, d. The isothermal crystallization rate can written in the following from:

$$\frac{d\alpha}{dt} = k(T)f(\alpha) \tag{16.2}$$

where T is the temperature and $f(\alpha)$ is a function describing the kinetic model. Usually the best fit to the experimental values is obtained by the Jhonson-Mehl-Avrami-Erofe'ev (JMAE) model [16,31,32] as given by (16.1). Then, the function $f(\alpha)$ obeys the following equation:

$$f(\alpha) = n(1 - \alpha) \left(ln\frac{1}{1 - \alpha} \right)^{1-1/n} \tag{16.3}$$

where n is the JMAE exponent.

Sometimes the most suitable form for $f(\alpha)$ corresponds to the autocatalytic two-parameter Sesták-Berggen (SB) model [33].

$$f(\alpha) = \alpha^m (1 - \alpha)^n \tag{16.4}$$

where n and m are parameters depending on the mechanisms of nucleation and growth.

Moreover, it is usually assumed that the rate constant k has an Arrhenius equation behavior with respect to the temperature during the crystallization process:

$$k(T) = k_0 \exp(-\frac{E}{k_B T}) \tag{16.5}$$

where k_0 is the pre-exponential factor, E the apparent activation energy and k_B is the Boltzmann constant. There are various differential and integral methods to calculate these terms from continuous heating data as i.e. Kissinger's [34] and Ozawa's [35].

In the modeling of the crystallization process, we will distinguish cold crystallization, for which the analysis mentioned above is often performed, as against solidification from the melt. Nevertheless, a very similar procedure can be used to analyze solidification data. The main difference is that cold crystallization of a polymeric mixture may be promoted by pre-existing nuclei. However, solidification from the assumed isotropic melt is expected to be driven by nucleation. It has been long recognized that a certain amount of undercooling $\Delta T = T_m - T$, with T_m the melting temperature, is necessary to start solidification [36] because nuclei have to be created prior to any crystal growth. The driving force for nucleation is the Gibbs energy difference between the liquid and the crystal, ΔG per unit volume. The activation energy for homogeneous nucleation is of the form [37–38]

$$E = \frac{16\pi}{3} \frac{\sigma^3}{\Delta G^2} \tag{16.6}$$

where σ is the energy per surface unit necessary to form the interface between the liquid and the nucleus. The Gibbs energy at low undercooling is propotional to ΔT and to the melting entropy ΔS_m. Thus,

$$E \simeq \frac{\sigma^3}{\Delta S_m^2 \Delta T^2} \tag{16.7}$$

Therefore, at low undercooling, one expects a rate constant of the approximate form

$$k(T) = A \exp(-\frac{B}{T(\Delta T)^2}) \tag{16.8}$$

where A is a function that shows a slight dependence with the temperature in the interval: $\Delta T \leq 0.2 T_m$ while B is a constant proportional to $\sigma^3 / \Delta S_m^2$.

The knowledge of E or B value is necessary prior to the analysis of the kinetic process. One of the methods to obtain E can be deduced taking logarithms in (2) combined with the Arrhenius equation. The expression obtained is [39]

$$\ln(\frac{d\alpha}{dt}) = \ln(k_0 f(\alpha)) - \frac{E}{k_B T} \tag{16.9}$$

and, in the processes governed by nucleation a analogous expression [22] can be deduced

$$\ln(\frac{d\alpha}{dt}) = \ln(A f(\alpha)) - \frac{B}{T \Delta T^2} \tag{16.10}$$

A basic assumption in the DSC technique is that the heat flow, ϕ, is proportional to the conversion rate, $\phi = (d\alpha/dt)\Delta H$, where ΔH is the enthalpy of the crystallization process. On substitution into previous equation, we can rewrite it as

$$\ln(\frac{\phi}{\Delta H}) = \ln(A f(\alpha)) - \frac{B}{T \Delta T^2} \tag{16.11}$$

The slope of $\ln(\phi/\Delta H)$ versus $(1/T\Delta T^2)$ provides the constant B for a given α. If we repeat the procedure for different values of α, the invariance of E with respect to α (which is a basic assumption for the validity of the model) is checked in a simple and reliable manner. This operation allows testing of the T_m accuracy at same time [27]. A first estimation of T_m results from previous heating, where samples are completely melted. As an example, see Fig. 16.1. The so-called isoconversional method was applied to find the activation energy for a given value of the degree of conversion. Here, it is adapted to evaluate the

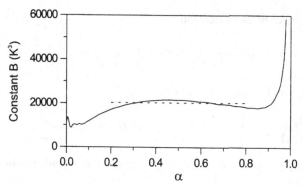

Fig. 16.1. Constant B obtained by the isoconversional method versus transformed fraction for a PEG 5600

constant B, of the constant rate equation, in the case of solidification from an isotropic melt.

Furthermore, the value of B can be determined by a procedure [40] similar to the Kissinger method, that is by imposing $(d^2\alpha/dt^2)_{T=T_p} = 0$ at a given cooling rate β. Thus,

$$(\frac{d^2\alpha}{dt^2})_{T_p} = \frac{dk(T)}{dt}f(\alpha) + \frac{df(\alpha)}{dt}k(T) \tag{16.12}$$

Taking into account the following equations:

$$\frac{dk(T)}{dt} = -\beta\left[A(\exp(-\frac{B}{T\Delta T^2}))\frac{B(\Delta T^2 - 2T\Delta T)}{T^2\Delta T^4}\right] = \beta\frac{k(T)B}{T^2\Delta T^3}(2T - \Delta T) \tag{16.13}$$

and

$$\frac{df(\alpha)}{dt} = \frac{df(\alpha)}{d\alpha}\frac{d\alpha}{dt} = \frac{df(\alpha)}{d\alpha}k(T)f(\alpha) \tag{16.14}$$

and substitution into (12), the following equation is obtained

$$(\frac{d^2\alpha}{dt^2})_{T_p} = k(T)f(\alpha)\beta B(2T_p - \Delta T_p)T_p^2\Delta T_p^3 + k^2(T)f(\alpha)\frac{df(\alpha)}{d\alpha} = 0 \tag{16.15}$$

from this we can obtain

$$\frac{\beta B(2T_p - \Delta T_p)}{T_p^2\Delta T_p^3} = -\frac{df(\alpha)}{d\alpha}k(T) \tag{16.16}$$

and in the logarithmic form

$$\ln B + \ln\left[\frac{\beta(2T_p - \Delta T_p)}{T_p^2\Delta T_p^3}\right] = \ln(-\frac{df(\alpha)}{d\alpha}) + \ln k(T) \tag{16.17}$$

and using the Arrhenius equation

$$\ln k(T) = \ln(A \exp(-\frac{B}{T_p \Delta T_p^2})) = \ln A - \frac{B}{T_p \Delta T_p^2} \qquad (16.18)$$

From the combination of (17) and (18)

$$\ln \left[\frac{\beta(2T_p - \Delta T_p)}{T_p^2 \Delta T_p^3} \right] = \ln(-\frac{df(\alpha)}{d\alpha}) - \ln B + \ln A - \frac{B}{T_p \Delta T_p^2} \qquad (16.19)$$

An approximation is to consider constant at the peak temperature the following relation

$$\ln(-\frac{df(\alpha)}{d\alpha}) - \ln B + \ln A = ct. \qquad (16.20)$$

Thus,

$$\ln \left[\frac{\beta(2T_p - \Delta T_p)}{T_p^2 \Delta T_p^3} \right] \approx -\frac{B}{T_p \Delta T_p^2} + ct \qquad (16.21)$$

The slope of $\ln(\beta(2T_p - \Delta T_p)/T_p^2 \Delta T_p^3)$ versus $(1/T_p \Delta T_p^2)$ provides the approximate value of constant B. Results obtained using this method and the isoconversional one are similar [41]. Nevertheless, with the isoconvertional method is observed the slight dependence of B on α, see Fig. 16.1. This confirms the validity of the rate constant model. As usual for low α values, ϕ tends to zero which explain the divergence of the results.

Once B and T_m are known, it is possible to evaluate the crystallization kinetics. The kinetic model and the coefficient A can readily be determined from a knowledge of two functions of the crystalline fraction, $y(\alpha)$ and $z(\alpha)$ defined by [42–44]

$$y(\alpha) = A\Delta H f(\alpha) \qquad (16.22)$$

and

$$z(\alpha) = \Delta H f(\alpha) g(\alpha) \qquad (16.23)$$

where $g(\alpha)$ is defined by

$$g(\alpha) = \int\limits_0^\alpha \frac{1}{f(\alpha')} d\alpha' \qquad (16.24)$$

From the heat flow equation and on substitution in the non logarithmic form of (11), we can rewrite the heat flow expression as

$$\phi = \frac{d\alpha}{dt} \Delta H = A \Delta H \exp(-\frac{B}{T \Delta T^2}) f(\alpha) \qquad (16.25)$$

and $g(\alpha)$ as

$$g(\alpha) = -\frac{1}{\beta} \int_{T_f}^{T} k(T)dT = -\frac{A}{\beta} \int_{T_f}^{T} \exp(-\frac{B}{T(\Delta T)^2})dT \qquad (16.26)$$

we can define $x = B/T\Delta T^2$ and

$$g(\alpha) = -\frac{A}{\beta}B\frac{\exp(-x)}{x}\left[\frac{\int_{T_f}^{T} B\exp(-x)dT}{\exp(-x)/x}\right] \qquad (16.27)$$

if we define the function $\tau(x)$ as

$$\tau(x) = \left[\frac{\int_{T_f}^{T} B\exp(-x)dT}{\exp(-x)/x}\right] \qquad (16.28)$$

the previous expression can be rewritten as

$$g(\alpha) = -\frac{A}{\beta}B\frac{\exp(-x)}{x}\tau(x) = -\frac{T\Delta T^2}{\beta}\tau(x)k(T) \qquad (16.29)$$

From the combination of the previous expressions, it is possible to evaluate the functions $y(\alpha)$ and $z(\alpha)$ directly from the experimental data once B and T_m are known

$$y(\alpha) = \phi\exp x \qquad (16.30)$$

$$z(\alpha) = -\frac{T\Delta T^2}{\beta}\tau(x)\phi \qquad (16.31)$$

The experimental functions $y(\alpha)$ and $z(\alpha)$, divided by $A\Delta H$ and ΔH, respectively, are plotted in Figs. 16.2 and 16.3 for a cooling rate of 10°C min^{-1} (symbols). The polymer is a PEG 5600. We can find the kinetic model which best describes the measured data, and its kinetics parameters [22,45]. In this case, the most suitable f(α) corresponds to the two-parameters Sesták-Berggen model [33]. Figures 16.2 and 16.3 demonstrate that the SB model gives a good description of the solidification process.

Once the best model is known, several experimental data can be compared directly with the "ideal" experimental behavior. As an example, the experimental DSC (symbols) and calculated (full lines) heat flow divided by enthalpy is shown in Fig. 16.4. for PEG 3400. In this case, the best fit model corresponds to the JMAE. Other example is shown in Fig. 16.5 where the experimental degree

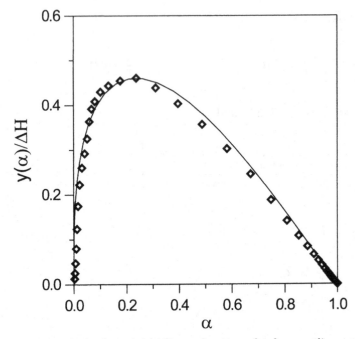

Fig. 16.2. Dimensionless value $y(\alpha)/\Delta H$ as a function of α for a cooling rate of 10 K min^{-1} for a PEG 5600.

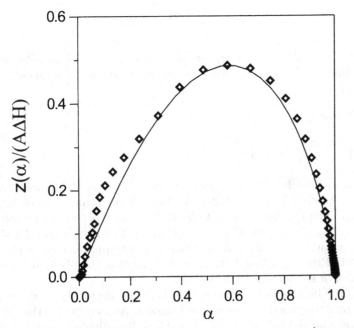

Fig. 16.3. Dimensionless $z(\alpha)/A\Delta H$ for a cooling rate of 10 K min^{-1} for a PEG 5600

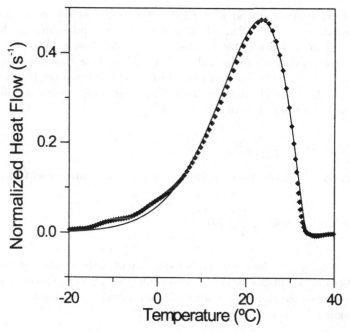

Fig. 16.4. Normalized heat flow versus temperature for a cooling rate of 10 K min^{-1}of a PEG 3400.

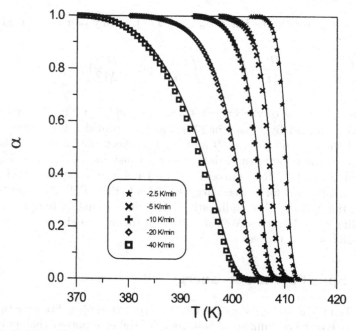

Fig. 16.5. Solidified fraction versus temperature at different cooling rates of a PP-PE copolymer.

of transformation (symbols) of a PP-PE copolymer at different cooling rates
is compared with the model. All figures verifies the reliability of the method
utilized and the validity of the constant rate model description.

From the previous analysis and the knowledge of the kinetic model it is possible to construct low temperature parts of the transformation diagrams [46,47],
in our case the T-CR-T diagram. Using the expressions deduced before, under
isothermal conditions the function $g(\alpha)$ is

$$g(\alpha) = A \exp(-\frac{B}{T \Delta T^2})t \tag{16.32}$$

Similarly, under constant cooling rate regime, the integrated form becomes

$$g(\alpha) = \frac{1}{\beta} \int\limits_o^{\Delta T} \exp(-\frac{B}{T \Delta T^2})dT \tag{16.33}$$

The quantity $g(\alpha)$ for a given α is independent of the mode of crystallization,
i.e. isothermal conditions or continuous cooling conditions. Once its value for a
given crystalline fraction α_o is known, the analytical expression for $\alpha_o = \alpha_o(T,t)$
for any pair (T,t) is given by

$$\exp(-\frac{B}{T \Delta T^2})t = \exp(-\frac{B}{T \Delta T_o^2})t_o \tag{16.34}$$

and the analytical expression of $\alpha_o = \alpha_o(T,\beta)$ for any pair (T,β) is given by

$$-\int\limits_o^{\Delta T} \frac{A}{\beta} \exp(-\frac{B}{T \Delta T^2})dT = -\int\limits_o^{\Delta T_o} \frac{A}{\beta_o} \exp(-\frac{B}{T \Delta T^2})dT \tag{16.35}$$

Thus, the forms of the curves $T = T(t)$ and $T = T(\beta)$ for a fixed value
of α, are obtained by means of (34) and (35) provides the values of B and
T_m are known. The values of B and T_m have been obtained from the best fit
approach to the experimental data under a constant cooling rate. Figure 16.6
shows calculated (T-CR-T) curves (full lines) and several experimental DSC data
for $\alpha = 0.1, 0.5, 0.9$ for a pharmaceutical blend (PEG-PD). As expected in any
solidification process, the solidification onset shifts to higher temperatures when
the cooling rates decreases. We can state that the models give a good description
of the solidification process.

16.3   Specific Surface Energy

In 1945 Bunn and Alcock were the first to report the spherulite structure formation on polymer crystallization [48]. In 1957 Keller reported that polyethylene
formed chain-folded lamellar crystals from solution [49]. Lamellar crystalls are
formed on crystallization, from both the solution and the melt for a wide variety

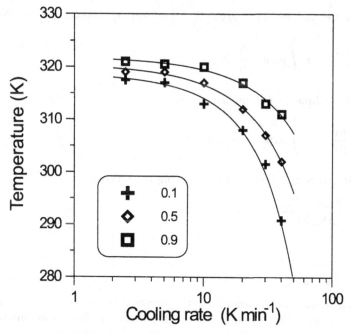

Fig. 16.6. Calculated T-CR-T curves (solid lines) and experimental DSC data (symbols) for a phermaceutical blend.

of polymers. Two of the most well known theories about lamellar formation are the Lauritzen-Hoffman surface nucleation theory [50,51] and the entropic barrier theory of Sadler and Gilmer [52,53]. The morphology of polymers during and after crystallization has been analyzed by different microscopy techniques [23,54]. Nevertheless, it is possible to obtain information about crystallization behavior only from calorimetric data [55,56].

The determination of the surface superficial energy is one of the most characteristic methods to describe the crystallization of previously molten polymers. Such energy, σ, is formed by lateral surface energy, σ_l, and by end surface energy, σ_e. The first corresponds to the non-deformed part of polymer's lamellae structure, while the latter is released by the ridged surface perpendicular to the axis of the polymeric chain [57,58]. Further on, an approximation is used according to which it is assumed that in the crystalline polymer nucleus the surface energy can be expressed as a function of lateral and end surface energy. That is to say, $\sigma^3 = \sigma_l^2 \sigma_e$ [59]. In this case σ has the meaning of an average specific surface energy.

It is possible to obtain surface energy information from DSC measurements. Taking into account that bi- or three dimensional nucleation is always present in crystalline polymer [60], in the bidimensional case, the function $g(\alpha)$ can be rewritten for the DSC peak crystalline fraction, α_p, and time, t_p, as follows [61]

$$\int_0^{\alpha_p} \frac{d\alpha}{f(\alpha)} = n \int_0^{t_p} \exp(-\frac{R}{\beta t})dt \tag{16.36}$$

In the three-dimensional one as:

$$\int_0^{\alpha_p} \frac{d\alpha}{f(\alpha)} = n \int_0^{t_p} \exp(-\frac{S}{\beta^2 t^2})dt \tag{16.37}$$

Terms R and S are given by the following expressions:

$$R = \frac{4a_o \sigma_l \sigma_e V_m}{\Delta S_m k T_m} \tag{16.38}$$

and

$$S = \frac{16\pi \sigma_l^2 \sigma_e V_m^2}{3\Delta S_m^2 k T_m n} \tag{16.39}$$

where V_m is the molar volume of the crystallizing material, ΔS_m is melting entropy and a_o is the crystalline network parameter. The approximate solution of previous equations results in the following relationships for the peak ΔT_p:

$$\log(\beta) = M - (R/2.3\Delta T_p) \tag{16.40}$$

and

$$\log(\beta) = N - (S/2.3\Delta T_p^2) \tag{16.41}$$

where M and N are constants. The graphical representation of $\log(\beta)$ versus $1/\Delta T_p$ or $1/\Delta T_p^2$ supplies us with parameters R and S from slope (as an example see Figs. 16.7 and 16.8 for several PP-PE copolymers with different colorant additives). Once these parameters are calculated their coefficient allows the calculation of σ_l; σ_e follows readily. The values obtained using this approach are similar to those reported by other techniques analysis [56,62].

16.4   Conclusions

The kinetic analysis of polymer crystallization was performed from calorimetric data. The polymers analyzed were: polyethylene glycol (PEG) with different mean molecular weight, blends of PEG with pharmaceutical drugs (PEG-PD) copolymers of polypropylene-polyethylene with different colorant additives (PP-PE) and blends with high density polyethylene (HDPE). A modified isoconversional is applied to evaluate the parameters governing the crystallization and the kinetic model. The reliability of the methods and models utilized was checked. The temperature - cooling rate - transformation diagram shows a good agreement between the experimental data and curves obtained by calculation. Furthermore the specific surface energies were calculated from calorimetry measurements.

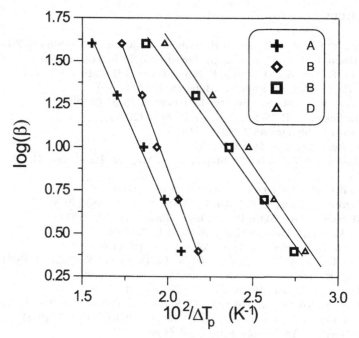

Fig. 16.7. $\ln(\beta)$ versus $1/\Delta T_p$ for the different samples.

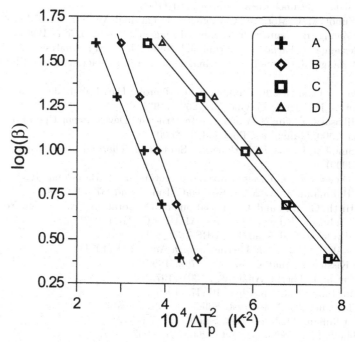

Fig. 16.8. $\ln(\beta)$ versus $1/\Delta T_p^2$ for the different samples.

References

[1] L. Zhu, Y. Chen, A. Zhang, B.H. Calhoun, M. Chun, R.P. Quirk, S.Z.D. Cheng, B.S. Hsiao, F. Yeh, T. Hashimoto: Phys.Rev.B.60-14, 0022 (1999)

[2] C.Y. Li, D. Yan, S.Z.D. Cheng, F. Bai, J.J. Ge, B.H. Calhoun, T. He, L-C. Chien, F.W. Harris, B. Lotz: Phys.Rev.B, 60-18, 12675 (1999)

[3] C. Basire, D.A. Ivanov: Phys.Rev.Letters:85-26, 5587 (2000)

[4] M. Muthukamar, P. Welch: Polymer 41, 8833 (2000)

[5] A. Ziabicki: Polymery:45-7, 520 (2000)

[6] J.P.K. Doye: Polymer 41, 8857 (2000)

[7] M. Mierzwa, G. Floudas, P. Stepanek, G. Wegner: Phys. Rev. B, 62-21, 14012 (2000)

[8] P. van der Schoot. Macromolecules, 33, 8497 (2000)

[9] A. Marotta, A. Buri, G.L. Valenti: J.Mater.Sci., 13, 2483 (1978)

[10] K. Matsumita, A. Sakka: Phys. Chem. Glasses, 20, 173 (1979)

[11] M.C. Weinberg: Thermochimica Acta, 194, 93 (1992)

[12] F. Carrasco, P. Pagès: J. Appl. Pol. Sci., 61, 187 (1996)

[13] B. Wunderlich, A. Boller, I. Okazaki, K. Ishikiriyama, W. Chen, M. Pyda, J. Pak, I. Moon, R. Androsch: Thermochimica Acta, 330, 21 (1999)

[14] W. Chen, I.K. Moon, B. Wunderlich: Polymer, 41, 4119 (2000)

[15] A. Toda, C. Tomita, M. Hikosaka, Y. Saruyama: Polymer 38, 231 (1997)

[16] M.J. Avrami: J.Chem.Phys.:7, 1103(1939);8, 212(1940);9, 177 (1941)

[17] C.D. Doyle: J. Appli. Poly. Sci., 5, 285 (1961)

[18] N. Clavaguera, M.T. Mora: Mater. Sci and Eng.A, 179, 288 (1994)

[19] J. Málek: Thermochimica Acta, 267, 61 (1995)

[20] J.J. Suñol, M.T. Mora, N. Clavaguera: J. Thermal Analysis, 52-853 (1998)

[21] D.L. Zhang, B. Cantor: Philosophycal Magazine A , 62-5, 557 (1990)

[22] R. Berlanga, J. Farjas, J. Saurina, J.J. Suñol: J. Therm. Analysis, 52, 765 (1998)

[23] R. Berlanga, J.J. Suñol, J. Saurina, J. Oliveira: J. Macromol. Sci.-Physics, B40, 327 (2001)

[24] C. Bas, A.C. grillet, F. Thimon, N.D. Albérola: Eur. Polym. J., 10, 911 (1995)

[25] A. Jonas, R. Legras: Polymer, 32, 2691 (1991)

[26] M. Hikosaka, F. Gu, Y. Yamazaki: In Structure Development Upon Polymer Processing 2000, Guimaraes, Portugal, 1 (2000)

[27] J.J. Suñol, J. Farjas, R. Berlanga, J. Saurina: J. Thermal Analysis and Calor., 61, 711 (2000)

[28] J.A. Diego: Ph. D. Thesis, Universitat Autònoma de Barcelona (1994)

[29] J.D. Hoffmann, J. Weeks: J. Res. Natl. Bul. Stand. 57, 217 (1962)

[30] J.P. Hirth, G.M. Pound: Condensation and Evaporation, Oxford, New York (1963)

[31] W.A. Johnson, R.F. Mehl: Trans. AIME, 135, 416 (1939)

[32] B.V. Erofe'ev: C.R.Acad.Sci., URSS, 52, 511 (1946)

[33] J. Sesták, G. Berggren: Thermochimica Acta, 3, 1 (1971)

[34] H.E. Kissinger: Anal. Chem., 29, 1702 (1957)

[35] T. Ozawa: J. Therm. Analysis, 2, 301 (1979)

[36] D. Turnbull: Contemp. Phys. 10, 473(1969)

[37] N. Clavaguera: J. Non-Cryst. Solids, 30, 301 (1970)

[38] D.R. Ulhmann: J. Non-Cryst. Solids, 7, 337 (1972)

[39] J. Málek: Thermochimica Acta, 200, 257 (1992)

[40] N. Clavaguera, J. Saurina, J. Lheritier, J. Masse, A. Chauvet, M.T. Mora: Thermochimica Acta, 290, 173 (1997)

[41] J.J. Suñol: Inter. J. Polym. Mater. 1, 8 (2001)
[42] J.M. Criado, J. Málek, A. Ortega: Thermochimica Acta, 147, 377 (1989)
[43] J. Málek: Thermochimica Acta, 138, 337 (1989)
[44] S. Montserrat, C. Flaqué, P. Pagés, J. Málek: J. Appl. Polym. Sci., 56, 1413 (1995)
[45] J.J. Suñol, J. Saurina, R. Berlanga, D. Herreros, P. Pagès, F. Carrasco: J. Thermal. Anal. and Calor., 55, 57 (1999)
[46] S. Suriñach, M.D. Baró, M.T. Mora, N. Clavaguera: J. Non-Cryst. Solids, 58, 209 (1983)
[47] M.T. Mora, S. Suriñach, M.D. Baró, N. Clavaguera: Solid State Ionics, 63, 268 (1993)
[48] C.W. Bunn, T.C. Alcock: Trans. Faraday Soc., 41, 317 (1945)
[49] A. Keller: Philos. Mag.: 2, 1171 (1957)
[50] J.I. Lauritzen, J.D. Hoffman: Res. Nat. Bur. Stds., 64, 73 (1960)
[51] J.D. Hoffman, R.L. Miller: Polymer, 38, 3151 (1997)
[52] D.M. Sadler, G.H. Gilmer: Polymer, 25, 1446 (1984)
[53] D.M. Sadler: Nature, 326, 174 (1987)
[54] G. Reiter, G. Castelein, P. Hoerner, G. Riess, J.U. Sommer, G. Floudas: Eur. Phys. J. E: 2, 319 (2000)
[55] H. Marand, A. Alizaheh, R. Farmer, R. Desai, V. Velikov: Macromolecules: 33, 3392 (2000)
[56] J.J. Suñol, J. Saurina, P. Pagès, F. Carrasco: J. Appl. Polym. Sci., 77, 1269 (2000)
[57] P.J. Flory: J. Chem. Phys., 17, 223 (1949)
[58] L. Mandelkern: J. Appl. Phys., 16, 1155 (1978)
[59] I. Gutzow, V. Dochev, E. Pancheva, K. Dimov: J. Polym. Sci. Polym. Phys., 16, 1155 (1978)
[60] F. Price: In Nucleation, A.C. Zettlemoyer, New York, (1969)
[61] A. Dobreva, I. Gutzow: Cryst. Res. Technol., 26-7, 863 (1991)
[62] B. Monasse, J.M. Haudin: Colloid Polym. Sci., 264, 117 (1986)

17   A Computational Model
for Processing of Semicrystalline Polymers:
The Effects of Flow-Induced Crystallization

Gerrit W.M. Peters

Materials Technology (www.mate.tue.nl), Dutch Polymer Institute, Eindhoven University of Technology, P.O. Box 513, 5600 MB Eindhoven, The Netherlands, g.w.m.peters@tue.nl

Abstract. A computational model for the combined processes of quiescent and flow-induced crystallization of polymers is presented. This modelling should provide the necessary input data, in terms of the structure distribution in a product, for the prediction of mechanical properties and shape- and dimensional-stability. Rather then the shear rate as the driving force, a viscoelastic approach is proposed, where the viscoelastic stress (or the equivalent recoverable strain) with the highest relaxation time, a measure for the molecular orientation and stretch of the high end tail molecules, is the driving force for flow induced crystallization. Thus, the focus is on the polymer that experiences the flow, rather then on the flow itself. Results are presented for shear flow, extensional flow and for injection moulding conditions of an isotactic Polypropylene (iPP).

17.1   Introduction

The resulting local properties of a product made of semicrystalline polymers strongly depend on the crystalline structure which is determined by both molecular properties and the processing conditions applied, i.e. the thermal-mechanical history experienced by the polymer in the process. Consequently, this thermal mechanical history (e.g. in injection moulding, film blowing or fiber spinning) has to be modelled in order to describe nucleation and crystallization kinetics and their dependence on flow-induced structure formation. Experimental results in the literature show that flow-induced crystallization correlates with the viscoelastic stresses, much more than with the macroscopic strain or strain rate, leading to the conclusion that chain orientation/extension is the governing phenomenon. The orientation of polymer chains can result in the development of anisotropic structures which influence the mechanical properties and the dimensional stability, i.e warpage and (anisotropic) shrinkage. A modelling tool is needed to prevent and correct for such unwanted properties and phenomena. The ultimate goal is to develop a tool that is able to improve or even optimize polymer synthesis and industrial processes. However, several of the underlying relations are still not clear due to the complex mutual interaction of many parameters. Therefore, optimization of properties in industry is still done by expensive and time consuming trial and error methods, which -at best- are based on cumulative experience.

The need for such a modelling tool became even more urgent since the development of novel metallocene-based catalysis. Compared to conventional Ziegler-Natta catalysis, metallocene-based polymers are much more well defined and possess a narrow molecular weight distribution. Moreover, differences in regularity of the polymer chains exist caused by sterical restrictions during polymer synthesis [1].

The relation between molecular properties, processing conditions and final properties that result from the micro-structure formed, can be determined once recent developments in detailed modelling are condensed into constitutive equations that are the input of continuum mechanical modelling. The relation between molecular weight distribution and linear viscoelastic properties [2,3] and that between rheology and flow-induced nucleation [4,5] has now become much more clear. Polymer rheology can at present quantitatively be described with the (extended) Pom–Pom model [6–8]. Consequently, a powerful tool could be obtained once these methods are combined to model the whole life cycle of a polymer: from the synthesis to properties via processing. However, much more advanced experimental methods and set-ups than used at present have to be developed to be able to validate the modelling and to obtain the required input parameters.

17.2   Modelling

17.2.1   Balance Equations

The basic equations of continuum mechanics are the balance equations. In their local form, the equations for mass, momentum, moment of momentum and energy read, respectively:

$$\dot{\rho} + \rho \nabla . v = 0 \tag{17.1}$$

$$\nabla . \sigma^c + \rho f = \rho \dot{v} \tag{17.2}$$

$$\sigma = \sigma^c \tag{17.3}$$

$$\rho \dot{e} = \sigma : \mathbf{D} - \nabla . h + \rho r_h \tag{17.4}$$

in which v the velocity, ρ the density, σ the Cauchy stress tensor, f the specific body force, e the internal energy, h the heat flux vector and r_h the specific heat source. In the following, constitutive relations will be presented for the stress and the density in terms of the kinetics of the different phases. Details on constitutive models for the other quantities, such as internal energy and heat flux can be found in literature, see for example [5].

17.2.2   Constitutive Equations

17.2.2.1   Viscoelastic Stress. A general expression for the evolution of the Cauchy stress tensor σ is given by [9]

$$\overset{\triangledown}{\sigma} + \mathbf{A} \cdot \sigma + \sigma \cdot \mathbf{A}^c = \mathbf{O} \tag{17.5}$$

where **A** is the slip tensor (also known as the plastic deformation rate tensor $\mathbf{D_p}$) which takes into account the non-affine deformation of the molecules and the upper convective derivative of the Cauchy stress tensor is defined by

$$\overset{\triangledown}{\sigma} = \dot{\sigma} - \mathbf{L} \cdot \sigma - \sigma \cdot \mathbf{L}^T \tag{17.6}$$

The Cauchy stress σ is related to the recoverable strain tensor $\mathbf{B}_e$ according to

$$\sigma = G\mathbf{B}_e \tag{17.7}$$

in which G is the elastic modulus. Once the viscoelastic stress is known, the recoverable strain is also defined by (17.7). With the definition of the extra stress tensor $\tau = G(\sigma - \mathbf{I})$ and the upper convective derivative of the unity tensor $\overset{\triangledown}{\mathbf{I}} = -2\mathbf{D}$, it follows

$$\overset{\triangledown}{\tau} + \mathbf{A} \cdot \tau + \tau \cdot \mathbf{A}^c + G(\mathbf{A} + \mathbf{A}^c) = 2G\mathbf{D} \tag{17.8}$$

For a realistic description of melt behavior a multi-mode model has to be used

$$\sigma = -p\mathbf{I} + \sum_{i=1}^{n} \tau_i \tag{17.9}$$

Most known models are captured within the following definition of the slip tensor

$$\mathbf{A} = \alpha_1 \sigma + \alpha_2 \sigma^{-1} + \alpha_3 \mathbf{I} \tag{17.10}$$

by a certain choice for the parameters α_1, α_2 and α_3. They are given for the Upper Convected Maxwell model (as an example: it is the most simple model for viscoelastic stress) the Leonov model and the extended Pom–Pom model in Table 17.1. The Extended Pom–Pom model has a generalized relaxation time λ_g defined by

$$\lambda_g^{-1} = \frac{1-\alpha}{\lambda_{0b}\Lambda^2} - \frac{\alpha \mathcal{I}_{\sigma \cdot \sigma}}{3G^2 \lambda_{0b}\Lambda^2} + \frac{2}{\lambda_s}\left(1 - \frac{1}{\Lambda}\right) \tag{17.11}$$

Table 17.1. Definition of the slip tensor for the upper convected Maxwell (UCM), the Leonov and the Extended Pom–Pom model (taken from [8,10]).

Model	α_1	α_2	α_3	Ref.
UCM	0	$-\frac{G}{2\lambda}$	$\frac{1}{2\lambda}$	[10]
Leonov	$\frac{1}{4G\lambda}$	$-\frac{G}{4\lambda}$	$-\frac{\mathcal{I}_\sigma - G^2 \mathcal{I}_{\sigma^{-1}}}{12G\lambda}$	[10]
Extended Pom–Pom	$\frac{\alpha}{2G\lambda}$	$-\frac{G(1-\alpha)}{2\lambda}$	$\frac{1}{2\lambda_g}$	[8]

with a material parameter α, which defines the amount of anisotropy in molecular friction and (Brownian) motion (i.e. reptation) and the molecular stretch Λ. Two different relaxation times are defined, namely the relaxation time of the backbone orientation λ_{0b} and the backbone stretch λ_s. Slip tensors for some other, well known constitutive equations can be found in [9,10].

The upper convected Maxwell (UCM) model can describe different aspects of linear and non-linear viscoelastic behavior of a polymeric fluid by using multi modes. Linear behavior can be predicted accurately, while the non-linear viscoelastic regime is only described qualitatively. It can predict a first normal stress difference in shear, strain hardening in elongation and stress relaxation after cessation of flow. However, shear viscosity and the first normal stress difference are predicted to be independent of shear rate, and extensional viscosity can become infinite for finite extensional rates [12]. The differential equation of the UCM model is given by

$$\overset{\triangledown}{\tau} + \frac{1}{\lambda}\tau = 2G\mathbf{D} \tag{17.12}$$

The UCM model is fully determined using linear viscoelastic data only.

The Leonov model can describe the (non)-linear viscoelastic behavior of a polymeric liquid in shear rather well by using multi modes. Elongation is poorly described [12]. The differential equation of the Leonov model is given by

$$\overset{\triangledown}{\tau} + \frac{1}{\lambda}\tau + \frac{1}{2G\lambda}\tau\cdot\tau - \underbrace{\frac{1}{6G\lambda}\left[\mathcal{I}_{(\tau+G\mathbf{I})} - G^2\mathcal{I}_{(\tau+G\mathbf{I})^{-1}}\right](\tau+G\mathbf{I})}_{(a)} = 2G\mathbf{D} \tag{17.13}$$

The Leonov model is, like the UCM model, determined by the linear viscoelastic data only. For incompressible, planar deformations part (a) of (17.13) is equal to zero. The Leonov model is in this case equivalent to the Giesekus model with the parameter α in the Giesekus model equal to 0.5.

The Pom–Pom model was proposed by McLeish and Larson [7]. It describes the strain hardening in elongation and thinning behavior in shear for branched polymers. They defined an idealized molecule, the 'Pom–Pom', which consists of a backbone with a number of branches at each end. A key feature in this model is the separation of stretch and orientation of a polymer molecule. Verbeeten et al. [8] proposed an extended version of the model which also predicts a second normal stress difference. Excellent quantitative agreement with measurements of branched (LDPE) and linear polymers (HDPE) was found by using a multi mode version. The extended Pom–Pom model is given by

$$\overset{\triangledown}{\tau} + \lambda(\tau)^{-1}\cdot\tau = 2G\mathbf{D} \tag{17.14}$$

with inverse relaxation tensor

$$\lambda(\tau)^{-1} = \frac{1}{\lambda_{0b}}\left[\frac{\alpha}{G}\tau + \frac{1}{f(\tau)}\mathbf{I} + \underbrace{G\left(\frac{1}{f(\tau)} - 1\right)\tau^{-1}}_{(a)}\right] \tag{17.15}$$

and extra function

$$\frac{1}{\lambda_{0b} f(\tau)} = \underbrace{\frac{2}{\lambda_s}\left(1 - \frac{1}{\Lambda}\right)}_{(b)} + \underbrace{\frac{1}{\lambda_{0b}\Lambda^2}\left(1 - \frac{\alpha \mathcal{I}_{\tau \cdot \tau}}{3G^2}\right)}_{(c)} \qquad (17.16)$$

Backbone stretch and stretch relaxation time are defined as

$$\Lambda = \sqrt{1 + \frac{\mathcal{I}_\tau}{3G_0}}, \qquad \lambda_s = \lambda_{0s}\exp(-\nu(\Lambda - 1)), \qquad \nu = \frac{2}{q} \qquad (17.17)$$

in which q is the number of branches. Parameter α ($\alpha \geq 0$) describes a Giesekus type of anisotropy [13]. This parameter influences the second normal stress difference only. The Extended Pom–Pom model is equivalent to the original (approximative) Pom–Pom model for $\alpha = 0$. The orientation relaxation times of the backbone λ_{0b} are obtained from linear viscoelastic data. The number of branches q the stretch relaxation times λ_s and the anisotropy parameter α have to be determined for each mode. This large number of parameters seems to be a drawback. A physical guideline related to the structure of the Pom–Pom molecule can be taken into account. The free ends of the molecule correspond to fast relaxation times and no or less branches. Towards the center of the molecule, relaxation and branches will increase. Moreover, the stretch relaxation time is constrained in the interval $\lambda_{0b,i-1} < \lambda_{0s,i} \leq \lambda_{0b,i}$.

Stretch and orientation are gathered in one equation in the extended Pom–Pom model. In case of $\alpha = 0$ two situations can be distinguished:

- Only orientation relaxation for low strains ($\Lambda \approx 1$; part (a) and (b) of (17.15) and (17.16) are equal to zero.)
- Only stretch relaxation for high strains ($\Lambda \gg 1$; part (c) of (17.16) is equal to zero.)

Verbeeten [8] showed that the described parameter trends are obtained automatically using a non-linear fitting procedure.

17.2.2.2 Crystallization. Non-isothermal crystallization of spherulites is described by the Schneider's rate equations [14,15], a set of differential equations which give the structure developing for quiescent conditions. Mean number of spherulites and their mean radius, surface and volume are calculated from:

$$\phi_3 = 8\pi\alpha \qquad (\phi_3 = 8\pi N) \qquad 'rate' \qquad (17.18)$$

$$\phi_2 = G\phi_3 \qquad (\phi_2 = 8\pi R_{tot}) \qquad 'radius' \qquad (17.19)$$

$$\phi_1 = G\phi_2 \qquad (\phi_1 = S_{tot}) \qquad 'surface' \qquad (17.20)$$

$$\phi_0 = G\phi_1 \qquad (\phi_0 = V_{tot}) \qquad 'volume' \qquad (17.21)$$

$$\phi_0 = -\ln(1 - \xi_g) \qquad 'spacefilling' \qquad (17.22)$$

with the nucleation rate α and the growth rate G. Impingement of the spherulites is captured by an Avrami model, (17.22). The morphology is described per uni volume by the total volume of spherulites V_{tot}, their total surface S_{tot}, the sum of their radii R_{tot} and the number of nuclei N. The relation of these parameters with ψ_i is given between brackets. The number of nuclei and the growth rate have to be measured as a function of temperature.

The very often used Avrami equation (17.22), which describes space filling in case of isothermal crystallization for which all nuclei appear at t_0 and where the growth rate is constant for $t > t_0$, is given by

$$\xi_g = 1 - \exp\left[-\frac{4\pi}{3}NG^3(t - t_0)^3\right] \tag{17.23}$$

with t_0 the time that the crystallization temperature was reached. Notice that the Avrami equation is a special case of the Schneider rate equations as shown in [4].

For flow-induced nucleation and crystallization Zuidema et al. [5,16] proposed a modification (S_{J2}-model) for the shear-induced crystallization model of Eder and Janeschitz-Kriegl [4], which gave a good description of the flow-induced structures obtained in their experiments. Zuidema et al. used the recoverable strain modelled with a Leonov type of model as a driving force for flow-induced crystallization. It was demonstrated that the flow-induced structure correlated most strongly with the viscoelastic mode with the highest relaxation time. Therefore, only the second invariant of the deviatoric part of the recoverable strain ($J_2(\bar{\mathbf{B}}_e^d) = 1/2\ \bar{\mathbf{B}}_e^d : \bar{\mathbf{B}}_e^d$) corresponding with the maximum rheological relaxation time was used. This invariant is a measure for the molecular orientation [12]. With the S_{J2}-model, non-isothermal crystallization of cylindrical structures (which are for convenience called shish-kebabs) can be described. It has the same structure of differential equations as the Schneider rate equations. Mean number of shish-kebabs and their mean length, surface and volume are calculated from:

$$\dot{\psi}_3 + \frac{\psi_3}{\tau_n} = 8\pi J_2 g_n' \qquad (\psi_3 = 8\pi N_f) \qquad 'rate' \tag{17.24}$$

$$\dot{\psi}_2 + \frac{\psi_2}{\tau_l} = \psi_3 J_2 \frac{g_l'}{g_n'} \qquad (\psi_2 = 4\pi L_{tot}) \qquad 'length' \tag{17.25}$$

$$\dot{\psi}_1 \quad = G\psi_2 \qquad (\psi_1 = S_{tot}) \qquad 'surface' \tag{17.26}$$

$$\dot{\psi}_0 \quad = G\psi_1 \qquad (\psi_0 = V_{tot}) \qquad 'volume' \tag{17.27}$$

$$\psi_0 \quad = -\ln(1 - \xi_g) \qquad 'space filling' \tag{17.28}$$

with the driving force J_2, the growth rate G, the scaling factors g_n' and g_l' to describe the sensitivity of the flow-induced nuclei and length on J_2, and the characteristic times τ_n and τ_l to describe the relaxation behavior of the flow-induced nuclei and length. Impingement of the cylindrical structures is captured by an Avrami model, (17.28). The morphology is described per unit volume by the total volume of shish-kebabs V_{tot}, their total surface S_{tot}; the sum of their lengths L_{tot} and the number of flow-induced nuclei N_f. The relation of these parameters with ψ_i is given between brackets. Eder's rate equations are obtained by replacing J_2 by a scaled value of the shear rate squared.

Molecular orientation will generate extra nuclei, (17.24). When the orientation is strong enough, these nuclei can grow in one direction, (17.25). The radial growth rate of the cylindrical structures is taken equal to the spherulitical growth rate. Of course, other choices for the radial growth dependent on J_2 are possible. The relaxation time τ_l is in general taken very large, because reduction of length can only occur via melting.

Zuidema [5,16] considered nucleation to act as physical cross-linking. Consequently, an increased number of nuclei causes an increase in the rheological relaxation time. A simple, linear relationship between flow-induced nuclei and the highest rheological relaxation time was chosen.

$$\lambda_{max} = a_T(T)\lambda_{0,max}\left(1 + \frac{\beta N_f}{g_n'}\right) \tag{17.29}$$

with $\lambda_{0,max}$ the highest rheological relaxation time at a reference temperature, a_T the shift factor and β a scaling factor that describes the interaction between nuclei and rheology. Consequently, the scaling factors β, g_n' and g_l' should be measured as a function of flow conditions.

The concept of the equivalence of physical cross-linking and nucleation in the Leonov model can be used in the Pom–Pom model in rather natural way. Increase in the number of nuclei gives an increase in the number of branches and relaxation time. This means that the scaling factor β in (17.29) is related to the number of branches in the Pom–Pom model.

In the case of flow, both spherulitical and flow-induced structures contribute to the degree of space filling, depending on the influence of J_2. The Avrami model for impingement is then described by

$$\phi_0 + \psi_0 = -\ln(1 - \xi_g) \tag{17.30}$$

There are two big advantages when applying the Pom–Pom model to the linear polymers used in this study. First, both elongation and shear data can be described excellent with the same set of fitting parameters. Second, the physical description that serves as a basis for this model (i.e. the Pom–Pom molecule) results in a more transparent model. The physical cross-linking process, as proposed by Zuidema et al. [5,16], can be related to the increase in the number of branches during crystallization.

17.3 Results

The most important features of the model will be demonstrated by some illustrative examples. A more extensive treatment of these results can be found in [5,16] and [17].

Zuidema [5,16] predicted the position of the transitions of different flow induced layers in a slit flow as measured by Jerschow [18]. He considered only shear flows and, therefore, used the Leonov model for viscoelastic modelling since this model gives an excellent description of viscoelastic shear data. Postulating that

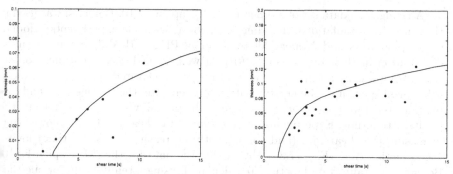

Fig. 17.1. Measured (*) thickness of the boundary of the flow-induced layer for different shear times (from [18]), together with the numerically determined position of this boundary (solid line) for a wall shear rate of 79 (left) and 115 $[s^{-1}]$ (right).

variables describing the flow induced structure should be the same at these transitions, and determining the values from one independent experiment, he could reproduce the transition position for a broad range of deformations (two different wall shear rates, many different shear times). Figure 17.1 shows the position of the transition between the highly oriented layer near the duct wall and the fine grained layer next to it. Predictions are within the experimental scatter.

Next, it was demonstrated that the model could predict qualitatively (for most materials used in literature the required data set is missing) a wide range of observations on flow induced crystallization as reported in literature. Two examples are given here. In Fig. 17.2 shows for a continuous, isothermal shear flow at different, constant shear rates the induction time, the time that a noticeable rise in the viscosity is found [19]. The induction time decreases dramatically if some critical shear rate is exceeded.

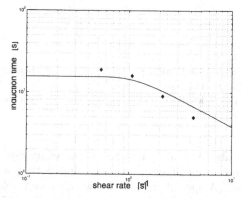

Fig. 17.2. The calculated induction time using parameters of the iPP Daplen KS10 (Borealis). Measurements (♦) after Lagasse [19].

Although the data set of a different iPP (Daplen KS10, Borealis) was used in the numerical simulations, the same behavior as with the experiments is found

Vleeshouwers and Meijer [20] used three iPP's with different molar mass distribution in their experiments. After a well defined shear treatment of the melt, using a cone and plate configuration, at an elevated temperature, and a subsequent quench to the crystallization temperature, the storage (G') and loss modulus (G'') were monitored. The rise in the moduli was used to determinethe the induction time. Experiments at a constant total shear treatment showed that increasing the shear rate (and consequently decreasing the shear time) lowers the induction time (Fig. 17.3, right). Using the data for the iPP Daplen KS10 Borealis, crystallization kinetics are calculated, using a temperature profile that resembles the one Vleeshouwers [20] used, and induction times are calculated for a constant total shear of $\gamma = 250[-]$. Results are given in Fig. 17.3. The model confirms the experimental observations in a qualitattive way. The measurements indicate the presence of a plateau region at low shear times. This probably is related to the difference in startup behavior of the materials. The experiments with a constant shear rate during various shear times show a strong decrease in induction time in the low total strain region (low shear times), while in the high total strain region (high shear times) the induction time decreases much more gradually. Again, the calculated induction times confirm this observation in a qualitative way.

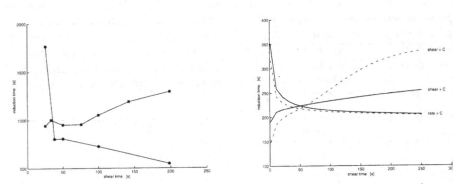

Fig. 17.3. Left: Measured induction time for a constant shear rate ($\dot{\gamma} = 5[s^{-1}]$) and different shear times (●), the induction time for a constant total shear ($\gamma = 250[-]$) and different shear rates (■). Right: The corresponding predicted induction times (—: Sg_2 model, – · –: Eder model).

Swartjes [17] studied extensional flows and, thus, he used also the Pom–Pom model since this model is superior in describing both, shear and extensional rheological behavior. A special cross-slot flow cell was developed enabling bire fringence and time resolved WAXS measurements, see Fig. 17.4. To create a flow an outer ring with two cams is rotated. Measurements and numerical results re late to the central part ($2x2\ mm$) of the flow cell where a planar elongational flow around the stagnation point is present.

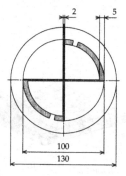

Fig. 17.4. Main dimensions in *mm* of the cross-slot flow cell.

One of the major outcomes, both, experimentally and numerically, was a highly oriented narrow strand ($\approx$ 80 μm) of a fibre-like crystalline structure around the outflow centerline that developed at a relatively high temperature of $\approx$ 149° C, a temperature at which noticeable quiescent crystallization effect takes hours. This is shown in Fig. 17.5, where the integrated WAXS intensity is given for the (110) refection as a function of the azimuth angle and the position perpendicular (y-coordinate) and parallel (x-coordinate) to the outflow direction. These results were obtained with the micro-focus-beam-line ID13 (beam size 30 μm) at the European Synchroton Radiation Facility (ESRF).

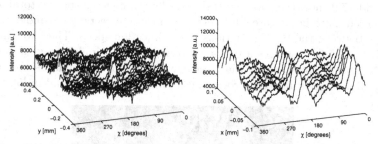

Fig. 17.5. WAXS intensity for the (110) reflection vs azimuthal angle and position on the inflow center line (left) and on the outflow center line (right).

From numerical simulations a qualitatively similar result was found, see Fig. 17.6, which shows the computed flow-induced mean space filling (the flow is essentially three dimensional) in the cross-slot flow cell. For symmetry reasons, only a quarter of the central area of the flow cell is shown. A very sharp, highly crystalline, narrow band is found also with the numerical simulations. Taking a level of 20% space filling as a threshold, the band has a width in the order of 100 μm.

Finally, some examples of predicted structure distributions for injection moulding flows are presented. With injection moulding a cold mould is filled at high speed during which a solidified layer is growing from the wall. At the

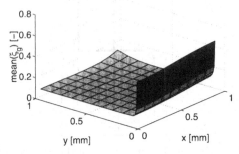

Fig. 17.6. Predicted mean flow induced space filling as a function of the position.

transition from this solidifying layer to the melt, the relaxation times of the melt become rather high, strongly increasing the effect of the flow gradients on the crystallization behavior. At the free surface of the progressing polymer (the flow front) hot material from the core moves towards the cold walls, experiences a complex deformation history and is cooled rapidly when it reaches the wall. These combined flow phenomena can create rather complex structure distributions in the sample. For example, sometimes a double oriented layer is found with an un-oriented, fine-grained spherulitical layer in between, see Fig. 17.7.

Results for two processing conditions of molding of iPP are compared. Figs. 17.8 and 17.9 show the specific total shish length and the specific total amount of oriented material at three different locations in a rectangular mold. The difference between the two simulations is the injection speed. The lower injection speed, Fig. 17.9, generates a two layer configuration, similar to what is seen in Fig. 17.7, that varies along the mold cavity. A more extended set of results based on a varying different processing parameters can be found in [5].

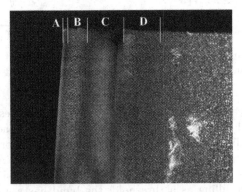

Fig. 17.7. Microscopic structure of cross-section near surface of a 1 *mm* injection molded plate. Showing the 'skin layers' (A), 'transition layer' (B), 'shear layer' (C) and 'fine grained layer' (D).

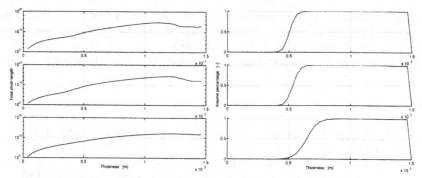

Fig. 17.8. The distribution of the flow-induced oriented structure across the thickness of the injection molded product close to the gate (top), far from the gate (bottom) and in between (middle). . Processing conditions: $T_{inj} = 430K, Q = 4.65\ 10^{-4}m^3s^{-1}, T_{wall} = 393K$. Left side: total shish-length, right side: volume percentage oriented material.

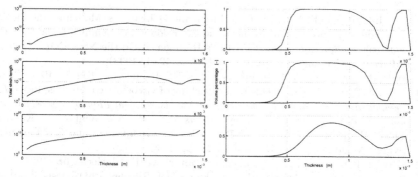

Fig. 17.9. The distribution of the flow-induced oriented structure across the thickness of the injection molded product close to the gate (top), far from the gate (bottom) and in between (middle). Processing conditions: $T_{inj} = 430K, Q = 1.16\ 10^{-4}m^3s^{-1}, T_{wall} = 393K$. Left side: total shish-length, right side: volume percentage oriented material.

17.4 Conclusions

A computational model for the simulation of flow induced nucleation and crystallization is presented. The model gives a detailed description of both, quiescent and flow-induced crystallization for any thermal-mechanical history. The advantage of the use of this model is the dependence of nucleation rate and growth on a (molecular) strain measure rather than the macroscopic shear rate. Improvements or other dependencies can be easily inserted (for example a deformation depended growth rate). The dependence of rheology on nucleation is described by physical cross-links (Leonov) or an enhanced number of branches (Pom–Pom), resulting in an increased relaxation time. The sensitivity of this relation between number of nuclei and rheological relaxation time has to be measured.

For comparison with experiments an extended set of material parameters is required. The lack of (part of) these data is the reason that until now most of the comparisons are still of qualitative nature. A more quantitative comparison is needed for validation and, if necessary, improvement of the model.

References

1. M. Gahleitner, C. Bachner, E. Ratajski and G. Rohaczek: J. Appl. Pol. Sci., **73** (1999)
2. D.W. Mead: J. Rheology, **38**, 6 (1994)
3. C. Pattamaprom, R.G. Larson and T.J. van Dyke: Rheol. Acta, **39** (2000)
4. G. Eder and H. Janeschitz-Kriegl, In: *Materials Science and Technology: A Comprehensive Treatment: Processing of Polymers*, **18**, ed. by H.E.H Meijer, Crystallization, (VCH, Weinheim 1997) pp 269–342
5. H. Zuidema, Flow induced crystallization of polymers. Application to injection moulding, PhD Thesis, Eindhoven University of Technology, The Netherlands (2000)
6. N.J. Inkson, T.C.B. McLeish, O.G Harlen and D.J. Groves: Modelling the rheology of low density polyethylene in shear and extension with the multi-modal Pom-Pom constitutive equation, Proc. PPS 15, 's Hertogenbosch, the Netherlands (1999)
7. T.C.B. McLeish and R.G. Larson: J. Rheol., **42**, 1 (1998)
8. W.H.M. Verbeeten, G.W.M. Peters and F.P.T. Baaijens: J. Rheol., **45**, 4, (2001)
9. G.W.M. Peters, Thermorheological modelling of viscoelastic materials, In: *IUTAM Symposium on Numerical Simulation of Non-Isothermal Flow of Viscoelastic Liquids: Fluid Mechanics and its Applications*, Proc. IUTAM Symposium, Kerkrade, The Netherlands, 1-3 November 1993, ed. by J.F. Dijksman and G.D.C. Kuiken (Kluwer Academic Publishers 1995), **28**, pp 21-35
10. G.W.M. Peters and F.P.T. Baaijens: J. Non-Newt. Fluid Mech., **68** (1997)
11. W.H.M. Verbeeten, Computational Polymer Melt Rheolgy, PhD Thesis, Eindhoven University of Technology, The Netherlands (2001)
12. R.G. Larson, *Constitutive Equations for Polymer Melts and Solutions* (Butterworths, London 1988)
13. R.B. Bird, C.F¿ Curtiss, R.C. Armstrong and O. Hassager, *Dynamics of Polymeric Liquids* (John Wiley and Sons, New York 1987)
14. W. Schneider, J. Berger and A. Köppl, In: *Physico-Chemical Issues in Polymers* Non-isothermal crystallization of polymers: Application of Rate Equations, (Technomic Publ. Co 1993), pp 1043-1054
15. W. Schneider, A. Köppl and J. Berger: Int. Pol. Proc. II, **3** (1988)
16. H. Zuidema, G.W.M. Peters and H.E.H. Meijer, Macromol.Theory and Simulations, **10**, 5, (2001)
17. F.H.M.S. Swartjes, Stress Induced Crystallization in Elongational Flow, PhD Thesis, Eindhoven University of Technology, The Netherlands (2001)
18. P. Jerschow, Crystallization of polypropylene. New experiments, evaluation methods and choice of material compositions, PhD Thesis, Johannes Kepler Universität Linz, Austria (1994)
19. R.R. Lagasse and B. Maxwell: Pol. Eng. Sci., **3**, 16 (1976)
20. S. Vleeshouwers and H.E.H. Meijer: Rheol. Acta, **35** (1996)

18 Physical Cross Links in Amorphous PET, Influence of Cooling Rate and Ageing

Stefano Piccarolo[1], Elena Vassileva[2], and Zebene Kiflie[1]

[1] Università di Palermo, Dipartimento di Ingegneria Chimica, Viale delle Scienze 90128 Palermo, Italy, piccarol@dicpm.unipa.it
[2] Universität Kaiserslautern, Institut für Verbundwerkstoffe GmbH, Erwin Schrödinger strasse Geb.58, D-67663 Kaiserslautern, Germany

Abstract. A Continuous Cooling Transformation (CCT) procedure can be used to distinguish the initial "state" of the amorphous PET samples produced upon solidification from the melt at different cooling rates. The material frozen at this stage behaves as a rubber when brought above the T_g due to the onset of physical cross links. The rubber is not a stable network, however, since physical cross links may eventually dissolve. Their size distribution, and possibly their number, depend on cooling rate and ageing. Some may be even stable above the glass transition and act as nuclei for further crystallization from the glass. Upon increasing cooling rate. size distribution becomes smaller and stability of physical cross links decreases. Ageing may give rise to the onset of further physical cross links or may affect the stability of the ones already set in depending on cooling rate, i.e. on how is constrained the initial network structure.

18.1 Introduction

Poly (ethylene terephthalate) (PET) is usually obtained as an amorphous polymer when solidified from the melt under most of the processing routes with a possible exception of high speed spinning. However, unlike completely amorphous polymers, PET can also be obtained in a semicrystalline state when the cooling conditions match its slow crystallization kinetics. Polycarbonate (PC) also crystallizes, but to a very moderate scale that it is often considered as an amorphous polymer. Yet, there are some interesting properties that associate these two polymers. In the amorphous state, PET shows some peculiar mechanical properties such as high modulus and toughness that are quite uncommon to amorphous materials. PC also shows, although to a lesser extent, similar behavior. On the other hand, these two materials, though crystallizable under suitable conditions, show a significant departure from the properties of crystalline polymers. PET, for instance, which is tough in the amorphous state becomes more and more brittle with increasing crystallinity. Nevertheless, this apparently clear contradiction is not often recognized in the literature and PET is usually considered as a semicrystalline material while PC in contrast as an amorphous.

In transformation operations such as reheat stretch blow molding of bottles, heat setting, production of films and fibers, etc., processing of slowly crystallizing polymers like PET is usually accomplished starting from the amorphous state. Obviously, the physical and mechanical properties of such products will be directly or indirectly influenced by the properties, and therefore by the morphology, of the starting "amorphous" polymer. Besides, thermal gradients arising during cooling give rise to a layered structure and as a result the material experiences a wide range of cooling conditions during solidification. Yet, since PET is a slowly crystallizing polymer, it will remain practically amorphous under all such conditions although this does not imply that the amorphous structure is not affected by these conditions.

To date, the ageing and crystallization behavior of amorphous PET, its structure as well as its properties, have been studied by many resaerchers. In spite of the vast number of available works, neither the scientific approaches nor the technical practices give specifications on how the quench is performed to obtain the starting amorphous PET. Descriptions about the thermal history experienced by the material during quenching are not available. In other words, no attempt is made to describe the initial state of the amorphous material. Although it would seem unconceivable to define an initial state for a material far from equilibrium it is possible, however, to trace the initial state by adopting a similar cooling rate range upon quenching the material from the melt as proposed in this work.

18.2   Continuous Cooling Transformation (CCT) of PET

It has been recently shown that a Continuous Cooling Transformation (CCT) procedure can be applied, for quiescent conditions, to determine the solidification behavior of polymers under a wide spectrum of cooling rates matching those usually adopted in polymer processing. The characteristic description of this behavior is the density dependence on cooling rate, which when combined with other structural probes, may add information on the phases formed and their pertinent cooling rate window. Apart from time dependent ageing phenomena, stable phases form preferentially at low cooling rates with a slow rate of change of density with cooling rate. With increasing cooling rates, phases of decreasing stability occur first [1] followed by metastable phases [2] and eventually the amorphous phase at the highest cooling rates. Normally, the connections between the cooling rate window of each phase are characterized by a sharp drop in density.

In this regard, a previous work which has examined the CCT behavior of a PET (IV 0.62) with reference to the phases formed in detail may be cited [3] in which it was pointed out that (see Fig. 18.1):

- the stable triclinic phase is observed for cooling rates below $\sim 1^\circ$C/s and the density is seen to slowly decrease with increasing cooling rate down to 1.38 g/cm^3;
- an abrupt change in density from 1.38 to 1.34 g/cm^3 between ~ 1 and $\sim 3^\circ$C/s;
- in this interval the stable triclinic phase disappears well before the minimum density is reached as can be deduced from the broad halo typical of the amorphous phase observed in the WAXD patterns;

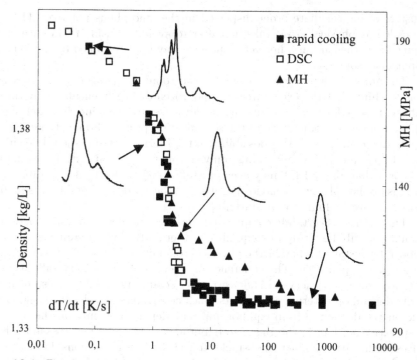

Fig. 18.1. Density, microhardness and some exemplyfying WAXD patterns of PET solidified by the CCT procedure and by DSC vs cooling rate.

- the linear relationship between latent heat of crystallization (obtained by DSC upon cooling from the melt) and density shows a discontinuity at a density of less than 1.35 g/cm^3 (at ~2°C/s) below which a "residual" very low and almost constant latent heat, incompatible with the onset of an amorphous phase, is observed;
- Lorentz corrected SAXS patterns show the existence of a long periodicity even above 2°C/s becoming very broad with increasing cooling rates;
- above 3°C/s the density decreases very slowly and reaches the value typical for the amorphous phase of PET (1.335 g/cm^3) at ~100°C/s which is maintained up to the maximum cooling rates achieved, ~7000°C/s;
- microhardness (MH), in contrast to density, shows a considerable sensitivity to cooling rate above 3°C/s and reveals a different morphological arrangement.

18.3 Physical Cross Links vs Entanglements

So based on these observations it has been assumed that a state of intermediate order exists in the amorphous phase of PET when solidified under quiescent conditions [3]. Microhardness (MH) measurements show that this state of intermediate order affects mechanical properties plausible if one thinks of crystalline clusters of intermediate order dispersed in the amorphous polymer. They will be called as physical cross links hereafter to avoid confusing them with entanglements as their nature, not yet identified, may be determined by different non topological sources.

The amorphous phase is indeed a glass at room temperature and it is reasonable to think that the excess free volume frozen during quenching may provide sufficient space for local mobility in such a way as to determine the onset of crystalline clusters at some stage of the solidification process. So far, there are only conflicting evidences on this possibility in the literature and it should be emphasized that these arguments are just only speculations. As an extreme example it is known that glassy PET may even crystallize below the T_g if a suitable stress state is applied [4]; this behavior is not only related to the mobility arising from the excess free volume frozen during quenching.

This subject is certainly complicated since the stability of physical cross links should not only be treated by equilibrium thermodynamics arguments. It is not known, for instance, whether the physical cross links could be identified as a true mesomorphic phase [5]. The coherence of the concept of paracrystallinity [6] is not clear since a paracrystal is defective per se and its stability does not necessarily depend on external constraints. Such constraints, which play a significant role on the departure from equilibrium, may depend not only on the frozen-in cooperative motions in the glassy state but also on the physical cross links themselves as the latter can act as obstacles to the elementary motions. Furthermore the relationship of physical cross links with the so called rigid amorphous phase frequently reported in semicrystalline PET [7-9], is not clear.

Physical cross links, as intended here, have also been inductively identified by other authors under different circumstances. Imai et al. [10] have given experimental evidence of a spontaneous nematic transition in PET already postulated by Doi et al. [11] on rigid chain polymers. On the basis of SANS experiments at 88°C a transition was observed to take place by spinodal decomposition at the early stages of crystallization due to the increase of trans conformations of the ethylene moieties. Fukao et al. [12] have shown the onset of an α' relaxation prior to the crystallization of PET by dielectric spectroscopy. Qian [13] has extensively studied the conformational changes occurring in different polymers by FTIR. He postulated the occurrence of cohesional entanglements, which according to him arise if sufficient free volume is frozen upon quenching from above the T_g. They are however unstable and sub-T_g annealing would cancel their effect. His viewpoint is different from the previous authors since it assumes that the cohesional entanglements arise due to the energy driven spontaneous self assembly of chain segments. The nematic transition observed by Imai and also the α' relaxation observed by Fukao may presumably arise simply due to short range interaction and therefore due to chain rigidity.

Other works have observed the influences of physical cross links on the crystallization behavior and on the properties of the amorphous phase. Ageing was shown to enhance crystallization above T_g thus invoking an increase of nucleation density in the amorphous polymer [14]. This behavior has been recently disputed since a maximum of crystallization rate vs. ageing time was observed and interpreted as the coexistence of two opposing mechanisms: a decrease of mobility should indeed take place for long ageing times [15].

The Continuous Cooling Transformation (CCT) approach seems however the most promising method to control the morphology of amorphous PET [2] giving rise to sensible structural variations as recently shown by a Small Angle X-Ray Scattering (SAXS) study [16]. With this method it is possible to continuously change a relevant processing variable, i.e. the cooling rate, with the constraint that the film obtained is homogeneous [17].

In this work the dependence of physical cross links on cooling rate will be reported on the basis of its influence on the crystallization kinetics from the glass [18]. Nevertheless, whether all physical cross links frozen during quenching act as nuclei is questionable as they may cancel out upon heating above the T_g and may even act as constraints for the subsequent crystallization. An attempt to analyze the stability of physical cross links is made with a rejuvenation procedure and comparing the results obtained on aged samples as well as on a non crystallizable amorphous polymer, polymethylmethacrylate (PMMA).

18.4 Sample Preparation and Test Procedure

The materials used were a PET resin (IV =0.62 dl/g, Mw=39000) kindly supplied by Mossi & Ghisolfi Group and a commercial extrusion grade PMMA. PET was dried at 170°C for 7 hrs in vacuum. Films of ca 200 μm were also prepared in vacuum by compression molding of the dried pellets.

Details on the experimental CCT approach have been reported previously [3]. A systematic analysis of the heat transfer conditions necessary for obtaining a homogeneous sample has also been examined [17]. In this approach samples of PET are first held in the melt at 280°C for 5 min in a nitrogen ambient and afterwards quenched to ca 0°C by spraying a cooling fluid while recording the temperature history. Desired cooling rates are obtained by a proper choice of the cooling fluid (water or air) and its flow rate. Cooling rates ranging from 2 to well above 1000°C/s were used to prepare apparently amorphous samples, (see Fig. 18.1). PMMA samples were also obtained with the same procedure although only one cooling rate, of ca 10°C/s was adopted. The comparison of the initial state of PET is made on samples with slightly different cooling rates since the results presented are collected from different preliminary works. The similarity of the initial state can however be immediately recognized once reference is made to Fig. 18.1 and to the features previously discussed.

A gradient column filled with a solution of n-heptane and carbon tetrachloride having a resolution of 0.0001 kg/L and a repeatability within 0.0002 kg/L was used for density measurement at 25°C. Each sample was properly checked against entrapped air bubbles using a microscope and afterwards degassed before being introduced into the column. An average of five samples were analysed for each test condition.

Microhardness measurements were carried out using an MHT-10 Vickers indenter by Anton Paar. Peak forces of 0.1, 0.15 and 0.25 N were used to correct for instantaneous elastic recovery [11] and an average of five measurements were done with each force. The microhardness, H_v was calculated from the residual projected indented area using the relation:

$$Hv = 2F \frac{sin(\alpha/2)}{d^2} = 1.854 \frac{F}{d^2} \tag{18.1}$$

where H_v is the microhardness in Pa, d is the mean diagonal length of the projected impression in meters, F is the peak force in Newton and α the included angle between two non adjacent faces (136° for the Vickers indenter).

Crystallization temperatures upon scanning at different heating rates and melting enthalpies were measured with a Perkin Elmer DSC 7 fluxed with dry nitrogen and calibrated for temperature with indium on heating at 10°C/min.

18.5   Isothermal Crystallization of Glassy PET

Isothermal crystallization of quenched samples was analysed as a function of initial cooling rate in order to determine its influence on nucleation density and therefore on the onset of stable crystal nuclei. The crystallization experiment was carried out by immersing the samples in an agitated thermal bath filled with silicon oil. Temperature was kept at 100°C±0.2°C. At the end of specified crystallization times, the samples were quenched (always at the same cooling rate) in an ethylene glycol bath maintained at -25°C. Analysis was made by measuring the density of these samples. This procedure was followed mainly because it was difficult to apply time resolved density measurements. Further experimental details are reported elsewhere [18].

Figure 18.2 shows that density increases with annealing time at 100°C and the final plateau density always attains the same value around 1.38 kg/L. Crystallization rate decreases with samples cooling rate although no difference is observed on further increasing cooling rate above ca 100°C/s. If this behavior is due to a difference of nucleation density associated with the occurrence of a precrystalline order, of what we have called physical cross links, then it is possible to deduce that the initial state of the amorphous quenched samples plays an important role on the density and size of the physical cross links. If one assume that the mechanism of crystal growth is the same for all of these samples and that only nucleation is affected by the initial state of the material, then the ratio of the nucleation density of the 2°C/s sample to that of the sample quenched at 100°C/s will be 2.3. In this case, however, it will be logical to ask if only such difference in nucleation density is responsible to produce the differences in density and microhardness (see Fig. 18.1) observed between these samples at the beginning of crystallization. In other words, if such nucleation density ratio represents the actual number of nuclei frozen in the samples during quenching or if it refers only to the stable nuclei at the isothermal crystallization temperature, supposing that those physical cross links smaller than the critical nuclei size should have rapidly dissolved.

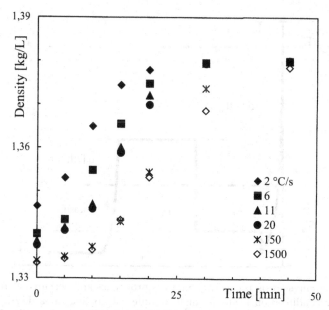

Fig. 18.2. Isothermal crystallization of PET amorphous samples quenched at different cooling rates. Density is measured on samples held for diferent times at 100°C.

18.6   Rejuvenation Procedure

Thermodynamics suggests that there exists a distribution of sizes of clusters (i.e. physical cross links) below the equilibrium melting temperature. This means when the sample is brought to the crystallization temperature, those clusters with sizes greater than the critical dimensions will remain stable upon temperature fluctuations, whereas the others may cancel out. However, the dynamics of this process may also be influenced reciprocally by the morphology of the amorphous phase itself constrained by the occurrence of physical cross links. So, it is attempted to clarify this role by making rejuvenation tests on these samples.

The rejuvenation procedure, schematically shown in Fig. 18.3, includes quenching the sample from the melt at a given cooling rate $dT/dt|_1$, and whenever ageing is involved, maintaining it at a constant ageing temperature, T_{ag}, for different times, t_{ag}, (ageing step), heating it rapidly (in about 10 s) to the rejuvenation temperature, T_{rej}, keeping it at this temperature for different times, t_{rej}, and finally cooling it to room temperature with a constant rate $dT/dt|_2$. The samples were then analyzed by microhardness measurements and DSC scans. A similar procedure was also applied to a PMMA sample. PMMA was chosen in order to compare the behavior of PET with that of a non crystallizable amorphous material. T_{ag} was chosen to be about 35°C below T_g while T_{rej} was chosen 10°C above T_g. The latter, as measured by DSC, on freshly quenched samples was 77 and 115°C for PET and PMMA respectively. At this point, however, it is good to note the experimental limitations of the whole procedure adopted which can

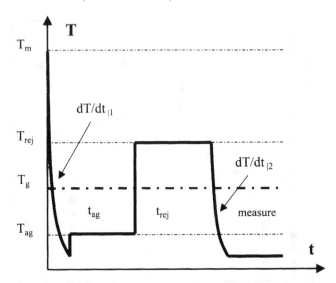

Fig. 18.3. Rejuvenation procedure. Samples properties are measured at room temperature after solidification from the melt at different cooling rates $dT/dt|_1$, ageing at $T_{ag} \sim T_g\text{-}35°C$, rejuvenation at $T_{rej} \sim T_g\text{+}10°C$ for different times t_{rej}.

be clearly understood from the dispersion of the data collected and reported in the sections below. They probably derive from the cumulative propagation of errors due to the complex thermal treatment the tested samples underwent.

18.7   Results

The dependence of the maximum crystallization rate temperature, T_c, measured by DSC at different heating rates is shown in Fig. 18.4 for a PET sample previously quenched at $100°C/s$. Under most circumstances a minimum of T_c with rejuvenation time at $87°C$, t_{rej}, is observed. It becomes more pronounced with increasing DSC heating rate and tends to cancel out at the minimum heating rate adopted ($2.5°C/min$). Since in most cases the minimum is observed at t_{rej} of ca 25 s, then it is clear that at a heating rate of $2.5°C/min$ the time spent above the T_g is relatively so high that it cancels out the observed phenomenon. Thus the minimum must be related to the occurrence of two competitive phenomena both with time scales of few tens of seconds. Noting that a decrease in T_c is related to a faster crystallization kinetics, it is possible to infer that the minimum may be related with an increase of physical cross links. In addition it can be observed from the same figure that such effect tends to cancel out on further increasing the time spent at $87°C$ since this temperature is above the T_g.

By T_g one understands that cooperative motions set in at this temperature in a time scale equivalent to that used for its determination. So for a T_g range of a few degrees, this time will be a fraction of a minute when using a DSC scanning rate of $10°C/min$. Therefore, at the rejuvenation temperature $10°C$ above T_g

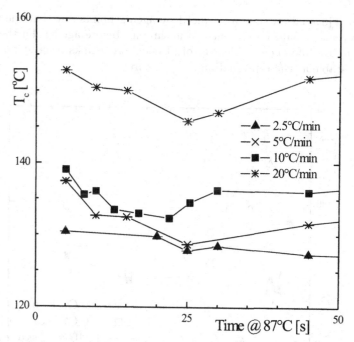

Fig. 18.4. Temperature of maximum crystallization rate, T_c, for PET rejuvenated samples vs rejuvenation time, t_{rej}, at a temperature of 87°C.

with an apparent activation energy of 330 KJ/mole for the cooperative relaxation process of PET [19], the characteristic time would be far lower than 25 s, thus implying that the occurrence of the minimum could not be related to the α relaxation.

Although it is clear that the quenched sample is characterized by an excess free volume which eventually cancels out even before the sample reaches 87°C, it is difficult however to speculate on the nature of the processes occurring. A number of questions have to be answered. Are there any physical cross links previously frozen in the 100°C/s sample during quenching from the melt? Do they constrain the material such that it behaves as a rubber even when cooperative motions sets in? Before attempting any further interpretation we shall try to examine other experimental evidence.

The crystallization temperature, T_c, of fresh and aged samples quenched at 100°C/s as well as that of sample quenched at 3800°C/s and aged subsequently, all three rejuvenated eventually, is compared in Fig. 18.5. Ageing was done at 60°C for 14 days and T_c was obtained using DSC by heating each rejuvenated sample at a rate of 10°C/min. The figure reveals that with ageing the dependence of T_c on rejuvenation time seems to vanish and the aged sample_100 assumes a constant value close to the plateau value observed for the fresh sample. The temperature at which T_c levels off in the case of the aged sample_3800, is slightly, although systematically, higher than that of aged sample_100. This independence of T_c on rejuvenation time introduced by ageing points out that the source of this difference should be present in the samples before ageing. Furthermore, if this is due to differences in density of physical cross links, then it implies that they are stable at the rejuvenation temperature.

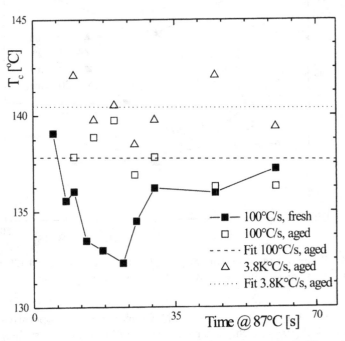

Fig. 18.5. Temperature of maximum crystallization rate, T_c, for PET aged and then rejuvenated samples vs rejuvenation time, t_{rej}, at a temperature of $87°C$.

The enthalpy of relaxation at the glass transition temperature of samples rejuvenated after ageing reported in Fig. 18.6 further confirms these observations. While in the case of a sample quenched at $3800°C/s$ the enthalpy cancels out immediately upon rejuvenation, similar to a PMMA aged sample, for an aged PET sample initially quenched at $100°C/s$ however, a residual relaxation enthalpy is observed. Even though this enthalpy appears to be small, it is nevertheless higher than those of PET $3800°C/s$ and PMMA aged samples and does not depend on rejuvenation time.

The source of a residual enthalpy upon rejuvenation above T_g might be related to the onset of a rubbery phase in which the excess enthalpy accumulated during ageing has not been released due to a constraint. Therefore the behavior of the PET quenched at $100°C/s$ and aged at $60°C$ should be related to the onset of a more constrained network as compared to the other samples.

The similarity between the $3800°C/s$ PET and the PMMA quenched and aged samples suggests that for a PET quenched at such a high cooling rate

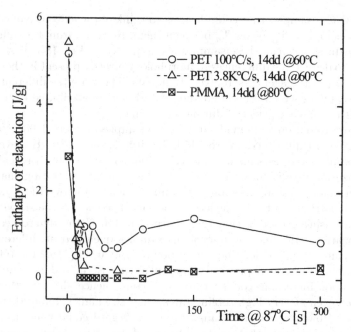

Fig. 18.6. Relaxation enthalpy at the glass transition of aged PET and PMMA samples afterwards rejuvenated at 87 and 125°C respectively vs rejuvenation time, t_{rej}.

(3800°C/s) there is a leveling off of the "effectiveness" of the physical cross links eventually formed.

PMMA was here used as a "blank" for which it is implicitly assumed that the phenomena observed in PET and discussed in the introduction, i.e., the onset of physical cross links, do not occur. In fact PMMA quenched samples do not show any time dependent phenomenon on rejuvenation as the comparison of the microhardness of rejuvenated fresh and aged samples could show. PMMA is therefore used in this context to compare the results obtained on rejuvenation of PET with a purely time dependent α relaxation process. This approach may look rather primitive, as it is difficult to exclude, even if the same degree of overheating (10°C above T_g) is used for the rejuvenation temperature, the possibility that these same processes could occur with a time scale outside the window of these experiments. To our knowledge however there are many instances justifying the assumption of PMMA as a blank in this context: it is non crystallizable although it is a flexible chain polymer, it has a bulky side group and it is a brittle material below T_g.

The results of Figs. 18.5 and 18.6 indicate that there are differences, upon rejuvenation, in T_c and relaxation enthalpy between samples 100 and 3800°C/s after ageing. At this stage it is plausible to assume that ageing could introduce physical cross links in favorable conditions. However, as can be clearly seen in Fig. 18.1, there is no significant density difference between the as-quenched 100 and 3800°C/s samples that could lead us to think of the existence of consequent differences in mobility below T_g between them. However, from the differences in microhardness between these samples also reported in Fig. 18.1, it shall not be difficult to deduce the differences in morphology already present in these samples prior to any of the ageing or rejuvenation tests. Therefore the different behavior of these samples seen in Figs. 18.5 and 18.6 should be attributed to the thermal history that they experienced during quenching.

The rejuvenation tests made at 87°C on samples 100 and 3800°C/s indicate that there exist differences in cross link density between them. However isothermal crystallization tests made at a higher temperature (100°C) reveal that the crystallization rate is constant for 100°C/s and above suggesting that the size distribution of physical cross links of samples quenched above 100°C/s does not add any contribution to the nucleation density. It could also be shown [18] that upon increasing crystallization temperature at 110°C the differences between crystallization half times of amorphous samples quenched at different cooling rates become even smaller than those pointed out in Fig. 18.2 for 100°C. This observation implies that physical cross links decrease in stability (i.e. decrease in size) upon increasing cooling rate. As a consequence although they may be generated even at very high cooling rates, their size and therefore their stability decreases. However, a definite answer on this regard could come from a more thorough investigation of their stability at different temperatures above T_g.

Thus in the case of the aged 100°C/s sample physical cross links stabilize the relaxation enthalpy through the onset of a network structure. The network is apparently stable at the rejuvenation temperature although larger temperatures may possibly affect its stability. Whether the network is generated by the ageing process or it is preliminary to it and therefore dependent on cooling rate can be judged by a comparison between fresh and aged samples on the basis of further experimental evidence.

Microhardness measurements of rejuvenated fresh and aged PET are reported in Figs. 18.7 and 18.8 respectively. In the case of the fresh ($t_{ag}=0$) PET sample quenched at 3800°C/s microhardness is not affected by rejuvenation time and shows the lowest asymptotic value with respect to the other two samples. Although the difference between 3800 and 100°C/s is small, the initial values of microhardness are however different as pointed out by Fig. 18.1 and decrease with increasing cooling rate. Furthermore a maximum of microhardness vs rejuvenation time is clearly observed for the 100°C/s fresh sample. Eventually a decrease of microhardness with rejuvenation time is also observed for the 5°C/s sample.

Figure 18.8 shows that in the case of aged samples the maxima disappear and the microhardness decreases from an initially much larger value with respect to the freshly quenched samples. Ageing has modified the morphology introducing new physical cross links: microhardness again decreases with initial cooling rate although it is now markedly higher for the 5 and 100°C/s samples with respect to the 3800°C/s sample. Upon rejuvenation there is a rather fast settling toward microhardness values close to the fresh samples. Asymptotic long time rejuve

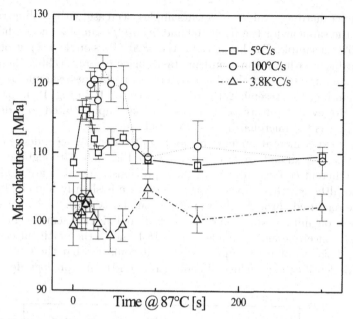

Fig. 18.7. Influence of cooling rate on time of rejuvenation dependence, t_{rej}, of micro-hardness of PET samples.

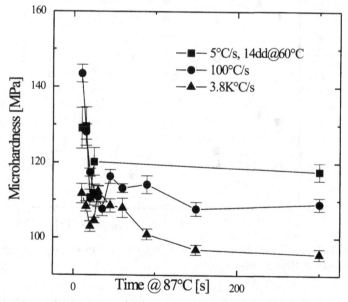

Fig. 18.8. Influence of cooling rate on time of rejuvenation dependence, t_{rej}, of micro-hardness of PET aged samples.

nation microhardness values of aged samples decrease with cooling rate. This value is the same as for the fresh 100 and 3800°C/s samples and slightly higher for the 5°C/s sample. Furthermore for the 3800°C/s sample a slight decrease of microhardness with rejuvenation time takes place before reaching the plateau.

The maximum in microhardness observed for the fresh sample quenched at 100°C/s in Fig. 18.7 resembles the one observed for T_c in Fig. 18.4, for the same sample. However the decrease of T_c cannot be simply related to a decrease of free volume as the microhardness data could imply.

A final experimental evidence comes from the density dependence on rejuvenation time of fresh and aged samples shown in Fig. 18.9. Although samples with slightly different cooling rates than those previously reported are analyzed this time and although the rejuvenation temperature is slightly higher (3°C higher than the one used through this work), the similarity of the behavior observed dependent on initial "state", i.e., on initial cooling rate, can be immediately recognized once reference is made to Fig. 18.1 and to the pertinent cooling rate window. This is worth to emphasize since it confirms that the CCT approach allows to identify the influence of cooling rate on the morphology developed.

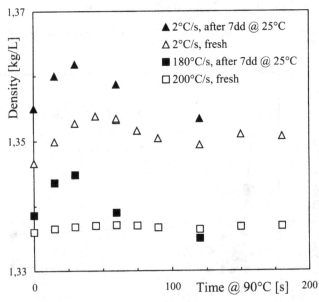

Fig. 18.9. Influence of cooling rate and ageing on time of rejuvenation dependence t_{rej}, of density of PET samples.

In Fig. 18.9 the density (measured at room temperature) dependence on rejuvenation time of two set of samples quenched at ca 2 and 200°C/s is compared with the behavior of the same samples after ageing at room temperature. Density of aged samples, after a first rapid increase with rejuvenation time eventually decreases with a slower pace to a final value smaller than the initial one. Because the samples are aged when cooperative motions set in the density should decrease and it does so although at long times. Comparison with fresh samples shows that the maximum is still observed only for the slowly quenched sample (2°C/s) while for the 200°C/s sample density immediately attains the equilibrium value at 90°C. This behavior when compared with the results already discussed makes the role of physical cross links still clearer. The physical cross links acting as nodes form a networked structure in the amorphous matrix. So, during rejuvenation a sudden entropic recoil on these nodes eventually accompanied by the onset of cooperative motions takes place. In other words, the material behaves as a rubber before the maximum and it collapses upon releasing the excess free volume. It is plausible that on further increasing the rejuvenation time, a slow process of dissolution of clusters, controlled by size distribution and thermodynamics, develops.

Hence, on this basis, it is possible to give a tentative interpretation about the behavior of T_c and microhardness, respectively shown in Fig. 18.4 and Fig. 18.7, for fresh sample quenched at 100°C/s. The minimum in T_c and the maximum in microhardness cannot be explained solely by a decrease of free volume. T_c decreases when the number of nuclei increases. Therefore, before the maximum occurs the onset of new physical cross links is favored by the decrease of constraints and by the local mobility associated with the release of the excess free volume frozen during quenching. In addition, the absence of the maximum in microhardness observed in Fig. 18.8 for the 100°C/s aged sample, confirms this interpretation. In this case, in contrast to the fresh sample, ageing has already given rise to a decrease of free volume and therefore a decrease of mobility. Therefore the behavior observed on the fresh 100°C/s sample in Fig. 18.7 for microhardness (maximum) and in Fig. 18.4 for T_c (minimum) is based on the same mechanism. However, it is also probable that a contribution to the maximum of microhardness may also be due to the decrease of free volume, as Fig. 18.9 suggests.

These results confirm that the size of physical cross links (and may be their number) is affected by cooling rate as well as by ageing. The relationship between cooling rate and ageing is also qualitatively clear. If cooling rate is small the network structure is immobilized by the bigger size (stability) and possibly the higher density of physical cross links. Ageing in this case will affect but very slightly the density of cross links. For intermediate cooling rates, the network is looser and new cross links can be generated. They are however smaller and unstable and hence will be cancelled at the rejuvenation temperature. At even larger cooling rates, for instance the fresh 3800°C/s sample, the initial density and size distribution of cross links is very small and the trend observed for the 100°C/s sample takes place in a much shorter time window not experimentally accessible. Therefore, for fresh samples in the cooling rate window of thousands of °C/s physical cross links are mostly generated during ageing. That this is indeed the case is shown by the similarity of the microhardness of the 3800°C/s aged sample where a faint maximum vs time is observed similar to the fresh 100°C/s sample. This is also confirmed by the influence of ageing on crystallization from the glass, already observed by other authors [14,15]. Although the cooling rate adopted for quenching was not specified, their samples should have been severely quenched with cooling rates above 1000°C/s.

18.8 Conclusions

Physical cross links arise in PET during cooling from the melt even if the sample is apparently amorphous. Whether they arise due to a fast nematic ordering process [10] or due to the onset of so called cohesional entanglements [13] is not clear. However, it is possible to freeze the morphology at the stage where the underlying mesomorphic structure, precursor of the crystallization process [20] is formed. The material frozen at this stage behaves as a rubber when brought above the T_g. However, the rubber is not a stable network since physical cross links may eventually dissolve due to the onset of cooperative motion. An entropic recoil takes place at short times upon rejuvenation followed by a slower trend towards equilibrium. In this light, it is plausible to think that a distribution of physical cross links arises and stability is determined by size effects and temperature. Yet, it is still unclear whether equilibrium thermodynamics determines the size of stable clusters even above the T_g since the underlying topology, determined by the cross links themselves, may affect the kinetics.

Density and size distribution of physical cross links depend on cooling rate. Some may be stable well above the glass transition and act as nuclei for further crystallization from the glass. For very high cooling rates, i.e. 3800°C/s, most are unstable at 100°C while upon decreasing cooling rate the stability of an increasing number of physical cross links is enhanced.

In an amorphous sample, excess free volume normally frozen during quenching favors local mobility. Moreover, local mobility is also affected by the departure from the temperature at which local elementary motions are frozen, which in the case of PET it is the β transition at 220°K [21]. Above this temperature ageing takes place and gives rise to further physical cross links. Possibly, ageing may also affect the stability of the ones already set in as can be seen in Fig. 18.8 in which rejuvenation of the aged sample 5°C/s clearly shows an increase of the asymptotic microhardness with respect to the fresh one.

On the other side, the network structure formed during quenching can act as a constraint for large scale motion so that only small new physical cross links that become less stable upon rejuvenation get formed. Therefore the stability of physical cross links formed during ageing depends in its turn on cooling rate: for high cooling rates constraints decrease and size distribution of physical cross links formed during ageing should increase.

If the mechanism invoked here for the onset of physical cross links can be considered as a prelude to the nucleation stage, i.e. if nucleation takes place by local molecular arrangements, similar to the onset of physical cross links described here for PET, then it is reasonable to ask as to what should be the lower thermal kinetic boundary for nucleation. Often, in crystallization kinetic modeling, the T_g has always been referred to as the limiting lower temperature boundary for molecular diffusion as it is associated with the freezing of the segmental chain mobility involved in cooperative motion. If a precrystalline order is first established in crystallization as suggested recently [20], one should rather consider that the glass transition temperature is not the lower boundary for the onset of such precrystalline order but rather the β transition temperature at which all the elementary motions get frozen.

The mechanism of dissolution of physical cross links is slow and it depends on the departure of the size distribution of the physical cross links from equilibrium. The dissolution process is also constrained by the physical cross links themselves. So, during processing the onset of physical cross links may significantly affect crystallization of PET from the glassy state. The influence of initial structure on crystallization from the glass cannot be explained only in terms of differences in nucleation density as this would be of minor entity as discussed previously. The network structure should have a major role, especially for fast processing, on the stress level the network is able to withstand, thus influencing significantly orientation and in turn flow induced crystallization.

Acknowledgements

This work has been supported by EU BRITE project "Decrypo" contract BRPR.CT96.0147 and by the Italian Ministry of University PRIN99. E.V. gratefully acknowledges a NATO-Consiglio Nazionale delle Ricerche (CNR), Italy, Outreach Fellowship and the hospitality of Dipartimento di Ingegneria Chimica, Università di Palermo, Italy where the fellowship was carried out.

References

1. T. Foresta, S. Piccarolo, G. Goldbeck-Wood: Polymer **42**, 1167 (2001)
2. S. Piccarolo: J. Macromol. Sci., Phys. **B 31**, 501 (1992)
3. S. Piccarolo, V. Brucato, Z. Kiflie: Polym. Eng. & Sci. **40**, 1263 (2000)
4. J.R.C. Pereira, R.S. Porter: J. Polym. Sci., Polym. Phys. **21**, 1133 (1983)
5. F. Auriemma, P. Corradini, C. De Rosa, G. Guerra, S. Petraccone, R. Bianchi, G. Di Dino: Macromolecules **25**, 2490 (1992)
6. G.S.Y. Yeh, P.H. Geil: J. Macromol. Sci., Phys. **B 1**, 235 (1967)
7. B. Wunderlich, I. Okazaki: J. Thermal Anal. **49**, 57 (1997)
8. M. Song, D. J. Hourston: J. of Thermal Anal. & Calorim. **54**, 651 (1998)
9. C. Schick, A. Wurm, A. Mohamed: Colloid & Polym. Sci. **279**, 800 (2001)
10. M. Imai, K. Kaji, T. Kanaya, Y. Sakai: Phys. Rev. **B 52**, 12696 (1995)
11. T. Shimada, M. Doi, K. Okano: J. Chem. Phys. **88**, 7181 (1988)
12. K. Fukao, Y. Miyamoto: Phys. Rev. Letts. **79**, 4613 (1997)
13. Y. Wang, D.Y. Shen, R.Y. Qian: J. Polym. Sci., Polym. Phys. **36**, 783 (1998)
14. L. Bove, C. DAniello, G. Corrasi, L. Guadagno, V. Vittoria: Polym. Bull. **38**, 579 (1997)
15. E.A. McGonigle, J.H. Daly, S. Gallagher, S.D. Jenkins, J.J. Liggat, I. Olsson, R.A. Pethrick: Polymer **40**, 4977 (1999)
16. F.J. Baltà-Calleja, M. Cruz Garcia, D.R. Rueda, S. Piccarolo: Polymer **41**, 4143 (2000)

17. V. Brucato, S. Piccarolo, V. La Carrubba: Chem. Eng. Sci., in press (2002)
18. Z. Kiflie, S. Piccarolo, V. Brucato, F.J. Baltà-Calleja: Polymer **43**, 4487 (2002)
19. A. Boller, I. Okazaki, B. Wunderlich: Thermochimica Acta **284**, 1 (1996)
20. G. Strobl: Eur. Phys. J. E **3**, 165 (2000)
21. V.A. Bershtein, V. M. Egorov: *Differential Scanning Calorimetry of Polymers* (Ellis Horwood PTR Prentice Hall Pub., 1994, p. 37)

19   Maximum Crystal Growth Rate and Its Corresponding State in Polymeric Materials

Norimasa Okui and Susumu Umemoto

Department of Organic and Polymeric Materials, International Research Center of Macromolecular Science, Tokyo Institute of Technology, Ookayama, Meguroku, Tokyo, Japan, nokui@o.cc.titech.ac.jp

Abstract. The temperature dependence of crystal growth rate (G) shows a bell shape with the maximum growth rate G_{max}. The activation energy for the molecular transport in G could be expressed in terms of equation of either Arrhenius or WLF. The G_{max} showed remarkable molecular weight dependence. The plots of G/G_{max} against T/T_{cmax} showed a single master curve without molecular weight dependence. The ratio of G_o/G_{max} gave a constant value for each polymer. Plots of $\ln(G/G_{max})/\ln(G_o/G_{max})$ against T/T_{cmax} for various polymers showed the universal curve. The molecular weight dependence of G_{max} was expressed as $G_{max} \propto MW^\alpha$, α was a constant but depending on the morphological features on the crystallization. The value of α was a function of the adsorption mechanism of polymer molecules on the crystal growth front and its diffusion mechanism. The ratio of T_{cmax}/T_m^o was formulated. T_{cmax} was also correlated to many other thermodynamic transition temperatures.

19.1   Introduction

According to a classical crystallization theory, a temperature dependence of linear crystal growth rate (G) can be expressed by two exponent factors, such as the molecular transport term (ΔE) and the nucleation term (ΔF) [1-3]. These two terms have opposing temperature dependence thereby bring about a maximum growth rate (G_{max}). In fact, many polymeric materials show a bell shape temperature dependence of crystal growth rate, showing the maximum growth rate at T_{cmax} [3-18]. This equation has been applied frequently to data of spherulitic growth rate. Application to the polymer crystallization leads to that the molecular transport term is considerably important in the lower temperature ranges. The transport term can be expressed by either WLF or Arrhenius type. In analyzing the crystallization data in bulk polymers, the WLF expression has been used much familiar than the Arrhenius type, since it has been believed that the former expression fits the data better than the later one. However, the transport term could be sufficiently expressed by the Arrhenius type in polymer crystallization [18-21]. It is worth to recheck which transport term equation is better describing the temperature dependence of the linear crystal growth rate.

On the basis of G_{max} for many crystalline materials, a universal master curve of temperature dependence of crystal growth rate has been proposed by Magill et al. [22,23]. This universal curve comes from a phenomenological basis. The universal master curve is derived from the corresponding state in the growth rate on the basis of a theoretical background [18,20]. G_{max} shows strong molecular weight dependence likewise the pre-exponential factor of G_o. Also, a polymer with a different chemical structure exhibits the big difference in G_{max}. Universal master curve for crystal growth rate in polymer is discussed here.

One of the characteristic features of polymeric materials is the molecular weight dependence of many of their material properties. Most of thermodynamic, spectral and mechanical properties of polymers are dependent on molecular weight (MW), for instance the viscosity of polymer solutions and melts. The influence of MW on the crystallization rate of polymers has been the subject of various publications, see [3-18,24-30]. Data existing in the literature for spherulite growth rate and the overall crystallization rate of several polymers have been analyzed as a function of MW expressed as a power law of MW^{α}. Large differences in the value of α have been reported for a large number of polymers. Here, we reinvestigate the power law for the molecular weight dependence of crystal growth rate on the basis of G_{max}.

For polymer crystallization many characteristic temperatures have to be considered, such as T_m^o, T_{cmax}, T_g and T_o. Here, T_m^o is a equilibrium melting temperature, T_g is a glass transition temperature and T_o is a hypothetical temperature at which the macro-Brownian motion of a polymer molecule ceases. It has been well known that a ratio of T_{cmax}/T_m^o is a nearly constant for a wide variety of materials [31-34] including metals, inorganic substances, organic compounds and polymers. For example, the ratio of T_{cmax}/T_m^o for polymeric materials is about 5/6 [33,34]. Constancy of the ratio of T_g/T_m^o is also widely known as Boyer Beaman rule [35,36]. We discuss these relationship based on the corresponding state of crystallization behavior.

19.2   Test of WLF and Arrhenius Expressions

Crystal growth data are often analyzed with a classical crystallization theory which is nucleation controlled, proposed by Turnbull and Fisher [1] as given by (19.1),

$$G = G_o \exp\left(-\frac{\Delta E}{RT} - \frac{\Delta F}{RT} \right) \tag{19.1}$$

where G_o is a constant for a given molecular weight and R is the gas constant. ΔE is the activation energy for the transport process at the interface between the melt and the crystal surface. ΔF is the work required to form a nucleus of critical size and commonly expressed as $\Delta F = KT_m^o/(T_m^o - T)$ [2], where K is a nucleation parameter. This equation has been applied frequently to data of spherulitic growth rate for polymeric materials. In a polymer crystallization from the melt, K is expressed as $K = nb_o\sigma_e\sigma_s/\Delta H_m$, where n is a parameter of a mode of nucleation, b_o is the thickness of the depositing growth layer and σ_e and σ_s are the end and the lateral surface energies, respectively, and ΔH_m is the heat of fusion. In (19.1), the term of ΔE and ΔF have opposing temperature dependence thereby bring about a maximum in the growth rate (G_{max})

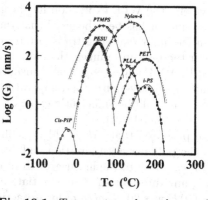

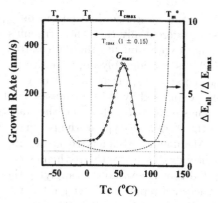

Fig. 19.1. Temperature dependence of crystal growth rate from the melt for a various polymers: PTMPS [6], Nylon-6 [7], PET [8], i-PS [6], PLLA [10], cis-PIP [12], PESU [18], Solid and broken lines are best fitting by Arrhenius and WLF expressions, respectively.

Fig. 19.2. Temperature dependence of linear crystal growth rate and overall activation energy (ΔE_{all}) for molecular transport and nucleation terms for PESU. ΔE_{max} is the activation energy at T_{cmax}. Solid and broken lines in linear crystal growth rate are best fitting by Arrhenius and WLF expressions, respectively.

Figure 19.1 shows spherulite growth data in a wide range of crystallization temperature for a wide variety of polymers as listed in Table 19.1. Data plots show a bell shape temperature dependence of crystal growth rate, giving the maximum crystal growth rate. Here, an application to the polymer crystallization leads to that the transport term of ΔE is considerably important in the lower temperature ranges. The transport term can be expressed in terms of the equation of either Arrhenius type ($\Delta E_{ARH}/RT$) or WLF type ($\Delta E_{WLF}/R(T - T_o)$ or $C_1 C_2/R(T - T_g + C_2)$) where C_1 and C_2 are adjustable parameters. In ana-

Table 19.1. Calculated values of the parameters characterizing the best fit to the data or various polymers by WLF and Arrhenius expressions in the molecular transport erm.

Polymer	Sample MW ($\times 10^4$)	T_m^o ($^\circ C$)	T_g	Arrhenius ΔE_{ARH} (kcal/mol)	K_{ARH}	$\langle R \rangle$ -	WLF ΔE_{WLF} (kcal/mol)	K_{WLF}	$\langle R \rangle$ -	Ref.
cis − PIP	89.70	30.0	-61.0	18.03	0.77	0.997	20.70	0.60	0.997	12
PESU	2.16	131.0	-10.0	29.96	1.25	0.999	3.16	0.90	0.999	18
PTMPS	14.30	160.0	-26.0	15.79	1.09	0.997	1.44	0.98	0.998	6
PLLA	10.00	203.0	54.0	49.79	2.11	0.999	6.22	1.67	0.999	10
Nylon−6	2.47	232.0	40.0	20.00	0.92	0.995	2.00	0.69	0.999	7
i − PS	29.00	261.0	100.0	49.94	1.69	0.995	4.89	2.02	0.996	16
PET	2.74	280.0	67.0	39.10	2.10	0.999	3.79	1.42	0.999	8

lyzing the crystallization data in bulk polymers, the WLF expression has been used preferentially than the Arrhenius-type, since it has been believed that the former expression fits the data better than the later one. The transport term of ΔE is considerably important in the lower temperature ranges, therefore the molecular transport term is often employed by WLF expression. The transport term expressed by WLF increases significantly when crystallization temperature reaches near to T_o. However, the transport term could be sufficiently expressed by the Arrhenius-type in the polymer crystallization [18-21].

Meanwhile, Mandelkern et al. [19] have proposed that ΔE could be suffi- ciently expressed by the Arrhenius-type in the polymer crystallization and also stated that the validity of applying the WLF equation to the linear growth rate is merely a repetitive assertion not involving any direct proof of substantiation Mandelkern et al. reported that the value of ΔE was $20.2 kcal/mol$ for isotactic polystyrene (i-PS). On the other hand, Suzuki and Kovacs [17] and Hoffman et al. [37] have claimed that the Arrhenius-type expression yielded a far inferior fit to the data of i-PS when $20.2 kcal/mol$ of ΔE was used (Hoffman et al. used actually $21 kcal/mol$ in their calculation). However, Hoffman et al. reported that both WLF and Arrhenius expressions could fit the data for many polymers with the high correlation coefficient, only except for i-PS. Here the question is raised as to why the i-PS data cannot be fitted by the Arrhenius. It is worth to re- examine the transport term expressed by WLF or Arrhenius in order to obtain an acceptable fit to the data for i-PS and other common polymers. In the ex- pression of Arrhenius, ΔE must be treated as an adjustable parameter to get the best fit to the data by a linear least squares procedure. A correlation coefficient $\langle R \rangle$ is strongly depended on a choice of ΔE value and a set of crystallization temperature range. In fact, Mandelkern et al. used the i-PS data at temperature ranges in the vicinity of T_{cmax} and above T_{cmax} (here, the data set is defined as A), while Hoffman et al. used the data with much wider temperature range especially at the lower temperature regions (here, the data set is defined as B) It is true that the Arrhenius yields a good fit to the data set A, but an inferior fit to the data set B when 20.2 or $21 kcal/mol$ is used. However, the data set B can be fitted with reasonable high correlation coefficient when $50 kcal/mol$ of ΔE is used as listed in Table 19.1 with the other several polymers, which are crystallized in the wide temperatures range encompassed through T_{cmax}. I is clear in Table 19.1 that both expressions of WLF and Arrhenius can fit the data with the sufficiently high correlation coefficient. Figure 19.1 shows both fittings results for various polymers. The values of $\langle R \rangle$ for the WLF-type (two adjustable parameters) are slightly higher than those for the Arrhenius-type (a single adjustable parameter). In principal, the usage of two adjustable param- eters in the data analysis yields the better fit than that of a single parameter The discrepancy between them could be thought within an acceptable error. In addition to the best fitting, a setting of T_m^o is one of the most important factor since the best fitting affects very much by its selection.

It is worth to recheck which transport term equation is better describing the temperature dependence of the linear crystal growth rate. We employ a conventional activation energy expressed as $\Delta E_D/(T - T_o)$ for the molecular transport term and $\Delta E_K/(T_m^o - T)$ for the nucleation term. Thus, the overall activation energy of the crystal growth process is given by the following (19.2),

$$\Delta E_{all} = \frac{\Delta E_D}{T - T_o} + \frac{\Delta E_K}{T_m^o - T} \tag{19.2}$$

Figure 19.2 shows the temperature dependence of ΔE_{all} and the linear crystal growth rate for poly(ethylene succinate) ($PESU$). The growth rate shows a typical bell shape curve. The observed bell-shape curve locates in the temperature range of $T_{cmax}(1 \pm 0.15)$, whereas the growth rate in the external temperature regions of the curve is extremely slow. No crystal growth data in the vicinity of T_g has been reported in polymeric materials. However, the crystal growth rate is comparable to/or slower than that of the rate of molecular transport via reptation process in the vicinity of T_g [21]. It is interesting to note that some organic [39] and inorganic [40] materials can crystallize even below T_g. The overall activation energy increases significantly when the crystallization temperature reaches near to T_o and T_m^o. However, the temperature dependence of ΔE_{all} becomes very small in the vicinity of T_{cmax}. This indicates that Arrhenius law can be used sufficiently for the temperature dependence of molecular transport term. In conclusion, the activation energy for the molecular transport is expressed by either WLF or Arrhenius type in a wide crystallization temperature range encompassed through T_{cmax}.

19.3   Master Curve for Crystal Growth Rate

Maximum growth rate (G_{max}) can be observed by equating to zero the derivative of (19.1) with respect to the temperature. The relations for the Arrhenius expression [33] and for WLF expression [18] in the molecular transport term are formulated as follows. For Arrhenius expression in the molecular transport term,

$$G_{max} = G_o \exp\left[-\frac{\Delta E_{ARH}}{R\left(2T_{cmax} - T_m^o\right)} \right] \tag{19.3}$$

$$\ln\left(\frac{G}{G_{max}}\right) = \left[\ln\left(\frac{G_{max}}{G_o}\right)\right]\left[\frac{(1-X)^2}{X(A-X)}\right] \tag{19.4}$$

For WLF expression in the molecular transport term,

$$G_{max} = G_o \exp\left[-\left(\frac{\Delta E_{WLF}}{R(2T_{cmax} - T_m^o)}\right)\left(1 + \frac{T_o(T_m^o - T_o)}{(T_{cmax} - T_o)^2}\right) \right] \tag{19.5}$$

$$\ln\left(\frac{G}{G_{max}}\right) = \left[\ln\frac{G_{max}}{G_o}\right]\left[\frac{(1-X)^2}{X(A-X)}\right]$$
$$\times \left[\frac{X(1 - 2B + AB) - B(A-1)^2}{(X-B)(1 - 2B + AB)}\right] \tag{19.6}$$

where X is the reduced crystallization temperature of T/T_{cmax} and A and B are the ratios of T_m^o/T_{cmax} and T_o/T_{cmax}, respectively. In polymer crystallization data obtained over a wide range of temperature through a maximum crystal growth rate, data sets of G_{max}, T_{cmax} and T_m^o for many polymers are available in the referenced literature. G_{max}, T_{cmax}, and T_m^o show remarkable molecular weight dependence. For example, Fig. 19.3 shows the temperature dependence of the linear crystal growth rate for *PESU* with various molecular weights [18] Each molecular weight fraction shows bell shape crystal growth rate behavior The G_{max} increases with the molecular weight up to 3, 200 and then decreases T_{cmax} increases with molecular weight similar to the molecular weight dependence of T_m^o. The ratio of T_{cmax}/T_m^o is almost independent of the molecular weight, yielding the constant value of 0.83. In fact, values of T_{cmax}/T_m^o lie between 0.8 and 0.9 for most polymers [31-34], which are discussed in details in the later section. When the crystal growth data are plotted according to (19.4) (Arrhenius expression), the reduced growth rate (G/G_{max}) shows a linear relationship with the reduced temperature of the second term $(1 - X)^2/X(A - X)$ in the right hand side of (19.4) as shown in Fig. 19.4, indicating no molecular weight dependence in growth rate. The linear relation is true for other various polymers [20], such as poly(ethylene terephthalate) (*PET*), *i-PS*, nylon-6 (*N-6*), poly(tetramethyl-p-silphenylene siloxane) (*PTMPS*), and poly(L-lactide) (*PLLA*) as listed in Table 19.1.

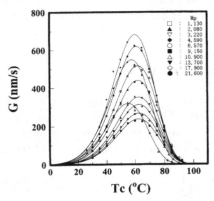

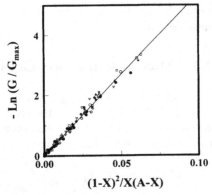

Fig. 19.3. Temperature dependence of linear crystal growth rate for *PESU* with various molecular weights. Indicated molecular weights in the figure are the peak molecular weight measured by GPC [18].

Fig. 19.4. Plots of the ratio of logarithm of the growth rate (G) for *PESU* at any temperature (T) to the value of G_{max} versus a reduced temperature $[(1 - X)^2/X(A - X)]$ based on Arrhenius expression of (19.4). Symbols in the figure are the same in those in Fig. 19.3. $X = T/T_{cmax}$, $A = T_m^o/T_{cmax}$.

WLF expression also shows the linear relationship plotting $\ln(G/G_{max})$ against the reduced temperature of $[(1 - X)^2/X(A - X)][(X(1 - 2B + AB) - B(A - 1)^2)/(X - B)(1 - 2B + AB)]$ in the right hand side of (19.6) as shown in Fig. 19.5, indicating also no molecular weight dependence. Here, parameter

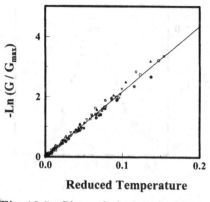

Fig. 19.5. Plots of the ratio of logarithm of the growth rate (G) for *PESU* at any temperature (T) to the value of G_{max} versus a reduced temperature $(1-X)^2/X(A-X)][(X(1-2B+AB)-B(A-1)^2)/(X-B)(1-2B+AB)]$ based on WLF expression of (19.15). Symbols in the figure are the same in those in Fig. 19.3. $X = T/T_{cmax}$, $A = T_m^o/T_{cmax}$, $B = T_o/T_{cmax}$

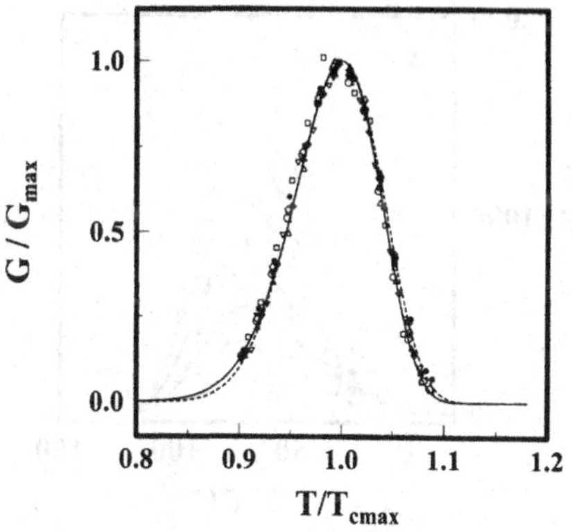

Fig. 19.6. Master curve of crystal growth rate for *PESU*. Symbols in the figure are the same in those in Fig. 19.3. Solid and dotted lines are calculated from equations 4 (Arrhenius) and 6 (WLF), respectively.

of A and B are assumed to be constants, which are discussed in details in the later section. The linear relation based on WLF expression was also true for other various polymers as listed in Table 19.1. These linear relationships indicate that ratios of G_{max}/G_o (the slope in Figs. 19.4 and 19.5) are independent of molecular weight. In other words, the molecular weight dependence of G_{max} is the mainly a consequence of the molecular weight dependence of G_o. It is much advantageous to use G_{max} rather than G_o for studying molecular weight dependence, because G_{max} can be observed by experimentally but not G_o.

Figure 19.6 shows the plots of the reduced crystal growth rate (G/G_{max}) against the reduced crystallization temperature of $(X = T/T_{cmax})$ for *PESU* 18]. Master curve of crystal growth rate is observed showing no molecular weight dependence of the crystal growth rate. The master curve can be fitted either by equation of (19.4) (Arrhenius ; solid line) or (19.6) (WLF ; dotted line). The master curve was also true for other polymers as listed in Table 19.1. These polymers were drawn in the master curve also as shown in Fig. 19.7 for *PTMPS* and in Fig. 19.8 for *i-PS*. However, each polymer gave a different value to the ratio of G_{max}/G_o indicating a material constant. The value of G_{max}/G_o could be dependent on characteristics of polymer molecules, such as molecular flexibility and/or crystallizability based on a chemical structure. For an example, Fig. 19.9 shows a relationship between $\ln(G_o/G_{max})$ and $R/\Delta S_m$, where ΔS_m is the entropy of fusion for polymer crystal. Therefore, the reduced growth rate (G/G_{max}) in each polymer should be normalized by the value of G_{max}/G_o. All

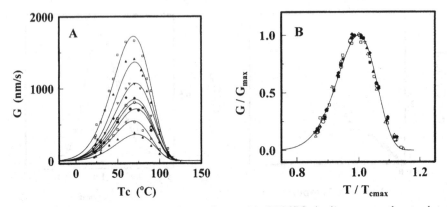

Fig. 19.7. Master curve of crystal growth rate for *PTMPS*. A : linear growth rate data, B : master curve. Solid and dotted lines are calculated from (19.4) (Arrhenius) and (19.6) (WLF), respectively. Data by Magill [6]

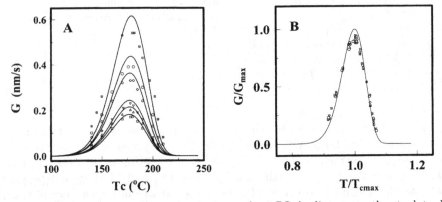

Fig. 19.8. Master curve of crystal growth rate for *i-PS*. A : linear growth rate data, B : master curve. Solid and dotted lines are calculated from equations (19.4) (Arrhenius) and (19.6) (WLF), respectively. Data by Lemstra *et al.* [16]

experimental crystal growth data (total numbers of data points are over 400) are drawn in to a single master curve as seen in Fig. 19.10.

19.4 Molecular Weight Dependence of Crystal Growth Rate

The influence of MW on the crystallization rate of polymers has been the subject of various studies [3-18,24-30]. For examples, the growth rate of *PTMPS* [6] decreases sharply with an increase in MW up to 40,000 and then becomes relatively insensitive at higher MW at a constant crystallization temperature. The growth rate of poly(ϵ-caprolactone) (*PCA*) [30] increases with an increase in MW and then decreases at a constant crystallization temperature, showing the maximum but the maximum depends on the crystallization temperature. The similar

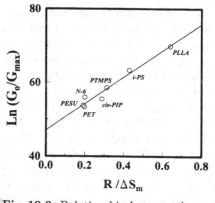

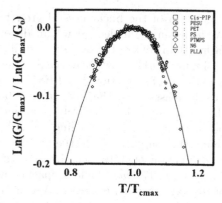

Fig. 19.9. Relationship between the ratio of G_o/G_{max} and the entropy of fusion (ΔS_m) for various polymers.

Fig. 19.10. Universal curve for crystal growth rate for various polymers. $\Diamond$:$PTMPS$ [6], $\triangle$:$Nylon$-6 [7], $\bigcirc$:PET [8], ∇:$PLLA$ [9,10], $\square$:cis-PIP [11,12], $\boxdot$:i-PS [15-17], $\odot$:$PESU$ [18]

maximum molecular weight dependence is found in poly(3,3-diethyl oxetane) [27] crystallized at the constant super cooling. The molecular weight dependence of the linear growth rate and the overall crystallization rate of poly(ethylene oxide) (PEO) [25] show a complex behavior. The growth rate decreases first with an increase in MW and then increases showing the minimum at MW about 4,000 and decreases again with MW yielding the maximum. The minimum crystallization could be related to the transition from extended chain crystallization (ECC) to folded chain crystallization (FCC); however, the minimum depends on the degree of super cooling. The growth rate at a given super cooling depends on MW.

Data existing in the literature for spherulite growth rate and the overall crystallization rate of several polymers have been analyzed as a function of MW expressed as a power law of MW^α. The large differences in value of α have been reported for a large number of polymers. Hoffman et al. [28] and Hikosaka et al. [29] have been reported for polyethylene (PE) that α lies in the range of -1.3 to -1.8 at a relatively small super cooling. At a relatively large super cooling, α shows nearly -0.5 [10,11,26,27]. The molecular weight dependence of G_o (preexponential factor of crystal growth rate) shows α to be -1 for PET [8] and -0.25 for i-PS [16]. The α value of -0.5 is also reported in the overall crystallization rate of $PESU$ [5]. The molecular weight dependence of the overall crystallization rate in PE [41,42] goes through a maximum and its location of the maximum is dependent on the super cooling. On the other hand, the linear growth rate is a function of MW but does not exhibit a maximum at a constant crystallization temperature in the case of $PTMPS$, whereas the overall crystallization rate shows the maximum [43]. Magill has pointed out that the overall crystallization rate will not provide detailed insight into the kinetics of polymer crystallization nor will universally applicable relationships, since the overall crystallization rate is a function of the linear crystal growth rate and the primary nucleation rate. We focus on the molecular weight dependence of the linear crystal growth rate rather than that of the overall crystallization rate.

Figure 19.11 shows a molecular weight dependence of a maximum crystal growth rate for fractionated $PESU$ samples [44]. The maximum crystal growth rate increases with the MW up to about $3,000$ and then decreases, yielding the maximum. This relatively small molecular weight of about $3,000$ seems to be an onset of a chain folding crystallization of $PESU$ at T_{cmax}. Figure 19.12 shows the long spacing of the fractionated samples crystallized at their T_{cmax} as a function of MW [44]. In the region below MW about 3,000 (the chain length is $ca.$ 17 nm), the long spacing increases with MW, indicating formation of the extended chain crystal. Above MW of $ca.$ 3,000, the long spacing keeps almost at a constant value of about 8 nm, however, data are scattered a little. The constant spacing indicates that polymer chain tends to form a folded chain crystal, yielding the onset of chain folding at about $ca.$ 17 nm for $PESU$ at T_{cmax}. This transition of crystal growth rate is just the opposite result in comparison to PEO crystallization rate at the constant super cooling, which shows the minimum [25]. The minimum crystallization of PEO occurs at about MW of 4,000, which is the onset of chain folding. The discrepancy between PEO and the present result of $PESU$ is not clear. However, the minimum point in PEO strongly depends on the degree of super cooling and its result is based on the study of the crystallization behavior over the limited ranges of crystallization temperature. On the other hand, $PESU$ can be crystallized over a wide temperature range that approaches to both the glass and melting transitions. This wide crystallization temperature is of primary importance for a reliable analysis of growth data based on kinetic nucleation theory. The molecular weight dependence of G_{max} in $PESU$ exhibits only one characteristic point, which is not influenced by the super cooling nor the crystallization temperature since G_{max} is a consequence of comparable weight between two factors of nucleation (ΔF) and transportation (ΔE). Therefore, the molecular weight dependence of G_{max} is much reliable data than that of G at a constant temperature or at a constant super cooling. The effect of super cooling on the molecular weight dependence of the crystal growth rate is discussed in details in the later section.

The logarithm of G_{max} increases linearly with the logarithm of MW in the lower molecular weight regions up to 3,000 with a slope of $ca.$ 1 and then decreases lineally with a slope of $ca.$ -0.5 in the higher molecular weight regions. The molecular weight dependence can be expressed as the following power law

$$G_{max} \propto G_o \propto MW^{\alpha} \tag{19.7}$$

where α is positive value of 1 in the lower molecular weight region for ECC and negative value of 0.5 in the higher molecular weight region for FCC. The detail in the value of α is discussed in the later section.

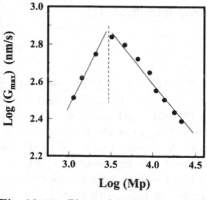

Fig. 19.11. Plots of logarithm of the maximum crystal growth rate against logarithm of molecular weight for *PESU*

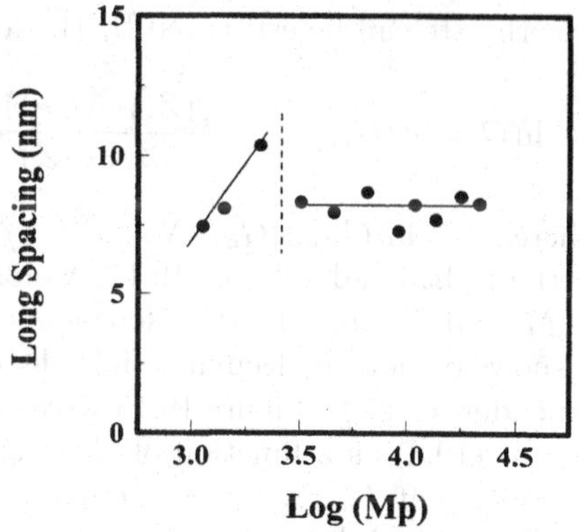

Fig. 19.12. Long spacing of *PESU* crystallized isothermally at T_{cmax} as a function of molecular weight

19.4.1   Effect of Super Cooling on the Molecular Weight Dependence of Growth Rate

Figure 19.13 shows molecular weight dependence of the crystal growth rate of *PESU* at a given super cooling ($\Delta T = T_m^o - T$). The value of α, the slope in Fig. 19.13, increases with decreasing in the super cooling. Such super cooling effect on α is shown in Fig. 19.14 plotted together with the reported reference data, such as *PTMPS* [6], trans-1,4-polyisoprene (*t-PIP*)[24], poly(1-butene) (*PB-1*)[26], *POX* [27] and *PE* [28,29]. α decreases clearly with an increase in ΔT. The effect of ΔT on the molecular weight dependence of the linear crystal

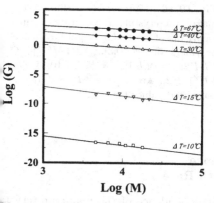

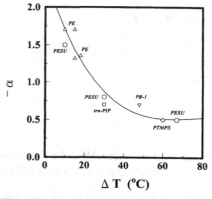

Fig. 19.13. Plots of logarithm of linear growth rate (G) at a given supercooling against logarithm of molecular weight for *PESU*

Fig. 19.14. Plots of the reported data for the exponent of power law (α) against supercooling for a various polymers. ◇ PTMPS[6], ○ PESU [18], □ t-PIP[24], ▽ PB-1[26], △ PE [28,29]

growth rate can be estimated by the following equation [20,44],

$$\ln G = \ln(G_{max}) + M\frac{(Z+N-1)^2}{Z(1-Z)} \qquad (19.8)$$

where $M = \ln(G_{max}/G_o)$, $N = T_{cmax}/T_m^o$, $Z = (T_m^o - T)/T_m^o$. The second term in the right hand side of (19.17) yields to zero at $T = T_{cmax}$. The parameters of M and N are almost independent of MW discussed previous section but Z shows obvious molecular weight dependence, because of the molecular weight dependence of T_m^o. Figure 19.15 shows plots of the second term in the right hand side of (19.8) as a function of Z. It is clear that the super cooling effect on c disappears at T_{cmax}. We can conclude that the molecular weight dependence of the linear crystal growth rate must be evaluated at T_{cmax}, otherwise α depends strongly on ΔT.

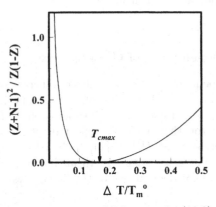

Fig. 19.15. Plots of Z function in (19.8) against $(Z = \Delta T/T_m^o)$

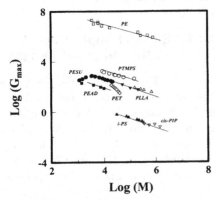

Fig. 19.16. Plots of logarithm of the maximum crystal growth rate against logarithm of molecular weight for various polymers. ■ *PEAD*[3], ○ *PTMPS*[6], ◇ *PET*[8], △ *PLLA*[9], ▼ *PLLA*[10], ▽ *cis-PIP*[11,12], ▲:*i-PS*[16], ●*PESU* [18], □ estimated values for *PE*[28]

19.4.2   Molecular Weight Dependence
of Maximum Crystal Growth Rate

Figure 19.16 shows the molecular weight dependence of the maximum crystal growth rate for various polymers, which data are re-plotted for polymers listed in Table 19.1. In the higher molecular weight regions, almost all polymers show a good linear relationship with a slope of about -0.5, except *PET*. Data of *PE* are in the limited small range of MW. Such limited molecular weight range could not be drawn reliable relationship. On the other hand, many polymer can be crystallized over only limited regions of their full growth rate curves. Even for those polymers, the maximum growth rate can be estimated by best fitting to the data by a linear least square method for (19.1) or (19.8). Thus calculated maximum crystal growth rate for PE using the referenced data [28] exhibits the similar molecular weight dependence with the slope of about -0.5 as plotted together in Fig. 19.16. According to the molecular weight dependence of the maximum growth rate and the long spacing for the fractionated $PESU$ samples, the positive slope of α with a value of about 1 can be associated with the mechanism of ECC and the negative slope of α with a value of 0.5 can be related to the mechanism of FCC. The details in molecular mechanism are not clear. We try to explain the molecular weight dependence on the basis of molecular levels. Transport process in polymer melts are sensitive to molecular weight and its distribution. DiMarizio et al. have proposed a negative value of 0.5 on the basis of sea-snake model [45]. In a model of the reptation proposed by de Gennes [46], a molecule is imagined to move only its worm-hole (or tube). In a sea-snake model, a molecule allows motion perpendicular to the worm-hole, as well as along it. The sea-snake model can be applied to crystallization from solution. Hoffman et al. have discussed the large differences between the experimental data and the theoretical prediction in α value by means of the presence of low molecular weight fractions, which acts as a solvent molecule in a general polymer sample [28]. Takayanagi has also proposed the value of -0.5 by means of activation entropy of molecular transport term [5]. That is, the free energy for the molecular transport term (ΔF_D) can be expressed by the enthalpy (ΔE_D) and entropy (ΔS_D) contributions. The entropy contribution to the molecular transport can be related to a change of molecular conformation during crystal growth. Such molecular transformation is associated with the entropy change from the molten state to crystal state, which entropy change can be proportional to $MW^{-0.5}$. However, these two molecular mechanisms cannot be applied to the lower molecular weight regions for ECC, where the slope gives a positive value of 1.

First, we have to consider the mechanism in molecular transport during crystal growth, such as a mass diffusion from the molten state to the crystal surface, adsorption of molecules on to the crystal surface, reel-in diffusion of molecules to the crystal surface from the melt, surface diffusion of admolecules on the crystal surface, migration and rearrangement with sliding diffusion prior to actual crystallographic attachment and etc. Figure 19.17 shows a schematic illustration for an activation process during polymer crystallization from the melt. Among these molecular diffusion processes, the rate determined process might be a process of surface diffusion of adsorbed molecules (admolecules) on the crystal surface, which may migrate on the crystal surface with the reptation mechanism.

In addition to the molecular diffusion on the crystal surface, a number of factors must be taken into account. One of those factors is the number of effective admolecules (N_{ad}) on the growth front, which are associated to form a critical nucleus. Lauritzen and Hoffman have introduced the probability parameter of admolecules on the crystal growth surface prior to surface nucleation but not

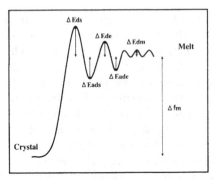

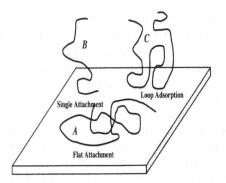

Fig. 19.17. Schematic illustration for activation process of molecular transport from molten state to crystal surface. ΔE_{dm} is activation energy of molecular mass transfer in the melt. ΔE_{ade} and ΔE_{de} are adsorption energy and diffusion energy of admolecule on the chain folding crystal surface, respectively. ΔE_{ads} and ΔE_{ds} are adsorption energy and diffusion energy of admolecue on the lateral surface of the crystal growth.

Fig. 19.18. Schematic illustration of the possible conformation of adsorbed polymer molecules: (A) flat multiple site attachment (pancake-like), (B) single point attachment (brush-like), (C) loop adsorption (loop-train)

considered the molecular weight dependence on it [47]. The polymer molecule goes down directly from the solution or sub-cooled melt on to the crystal growth front. At the crystal growth front adsorbed segments can migrate and rotate on the surface prior to crystallographic attachment. The admolecules will resemble a two-dimensional random walk with a number of contacts on the crystal surface. Polymer molecules can be adsorbed on the fold surface as well, however these admolecules are less important on the lateral crystal growth rate.

Taking into account of these factors into the polymer crystallization kinetics, the growth rate can be written as follow,

$$G \propto N_{ad}(T)D(T)\exp\left[-\frac{KT_m^o}{RT\Delta T}\right] \tag{19.9}$$

where $D(T)$ is a transport term for admolecules on the crystal growth from and $N_{ad}(T)$ is a number of effective admolecules prior to crystallographic attachment. The concept of admolecules on the growth front is not new but is well known for ordinary crystallization processes, such as for metals or small organic molecules, from vapor or solution [48]. Both $D(T)$ and $N_{ad}(T)$ depend strongly on MW. In fact, the molecular weight dependence of the adsorption of polymer molecules on the surface can be explained as MW^{β} [49]. The value of β is a function of the conformation of the adsorbate polymer molecules at the surface. The conformation of admolecules can be classified into following models as illustrated schematically in Fig. 19.18. (A) If all polymer segment lie in the surface, β equals to zero. These admolecules are difficult to move on the surface. (B) If all polymer molecules are attached to the surface by a single segment with other tail segments extending into the sub-cooled melt, β is unity. The single attachment occurs generally in a short chain molecule or a rigid chain molecule. The single attachment molecules can be migrated easily on the surface and might be independent of MW. (C) If all polymer molecules are attached with a number of contacts on the surface, β is 0.5. The point of attachment is called anchor segments or train segments. The molecule between the anchor segments is called loop adsorption or loop segments. These loop-train adsorptions of polymer molecules occur in polymers with relatively high molecular weight. Such loop-train molecules might be migrated on the surface based on the reptation mechanism without chain entanglements. Here, the molecular weight dependence of $D(T)$ can be simply expressed as $MW^{-\gamma}$.

Taking account of these factors into of (19.7) and (19.9), the parameter of α can be written as $\alpha = \beta - \gamma$. In the region of low molecular weight, polymer molecules may be adsorbed with a single attachment on the surface (β is unity) and the admolecules can be easily migrated on the surface without molecular weight dependence ($\gamma = 0$), then α yields to unity, which may be the case in ECC. On the other hand, high molecular weight polymers exhibit their molecular adsorption with a train-loop conformation (β equals to 0.5) and their molecular movements are controlled with the reptation mode ($\gamma = 1$), consequently α yields to -0.5, which may be the case for FCC.

19.5 Relationship Between T_m^o and T_{cmax}

It has been well known that a ratio of T_{cmax}/T_m^o is a nearly constant for a wide variety of materials [31-34] including metals, inorganic substances, organic compounds and polymers. Experimental results of T_{cmax}/T_m^o reported in the literature for many polymers have been summarized by Godovskii [31], where T_{cmax}/T_m^o lies about 5/6 as shown in Fig. 19.19. The ratio of T_{cmax}/T_m^o can be formulated from (19.1).

For Arrhenius expression in the molecular transport term [33,34],

$$\frac{T_{cmax}}{T_m^o} = \frac{C_{ARH}}{1 + C_{ARH}} \tag{19.10}$$

$$C_{ARH} = \sqrt{1 + \left(\frac{\Delta E_{ARH}}{K_{ARH}}\right)} \tag{19.11}$$

For WLF expression in the molecular transport term [18],

$$\frac{T_{cmax}}{T_m^o} = \frac{C_{WLF}}{1 + C_{WLF}} \tag{19.12}$$

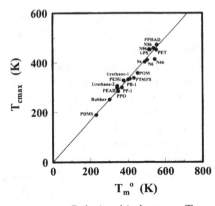

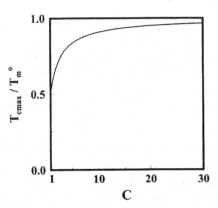

Fig. 19.19. Relationship between T_{cmax} and T_m^o for various polymers with the slope of 5/6.

Fig. 19.20. Plots of the ratio of T_{cmax}/T_m^o against C factor. C is a function of the ratio of the activation energy for molecular transport (ΔE) to the nucleation parameter (K).

$$C_{WLF} = \sqrt{1 + \left(\frac{\Delta E_{WLF}}{D K_{WLF}}\right)} \qquad (19.13)$$

$$D = \left(\frac{T_o}{T_{cmax}}\right)^2 \qquad (19.14)$$

Figure 19.20 shows the ratio of T_{cmax}/T_m^o vs. C (either C_{ARH} or C_{WLF}) according to (19.10) or (19.12). T_{cmax}/T_m^o increases with an increase in C value and its value saturates to 1. In general, the mean value of C lies in about 5 [33], yielding that T_{cmax}/T_m^o is about 5/6. T_m^o and T_{cmax} show strong molecular weight dependence, however the ratio of T_{cmax}/T_m^o yields almost a constant value of 0.83. The constant value of T_{cmax}/T_m^o indicates that the ratio of ($\Delta E/K$) remains invariant against MW. In fact, the values of ΔE and K show almost no molecular weight dependence [18,20], when ΔE and K were estimated by best fitting the growth data for (19.1). Figure 19.21 shows the plots of $\ln(G) + \Delta E/RT$ against $T_m^o/T\Delta T$ based on (19.1) for the fractionated $PESU$, whereas the straight line were calculated to get the best fit to the data by a linear least square procedure. A set of parameters of ΔE and K are obtained to that $\Delta E_{ARH}/K_{ARH}$ give 24 for Arrhenius expression and $\Delta E_{WLF}/K_{WLF}$ gives 3.8 for WLF expression, which values are almost independent of MW whereas G_o depends remarkably on MW. Therefore, these constant ratios with no molecular weight dependence give rise to the master curve of crystal growth rate in polymeric materials as discussed the former section. Here, the ratio of $\Delta E/K$ is rewritten as the following equation,

$$\frac{\Delta E}{K} = \frac{1}{n}\left(\frac{\bar{\sigma}}{\Delta H_m}\right)^{-2}\left(\frac{\Delta E}{\Delta H_m}\right) \qquad (19.15)$$

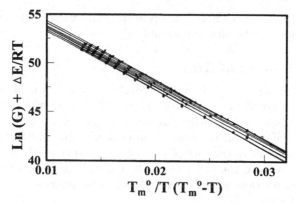

Fig. 19.21. Plots of $\ln(G)+\Delta E/RT$ against $T_m^o/T\Delta T$ for *PESU* with various molecular weight.

where $\bar{\sigma}$ is the mean molar surface free energy denoted as $\bar{\sigma} = \sqrt{b_o\sigma_e\sigma_s}$. It is interesting to note that the ratio of the mean molar surface energy and the heat of fusion $(\bar{\sigma}/\Delta H_m)$ and the ratio of the activation energy for molecular transport to the heat of fusion $(\Delta E/\Delta H_m)$ are generally found to be constant for a given type of crystalline materials. The constancy of these values gives rise to the constant value of T_{cmax}/T_m^o.

19.5.1   Ratio of ΔE and ΔH_m

The crystal growth data for many polymers have been analyzed by Mandelkern *et al.* according to possible nucleation mechanism and they have estimated ΔE by assuming Arrhenius expression in the molecular transport term [19]. Here ΔE is compared with the activation energy for viscous flow or for self-diffusion of polymer chains that may be related to the reptation mechanism. In other words, ΔE can be regarded as being equal to the activation energy for sliding diffusion of molecules on the growing crystal surface [50]. In the chain folding mechanism for polymer crystallization, a part of a polymer molecule adsorbs a preexisting crystal surface and subsequently some other parts of the molecule deposit adjacently on the same crystal surface. Translational shift along the chain axis needs the self-diffusion energy to generate crystalline packing finding a set of nearest lattice points. In fact, molecules cannot jump directly into the lattice points from the molten state. Such diffusion energy must be associated with the reptation energy of polymer molecules. For example, the reptation energy is close to $5.5\,Kcal/mol$ for n-paraffin [51]. This suggests that the ratio of $(\Delta E/\Delta H_m)$ yield about 5.6, which is satisfactory agreement with results for polymers [33]. In metals, ΔE is available experimentally from self-diffusion study both in the melt and in the solid. The ratio of $(\Delta E/\Delta H_m)$ in metals is also found to be fairly constant both for melts and solid [34,52] as shown in Fig. 19.22. These two thermodynamic energies are supposed to be a function of cohesive energy.

The ratio of the molar cohesive energy in polymer melts to the heat of fusion has been reported to be almost constant [32].

19.5.2   Ratio of $\bar{\sigma}$ and ΔH_m

An empirical relationship between $\bar{\sigma}$ and ΔH_m in most metals and molecular liquids has been noted by Turnbull [1]. Figure 19.23 shows the ratio of $(\bar{\sigma}/\Delta H_m)$ for metals is about 0.45, which value has been predicted to be $0.46 - 0.48$ by theoretical calculation on the basis of nearest neighbor cluster approximation [53]. Since a molecule in the surface may be regarded as partly in the liquid, one might expect to be of the order of one-half the ΔH_m. It has been also found that analogous relations exist between liquid vapor interfacial energies and the heat of vaporization [54].

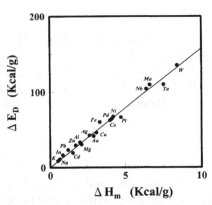

Fig. 19.22. Relationship between the activation energy for self-diffusion (ΔE_D) and the heat of fusion (ΔH_m) for various metals [34].

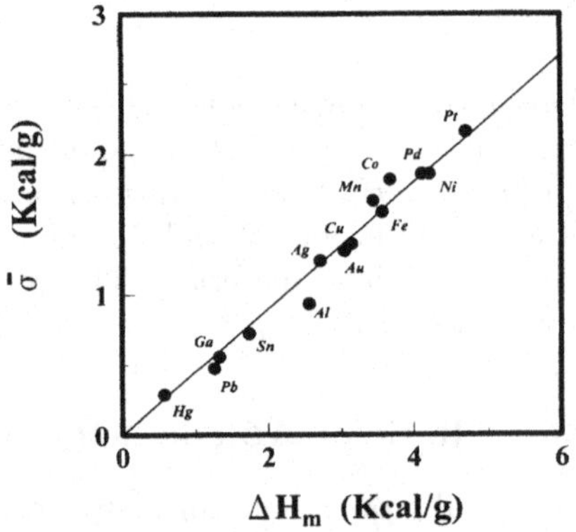

Fig. 19.23. Relationship between the mean molar surface energy $(\bar{\sigma})$ and the heat of fusion (ΔH_m) for various metals [52].

Figure 19.24 shows the chain length dependence of the ratio of $(\bar{\sigma}/\Delta H_m)$ for short chain paraffin to PE. For paraffin with about 30 carbon atoms, $(\bar{\sigma}/\Delta H_m)$ is about 0.05, which is comparatively small to that for PE [55] with the value of 0.3. The increase in $(\bar{\sigma}/\Delta H_m)$ with the chain length is associated with the onset of chain folding. Below carbon number of 30, $(\bar{\sigma}/\Delta H_m)$ increases with a decrease in the chain length, which indicates the surface roughness at the chain ends becomes to the contribution of surface entropy.

19.6   Relationship Among Thermodynamic Transition Temperatures

There are many other characteristic temperatures commonly observed on a mechanical relaxation spectrum, DSC and the other methods. Such characteristic

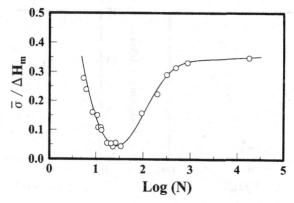

Fig. 19.24. Relationship between the ratio of the mean surface energy $(\bar{\sigma})$ and the heat of fusion (ΔH_m) and the logarithm of the number of carbon atoms in paraffin and PE [55].

are glass transition temperature (T_g), β-relaxation temperature (T_β) at which polymer segmental motion is ceased, liquid-liquid relaxation temperature (T_{ll}) and crystalline dispersion temperature $(T_{\alpha c})$. Also, there are second order transition (T_2) predicted by Gibbs-DiMarzio theory [56] and the temperature at which a free volume is zero (T_o), although hypothetical temperatures. T_o, however, is identical with T_2. Among these characteristics, there are several empirical rules. Constancy of the ratio of T_g/T_m^o is also widely known as Boyer-Beaman rule [35,36] and Boyer classified polymers into two groups as a symmetrical and an unsymmetrical. However, in an extensive study for 132 polymers, there is no sharp division between the ratios of T_g/T_m^o for symmetrical polymers and that for unsymmetrical polymers and the average value for whole polymers is about 2/3 [57] as shown in Fig. 19.25. It is interesting to note that the ratio of T_g/T_m^o is found to be about 2/3 not only for polymers but also for inorganic substances [36,58] and organic compounds [36,59]. According to WLF relationship, $T_o = T_g - f_g/\Delta\alpha$, where f_g is a free volume at T_g and $\Delta\alpha$ is the difference of the thermal expansivity between a glass and a super-cooled liquid [60]. This equation is rewritten as follows,

$$\frac{T_o}{T_g} = 1 - \frac{f_g}{\Delta\alpha T_g} \qquad (19.16)$$

The mean values of $\Delta\alpha T_g$ and f_g are approximated to be 0.1 and 0.025 [59], respectively. Thus the ratio of T_o/T_g yields to be 3/4, which coincides with the ratio reported by Gibbs-DiMarizio theory [56]. In other words, the free volume f_g is expressed as $\Delta\alpha T_g/4$. We can obtain the values for $\Delta\alpha$ and T_g in many referenced data. Then the histogram of $4f_g$ can be drawn as shown in Fig. 19.26. The distribution curve is considerably wide with the maximum value of 0.1 (the most probable value), which coincides, with the value reported by Shimha et al. [60]. Such wide distribution of the free volume is attributed to the wide distribution of the ratio of T_g/T_m^o as seen in Fig. 19.25.

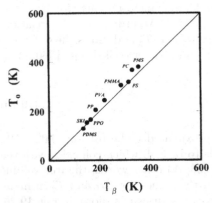

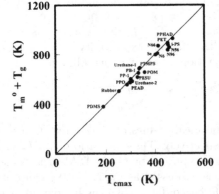

Fig. 19.25. Histogram for the ratio of T_g/T_m^o for various symmetrical ◯ and unsymmetrical (●) polymers [57].

Fig. 19.26. Histogram for the free volume defined as $\Delta \alpha T_g$ for various polymers [59].

It is also interesting to note that the ratio of T_β/T_g shows the constant value of 0.75 [61] and T_o/T_g is about $0.77 - 0.80$ [62,63]. It might be thought that T_β appears at almost the same temperature range with T_o. In fact, Fig. 19.27 shows the linear relationship between T_β and T_o [64]. Moreover, similar relationship between T_{cmax} and T_g is found as $T_{cmax}/T_g = 1.25 - 1.33$ [32,65]. In addition, the ratio of T_{ll}/T_g is also found to be about 1.2 [61], which value coincides with the value of $T_{\alpha c}/T_g$ or T_{cmax}/T_g. This might be suggested that there is some correlation between the molecular motion at T_{ll} in amorphous state and that at $T_{\alpha c}$ in crystalline state. The origin of these characteristic temperatures may be based on a similar mechanism associated with large molecular motions in the amorphous and crystalline states.

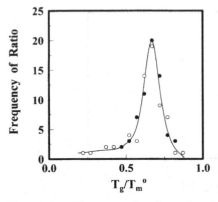

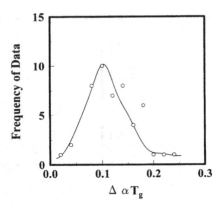

Fig. 19.27. Relationship between T_o and T_β for various polymers [64].

Fig. 19.28. Relationship between T_{cmax} and $T_m^o + T_g$ for various polymers with a slope of 2 [21].

According to the above empirical rules, $2T_{cmax} = T_g + T_m^o$ is found [59] as shown in Fig. 19.28. This means that T_{cmax} appears in the middle of T_m^o and T_g in polymeric materials. In addition, we can get interesting relationships as follows.

$$T_g - T_o = T_{cmax} - T_g = T_m^o - T_{cmax} \tag{19.17}$$

This means that each characteristic temperature appears at the same temperature intervals in polymer. These relationships come from totally empirical backgrounds, however those characteristic temperatures could be highly intercorrelated [66].

19.7 Conclusions

One of the characteristic behaviors of polymeric materials is the molecular weight dependence (MW) of many properties. The crystal growth rate (G) has a significant dependence on MW. However, its molecular weight dependence was strongly affected by the degree of supercooling when the growth rate was plotted against MW at a constant supercooling. Small supercooling gave the large molecular weight dependence. Therefore, we must employ the reference, standard or intrinsic states in a similar way for the MW dependence of intrinsic viscosity for polymer solution or zero shear viscosity for polymer melt.

1. The temperature dependence of G showed a bell shape with a maximum growth rate (G_{max}) that showed a remarkable MW dependence.
2. The G_{max} was formulated from a crystallization theory based on the molecular transport term expressed by Arrhenius or WLF expression.
3. The plots of the reduced growth rate (G/G_{max}) against the reduced temperature (T/T_{cmax}) showed a single master curve without MW dependence.
4. The plots of (G/G_{max}) normalized by (G_{max}/G_0) against (T/T_{cmax}) gave a single universal master curve for many polymers.
5. The G_{max} must be employed as a reference state in a similar way for the MW dependence of intrinsic viscosity for polymer solution or zero shear viscosity for polymer melt.
6. The MW dependence of G_{max} was expressed as the power law of $G_{max} \propto MW^{\alpha}$. The value of α was -0.5 for folded chain crystallization and α was unity for extended chain crystallization. The value of α was a function of the adsorption mechanism of polymer molecules on the growth front and its mechanism.
7. The ratio of T_{cmax}/T_m^0 was formulated from a crystallization theory and its value gave a constant value of $5/6$. T_{cmax} was correlated to many other thermodynamics transition temperatures.

References

[1] D.Turnbull, J.C.Fisher: J.Chem.Phys. **17**, 71 (1949)
[2] J.D.Hoffman, J.J.Weeks : J.Chem.Phys. **37**, 1723 (1962)
[3] M.Takayanagi, T.Yamashita: J.Polym.Sci. **22**, 552 (1956)
[4] K.Steiner, K.J.Lucas, K.Ueberreiter: Kolloid-Z.Z.Polym. **214**, 23 (1966)
[5] M.Takayanagi: J.Polym.Sci. **19**, 200 (1956)
[6] J.H.Magill: J.Appl.Phys. **35**, 3249 (1964)
[7] B.B.Burnett, W.F.McDevit: J.Appl.Phys. **28**, 1101 (1957)
[8] F. van Antwerpen, D.W.van Krevelen: J.Poly.Sci.,B,Polym.Phys. **10**, 2423 (1972)
[9] R.Vasanthakumari, A.J.Pennings: Polymer **24**, 175 (1983)
[10] T.Miyata, T.Masuko: Polymer **39**, 5515 (1998)
[11] P.J.Phillips, Vantansever, Macromolecules **20**, 2138 (1987)
[12] G.J.Rensch, P.J.Phillip, N.Vatanseyer: J.Polym.Sci. **B24**, 1943 (1986)
[13] H.Berghmans, E.Lanza,G.Smets: J.Polym.Sci. Phys.Ed. **11**, 87 (1973)
[14] A.J.Lovinger, D.D.Davis, F.J.Padden: Polymer **26**, 1595 (1985)
[15] J.Boon, G.Challa, D.W.van Krevelen: J.Poly.Sci. **A-2, 6**, 1791 (1968)
[16] P.J.Lemstra, J.Postma, G.Challa: Polymer **15**, 757 (1974)
[17] T.Suzuki, A.J.Kovacs: Polym.J. **1**, 82 (1970)
[18] S.Umemoto, N.Okui: Polymer **43**, 1423 (2002)
[19] L.Mandelkern, N.L.Jain, H.Kim: J.Polym.Sci. **A-2, 6**, 165 (1968)
[20] N.Okui: Polymer Bull. **23**, 111 (1990)
[21] N.Okui: Crystallization of Polymers, M.Dosiere Ed., Nato ASI Series **#405**, p-593 Kluwer Academic Pub., Netherlands (1993)
[22] A.Gandica, J.H.Magill: Polymer **13**, 595 (1972)
[23] J.H.Magill, H.M.Li, A.Gandica: J.Cryst.Growth **19**, 361 (1973)
[24] E.G.Lovering: J.Poly.Sci. **C-30**, 329 (1970)
[25] Y.K.Godovsky, G.L.Slonimsky, N.M.Garbar: J.Polym.Sci. **C-38**, 1-21 (1972)
[26] M.Cortazar, G.M.Guzman: Makromol.Chem. **183**, 721 (1982)
[27] M.A.Gomez, J.G.Fatou, A.Bello: Eur.Polym.J. **22**, 661 (1986)
[28] J.D.Hoffman, R.L.Miller: Macromolecules **21**, 3038 (1988)
[29] M.Okada, M.Nishi, M.Takahashi, H.Matsuda,A.Toda and M.Hikosaka: Polymer **39**, 4535 (1998)
[30] H.L.Chen, L.J.Li, W.C.Ouyang, J.C.Hwang, W.Y.Wong: Macromolecules **30** 1718 (1987)
[31] Y.K.Godovskii: Polym.Sci.USSR **11**, 2423 (1969)
[32] V.P.Privalko: Polymer **19**, 1019 (1978)
[33] N.Okui: Polymer J. **19**, 1309 (1987)
[34] N.Okui: J.Materils.Sci., **25**, 1623 (1990)
[35] R.F.Boyer: J.Appl.Phys. **25**, 825 (1954)
[36] R.G.Beaman: J.Polym.Sci. **9**, 470 (1953)
[37] J.D.Hoffman,T.Davis,J.I.Lauritzen: "Treaties on Solid State Chemistry", **Vol.3** Ed.,N.B.Hannay, Plenum Press, New York, p-555 (1976)
[38] E.Vrbanovici, H.A.Scheider, H.J.Cantow: J.Polym.Sci., B, Polym.Phys., **35**, 359 (1997)
[39] T.Hikima, Y.Adachi, M.Hanaya, M.Oguri: Physical Rev. **52**, 3900 (1995)
[40] Y.Abe, T.Arahori, A.Naruse: J.Am.Ceramic Soc. **59**, 487 (1976)
[41] E.Ergoz, J.G.Fatou, L.Madelkern: Macromolecules **5**, 147 (1972)
[42] J.G.Fatou, C.Marco, L.Mandelkern: Polymer **31**, 890 (1990)
[43] J.H.Magill: Polym.Lett. **6**, 853 (1968)

[44] S.Umemoto, N.Okui: J.Macromol.Sci. **B41**, 923 (2002)

[45] E.A.DiMarzio, C.M.Guttman, J.D.Hoffman: Faraday Disc.Chem.Soc. **68**, 210 (1979)

[46] P.G.de Gennes: J.Chem.Phys. **55**, 672, (1971)

[47] J.I.Lauritzen, J.D.Hoffman: J.Appl.Phys. **44**, 4340 (1973)

[48] W.A.Tiller: 'The Science of Crystallization', p-74 Cambridge University Press. (1991)

[49] Stablization of Colloidal Dispersions by Polymer Adsorption, Ed. T.Sato, R.Richard, Surface Science series **9**, p8 Marcel Dekker Inc. (1980)

[50] N.Hikosaka: Polymer **28**, 1257 (1987)

[51] J.H.Flynn: Polymer **23**, 1325 (1982)

[52] D.Turnbull: J.Appl.Phys. **21**, 1022 (1950)

[53] F.Spaepen: Acta Meta. **23**, 729 (1975)

[54] A.S.Skapski: J.Chem.Phys. **16**, 386 (1948)

[55] N.Okui: Polymer Crystal, Science One Point #4, Kyouritu Shuppan (1993)

[56] J.H.Gibbs, E.A.DiMarzio: J.Chem.Phys. **28**, 373 (1958)

[57] W.A.Lee, G.J.Knight: Br.Polym.J. **2**, 73 (1970)

[58] S.Sakka, J.D.Mackenzie: J.Non-Cryst.Solids. **6**, 145 (1971)

[59] D.W.van Krevelen: 'Properties of Polymers', Elsevier Sci.Pub.Co. Chap. 4, 16, (1976)

[60] R.Shimha, R.F.Boyer: J.Chem.Phys. **37**, 1003 (1962)

[61] R.F.Boyer: J.Polym.Sci.Symp. **50**, 189 (1975)

[62] A.B.Bestul, S.S.Chang: J.Chem.Phys. **40**, 3731 (1964)

[63] A.Adam, J.H.Gibbs: J.Chem.Phys. **43**, 139 (1965)

[64] V.A.Bershtein, V.M.Egrov: Differential Scanning Calorimetry of Polymers, Ellis Horwood Ltd. (1994)

[65] A.Utracki: J.Macromol.Sci.Phys. **B10**, 477 (1974)

[66] N.Okui: Polymer, **31**, 92 (1990)

20 Lamellar Growth in Melt-Crystallizing Polymers: Some Effect Related to a Nucleating Agent

Gaetano Di Marco and Marco Pieruccini

C.N.R. – Istituto per i Processi Chimico–Fisici, sez. Messina, Via La Farina 237, I-98123 Messina, Italy, pieruccini@me.cnr.it

Abstract. The influence of a nucleating agent on the free growth kinetics and on the final average thickness of lamellae grown from melts in isothermal conditions is considered. The growth kinetics in isotactic polypropylene, high density polyethylene and polyethylene oxide, is observed by means of calorimetry, whereas information on the long period and lamellar thickness at crystallization completed is drawn from SAXS measurements (on polyethylene oxide only). Both the observations are analyzed in terms of a decrease of the lamellar basal surface tension associated to the accumulation of the nucleant at the interface. This decrease is also discussed by means of a simple lattice model, where a finite compressibility of the system is accounted for. In light of the assumed mechanism, kinetic and morphological observations are interpreted consistently.

20.1 Introduction

The presence of additives (nucleating agents) is known to affect significantly both the growth kinetics and the morphology of structures developing in melt-crystallizing polymers. This circumstance is very important from the practical point of view, because it gives the possibility to control several properties of polymeric based materials in view of their application. On the other hand, still much has to be done in this area of research, both to improve the knowledge on the interplay between the nucleant and the polymer, and to help better understanding the complex mechanisms of the basic crystallization process as well.

This contribution is aimed at giving an heuristic interpretation of some experimental observations concerning the free growth kinetics and the morphology of laminar structures developing in moderately undercooled, nucleated polymeric melts. The following exposition substantially retraces the steps which led the authors to form a personal viewpoint on some aspects of the problem, on the basis of their own experience.

In this context, important suggestions came from the tentative application to nucleated systems, of a frequently used method suited to analyze the growth kinetics data from pure systems. A brief account of these first results is thus given in the following section. Within the framework of thermodynamics, a possible mechanism for the interpretation of the data is then presented, based on the idea that the nucleant molecules cause a local decrease of the basal surface tension. A lattice description of the basal region, which is outlined afterwards, supports the above argument and allows for the theoretical estimate of the effect in simple cases. Finally, SAXS data obtained from pure and nucleated samples of polyethylene oxide are reported, which confirm the predictions of the model.

20.2   Preliminary Analysis

In moderate undercooling conditions, the temperature dependence of the lamellar free growth velocity can be described by

$$v \propto \exp\{-\Delta G/kT\}, \tag{20.1}$$

with k the Boltzmann constant, T the temperature and ΔG, the secondary nucleation free enthalpy, given by

$$\Delta G = 4b_0 \sigma \sigma_e \frac{v_c T_m}{H_f \Delta T}, \tag{20.2}$$

where b_0 is the stem diameter, H_f/v_c is the enthalpy of fusion per unit volume, T_m is the equilibrium melting point and $\Delta T \equiv T_m - T$; σ and σ_e are the lateral and basal contributions to the surface tension [1] (see Fig. 20.3).

An estimate of $\sigma\sigma_e$, assumed to be independent of T, can be usually obtained by best fitting with (20.1) and (20.2) the experimental data of v as a function of T. This has been done with samples of isotactic polypropylene (iPP) nucleated with different amounts of synthetic indigo [2]. Isothermal crystallization

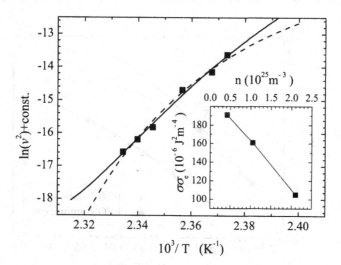

Fig. 20.1. $\ln(v^2)$ as a function of T^{-1} for an iPP sample with number density $n \cong 2 \times 10^{25}\mathrm{m}^{-3}$ of indigo (i.e. $\approx 1\%$ wt); the dashed line best fits the data assuming $\sigma\sigma_e$ independent of T; the solid line is obtained after inserting (20.5) into (20.2). The inset shows the values of $\sigma\sigma_e$ worked out from fittings like the dashed line in the main frame of this figure.

observed with a calorimeter allowed for the determination of v vs. T from the initial slope of the thermograms which, at suitably small to moderate ΔT's, were characterized by a linear behavior in the early stages of growth. The results are reported in the main frame of Fig. 20.1 for one of the samples. A rather marked dependence of the worked out *apparent* $\sigma\sigma_e$ on the nucleant number density, n is shown in the inset. On the other hand, there is a qualitative discrepancy between the adopted best fitting $v(T)$ and the data, which however is very weak. This discrepancy is systematic, and rather evident in Fig. 20.2 for the case of indigo-nucleated polyethylene oxide (PEO) [4]; nucleated high density polyethylene (HDPE) shows the same behavior [3]. In all these cases the nucleant species did not co-crystallize.

No dependence whatsoever of any of the parameters in (20.2) on just the nucleant number density n, can improve the situation, because it is inherent to a somewhat different T-dependence of ΔG [2][4]. Moreover, the possibility that either a regime I $\rightarrow$ regime II transition, or effects of chain segments' transport are being observed, has to be excluded [4]. Alternatively, the indications emerging from this preliminary analysis can be accepted, just as purely qualitative suggestions, as a basis for the development of a more adequate model.

20.3 Thermodynamic Viewpoint

In the spirit of a pure thermodynamic approach, i.e. disregarding the underlying microscopic mechanisms, deviations such as those pointed out by Figs. 20.1 and 20.2 are viewed as an effect of a *local* decrease of σ_e, due to the nucleant molecules

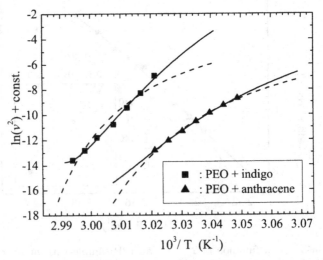

Fig. 20.2. $\ln(v^2)$ as a function of T^{-1} for PEO samples containing a number density $n \cong 2.6 \times 10^{25}\,\mathrm{m}^{-3}$ of either indigo or anthracene. Dashed lines are best fits obtained assuming $\sigma\sigma_e$ independent of T; solid lines are obtained using (20.5).

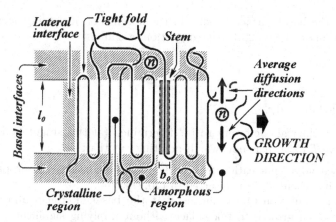

Fig. 20.3. Schematic representation of the chain configuration in a region of a growing lamella (the lamellar plane is perpendicular to the sheet). The nucleant molecules are indicated by circles labelled with an n.

which must leave the regions where crystallization is about to take place. More or less implicitly, this amounts to say that:

1. Secondary nucleation is still considered the controlling process of lamellar growth.
2. Nucleant molecules are not all involved in primary nucleation only, but in the growth process as well. In fact, in the opposite situation, free growth wouldn't be possible with the actual nucleant concentrations; moreover, extended lamellae wouldn't form, contrary also to the SAXS results below.
3. The nucleant molecules leaving the crystallizing regions remain close to the folding planes to keep the local σ_e low (see Fig. 20.3).
4. Making the hypothesis that the largest contribution to the change of σ_e comes from nucleant molecules which where previously in the crystallizing regions, means assuming that either lamellae are suitably thicker than the amorphous interlamellar regions, or the nucleant concentration in the amorphous bulk remains substantially unaltered.

In order to give an explicit expression for σ_e to be inserted in (20.2), consider the average number of nucleant molecules per stem, φ, which may reach *one* of the two basal interfaces upon secondary nucleation. This is given by:

$$\varphi = \frac{1}{2} n b_0^2 l, \tag{20.3}$$

where l is the stem length, to be approximated with the expression appropriate for pure systems (see 3rd item below):

$$l \approx l_0 \equiv 2\sigma_{e0} \frac{v_c T_m}{H_f \Delta T}, \tag{20.4}$$

being σ_{e0} the basal surface tension in the absence of nucleating agents. Then, taking the inset of Fig. 20.1 as a suggestion, assume:

$$\sigma_e = \sigma_{e0}(1 + \Gamma_e\varphi),$$ (20.5)

with

$$\Gamma_e = [d\ln(\sigma_e)/d\varphi]_{\varphi=0} < 0$$ (20.6)

a phenomenological coefficient to be determined (at this stage) by best fitting to the v vs. T data. It is now worth pointing out that:

1. The secondary nucleation free enthalpy ΔG maintains the property of extensivity when (20.5) is inserted in (20.2) [3] [6].
2. The accumulation of the nucleant at the basal interfaces allows for two-dimensional growth to take place without implying significant continuous increase of its concentration in the remaining *bulk* amorphous phase (compare "SAXS" and "Cal." columns in Table 20.2).
3. Equations (20.3)-(20.6) introduce a correction in the T *functional dependence* of ΔG, which is proportional to n; this is in fact a first order term in a Taylor expansion of $\Delta G \equiv \Delta G(n)$, consistent with the circumstance that the qualitative discrepancy pointed out above is weak.

The best fitting curves for nucleated iPP and PEO obtained after inserting (20.5) into (20.2) are shown in Figs. 20.1 and 20.2 as solid lines; the situation for HDPE is similar [3]. Table 20.1 reports the values of $\sigma\sigma_{e0}$ and of the (T-independent) coefficient

$$a \equiv \Gamma_e\varphi\Delta T$$ (20.7)

determined by best fitting.

It may be interesting to point out that the average $< \sigma\sigma_e >$ as estimated by (20.1) and (20.2) assuming $\sigma\sigma_e$ independent of T, differs in general significantly from $\sigma < \sigma_e >$. As an example consider iPP with 1% wt indigo for which $< \sigma\sigma_e > \cong 10^{-4}$ J^{2}m^{-2} and, from Table 20.1, $\sigma < \sigma_e > \cong \sigma\sigma_{e0}\left(1 + a_{fit}\Delta T^{-1}\right) \cong 1.6 \times 10^{-4}$ J^{2}m^{-2}.

Before proceeding further, notice that an entropy contribution S_n now appears to be associated to the formation of a basal surface of extension A, due to the presence of the nucleant. This follows from the general relationship [8]

$$\left(\frac{\partial S_n}{\partial A}\right)_T = -\left[\frac{\partial}{\partial T}(\sigma_e - \sigma_{e0})\right]_A.$$ (20.8)

Note, by inserting (20.3)-(20.5) into (20.8), that S_n increases with the interfacial extension only if $\Gamma_e < 0$, i.e. if σ_e is lowered by the nucleant as found (Table 20.1).

In conclusion, qualitative discrepancies such as those pointed out by Figs. 20.1 and 20.2, and the decreasing of the apparent $\sigma\sigma_e$ upon nucleant addition (e.g. the inset of Fig. 20.1 with the due interpretation), are in fact interpreted through (20.5) as the manifestation of the same mechanism: the nucleant concentration at the interface can be increased by either increasing n or decreasing ΔT.

Table 20.1. Nucleant number density n, equilibrium melting point T_m, best fit values for a and $\sigma\sigma_{e0}$, theoretical estimate of a, and average undercooling $\overline{\Delta T}$ at which the data are collected, for nucleated iPP, PE and PEO

System	n	T_m	$\sigma\sigma_{e0}$	a_{fit}	$a_{lattice}$	$\overline{\Delta T}$
	$(10^{25}\ \mathrm{m^{-3}})$	(K)	$(10^{-6}\mathrm{J^2m^{-4}})$	(K)	(K)	(K)
iPP/indigo [a]	2.1	448.5	220	-6.3	—	24
PE/indigo [b]	2.2	414	1900	-1.2	-1.12	15
PEO/indigo [c]	2.6	338	212	-1.83	-1.7	5.5
PEO/anthracene [c]	2.6	338	198	-2.09	-1.7	8.5

[a] [6]; see also [5] for T_m.
[b] [3], [4]; compare [7] for the value of $\sigma\sigma_e$.
[c] [4].

20.4 Lattice Model

The arguments of the preceding section find support in the lattice model of the basal transition region which is briefly described below (for a detailed account refer to [4]). For simple systems the model also allows for reasonable estimates of Γ_e.

The starting point is a cubic lattice model of the basal region which has been initially developed for pure, incompressible systems [9][10], and then improved to account for the density gradient's contribution to σ_e [11]. With this generalization, the chain segments' specific volume contraction in the basal interphase region, where crystalline order is still not established, is also accounted as a contribution to the overall σ_e, together with folding energy and configurational entropy change. This aspect can be important as a route through which the nucleating agent can be effective.

Although the model has been developed for pure systems, it can be easily adapted to account for the nucleant. To this aim, a fundamental role is played by item 3, preceding (20.3) above, which is now reformulated more clearly. If it is true that the nucleant molecules decrease σ_e locally, it means that these molecules tend to remain confined in the interfacial region, because their diffusion towards the amorphous bulk would require an additional energy to restore the local surface tension to the value σ_{e0}. Thus the calculation of the surface tension can be tentatively carried out assuming the nucleant molecules as being placed at fixed positions in the interfacial region (close to the folding plane). If the resulting σ_e is sufficiently lower than σ_{e0}, then the estimate is reliable. This in fact has been shown to be the case [4] for HDPE and PEO.

The surface tension σ_e is calculated from the change in the appropriate thermodynamic potential of a region undergoing the transition from the state of amorphous *bulk* to the state of amorphous *interphase*.

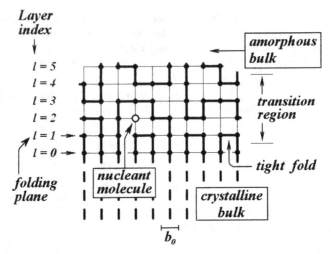

Fig. 20.4. Representation of part of a basal transition region in the cubic lattice for a compressible system in the presence of the nucleant. The l's label the lattice layers which are perpendicular to the sheet.

The chain configurational statistics is modelled by accommodating a virtually infinite linear sequence of freely joined segments in a cubic lattice (coordination $z = 6$); some sites are left empty to account for a finite compressibility (see Fig. 20.4).

Before crystallization takes place the fraction of the lattice sites occupied by the segments has everywhere the value $\phi_\infty = \rho_{\text{amorph.}}/\rho_{\text{cryst.}} < 1$, where ρ stands for the density, and bonds connecting segments are oriented randomly. This means that the conditional probabilities p_l^ν that a segment be placed in a neighboring site in the $l + \nu$ layer ($\nu = 0, \pm1$), given that the preceding one is in layer l, are all equal to $1/6$.

In the semicrystalline state, the situation is depicted in Fig. 20.4 (where the layer numbering is explicitly reported): the bond orientational distribution is now determined by the fact that i) $p_0^0 = 0$ and $p_0^{\pm1} = 1/2$ (i.e. bonds are all aligned along the same direction within the bulk crystalline phase) and ii) pairs of mutually bonded segments, both in $l = 1$, cannot be directly bonded with others placed in either $l = 1$ or $l = 2$ layers. Moreover, the fractional site occupation, ϕ_l, rises to unity in layers $l = 0$ and $l = 1$ (while remaining ϕ_∞ in the amorphous bulk), so the volume occupied by a given number of segments at the interface has changed against an elastic tension, τ, related to the segment chemical potential in the amorphous bulk, μ_∞, by the Gibbs-Duhem relation $\tau = -\mu_\infty \phi_\infty b_0^{-3}$, where b_0^3 is the volume of a lattice cell [4] [11].

In order to account for the confinement of the nucleant at the interface, the calculation of the configurational entropy $S \equiv S(\phi_l, p_l^\nu)$ is performed by first placing its molecules in the $l = 2$ layer (the closest one to the folding plane), and only afterwards counting the number of ways the chain can be accommodated in the sites left empty (with this approximation the nucleant confinement time is virtually infinite).

The relevant energy is in general expressed as a sum of three contributions, i.e. $E = E_{p0} + E_n + E_f$, where E_{p0} is a mixing energy proportional to a segment/vacuum interaction parameter χ_{p0}; E_n is another mixing contribution proportional to φ and to the quantity $\chi_s - \chi_{p0}$, with χ_s a "solvation" parameter related to the energy change of an isolated nucleant molecule when it is carried from vacuum into the polymer; $E_f \propto p_1^0 \mathcal{E}$ is the folding contribution, with $\mathcal{E}$ the energy of one tight fold in units of kT.

The work $\tau \Delta V$ associated to the volume change $\Delta V \propto (b_0^3/\phi_\infty) \sum_{l \geq 1}(\phi_l - \phi_\infty)$, must be finally accounted for.

The calculation of the surface tension is done by minimizing the Gibbs potential (associated to the entropy S, energy E and volume change above) under constraints associated to chain connectivity, namely, that given a segment placed in a site whatsoever, both the preceding and the subsequent ones can only be accommodated as nearest neighbors. These constraints allow for the interfacial region at equilibrium to be described by just two functions of the layer index, which are solutions of an appropriate boundary value problem. A possible choice is represented by the two functions of l: ϕ_l and p_l^0; their interfacial profiles are shown in Fig. 20.5 for the case of PEO at various φ and χ_s (for the calculation it is assumed $\sigma_{e0} = 2 \times 10^{-2}$ J m^{-2} [4][5]). The changes in the ϕ_l profile indicate that the contribution to σ_e associated to the specific volume variation of interfacial segments, decreases as an effect of the nucleant. This mechanism is accompanied by a reduction of the average bond orientational entropy in layer 2. This can be seen from the inset of Fig. 20.5, where it is shown that the pres-

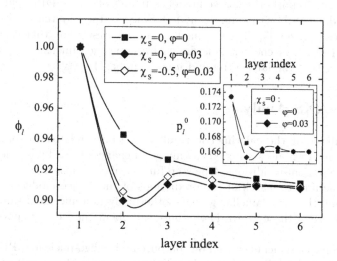

Fig. 20.5. Fractional site occupation ϕ_l (relative density) in semicrystalline PEO, for different values of χ_s and φ in the $l = 2$ layer. The inset shows the corresponding p_l^0 profiles for $\chi_s = 0$.

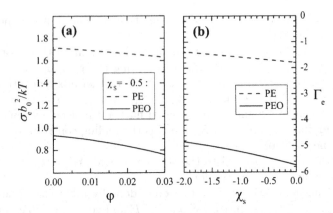

Fig. 20.6. Normalized interfacial free enthalpy σ_e as a function of φ for PE and PEO with a value $\chi_s = -0.5$ of the solvation parameter **(a)**; dependence of the derivative Γ_e on χ_s for the same polymers **(b)**.

ence of the nucleant doesn't affect the number of tight folds significantly (i.e. p_1^0), contrary to p_2^0. This means that the number of segment sequences emerging from the fold plane ($l = 1$) are forced not to bend in the $l = 2$ layer (the same topological constraint holds for sequences approaching the crystalline bulk from $l = 3$).

The overall effect on σ_e is reported in Fig. 20.6a for the cases of PE and PEO, with a parameter $\chi_s = -0.5$ corresponding to a ~ 1.3 kJ/mol solvation energy. Figure 20.6b shows that the derivative Γ_e is weakly dependent on χ_s.

Table 20.1 reports the lattice model estimates of a for HDPE and PEO (iPP is poorly described by this simple model). Since for these systems b_0 is ~ 4 Å and ~ 4.6 Å respectively [5], it is assumed, as an underestimate, that each molecule of indigo (or anthracene, which is similar in shape and dimensions) occupies two adjacent cells in the lattice (remind n is a number density).

20.5   SAXS Analysis

Up to this point the calorimetric results have been related to some assumed mechanism which at this stage has only good chances to really be the dominant one in the present conditions. One step further ahead consists in an alternative direct experimental check of the predictions of the lattice model. If (20.5) and (20.6) hold, then the lamellae grown isothermally in a nucleated sample should be thinner than those grown in the pure sample at the same undercooling [see (20.4)].

SAXS measurements carried out on pure and indigo-nucleated PEO indicate that this is in fact the case [12]. Here follows some further result on the same system (n as in the case of Fig. 20.2). The analysis has been carried out following [13] and [14] for the direct evaluation of the correlation function $K(x)$ and its second derivative $K''(x)$ [1] Especially in nucleated samples, however, PEO is characterized by a non negligible forward scattering contribution arising from local density fluctuations, which must be duly subtracted before Fourier analysis [this circumstance may be possibly related to an enhanced primary nucleation rate; cf. item 2 in the list preceding (20.3)].

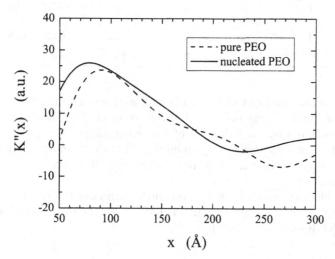

Fig. 20.7. Interface distance distribution function, $K''(x)$, obtained from SAXS analysis on pure and indigo nucleated PEO crystallized with an undercooling $\Delta T = 10$ K.

Figure 20.7 reports the second derivatives of the correlation function of pure and nucleated PEO crystallized at the same undercooling ($\Delta T = 10$ K). The worked out data are listed in Table 20.2, where the values of the ratio

$$\frac{\delta l}{l_0} \equiv \frac{l_0 - l}{l_0} = -\frac{a}{\Delta T} \tag{20.9}$$

predicted also by a as derived from calorimetry (Fig. 20.2 and Table 20.1), are reported for a comparison.

The lattice results in last column of Table 20.2 have been calculated assuming $\sigma_{e0} \cong 2 \times 10^{-2}$ J m^{-2} for consistency with [4], Table 20.1, and Figs. 20.5 and 20.6 above. On the other hand, the l_0 values measured by SAXS indicate $\sigma_{e0} \approx 4 \times 10^{-2}$ J m^{-2} as the correct value (see [7]). This inconsistency, however, is irrelevant: recalculation with the new value of σ_{e0} leads to $a \cong -1.5$ K, which is very close to $a_{lattice} \cong -1.7$ K in Table 20.1. So the change of Γ_e [(20.6)] with σ_{e0} is almost balanced by the change in the stem volume via the fold length (20.3) and (20.4); this means that $d\sigma_e/d\varphi$ is weakly dependent on σ_{e0}. In fact, the number of tight folds (i.e. p_1^0) is practically independent of φ in the range of

---

[1] See also Sect. 4.2.2 in the contribution by Al-Hussein and Strobl on page 50.

Table 20.2. SAXS results for long period L, interlamellar d_a, and lamellar thickness for pure (l_0) and nucleated (l) PEO, all in Å , at different undercoolings ΔT (with T_m and n as in Table 20.1). The estimates of $\delta l/l_0$ from SAXS, calorimetry (Cal.) and lattice calculations are also reported.

ΔT	pure			nucl.			$\delta l/l_0$		
	L	d_a	l_0	L	d_a	l	SAXS	Cal.	Lattice
7	315	95	220	245	95	150	0.32	0.26	0.24
10	270	90	180	230	80	150	0.17	0.18	0.17

interest (remind the inset of Fig. 20.5) and, as discussed in [12], $d\sigma_e/d\varphi$ mainly consists of the following two contributions connected to steric hindrance. The first is associated to the reduction of the orientational entropy of chain bonds connecting segments neighboring an interfacial nucleant molecule. The second is a local decrease of the $\tau\Delta V$ contribution (see Fig. 20.5) and mainly depends on $\rho_{amorph.}/\rho_{cryst.}.$

In conclusion, calorimetric analysis underestimates the ratio $\delta l/l_0$ obtained with SAXS by at most 30 %, which is however in the direction of mutual consistency [cf. item 2 after (20.6)].

20.6   Concluding Remarks

The idea that the surface tension characterizing crystal-to-amorphous interfaces can in general be affected by "impurities" is well established, for example as to what concernes primary nucleation [13]. Upon considering the implications that this hypothesis carries in, the same concept has been shown to be applicable to secondary nucleation in order to describe some effects of a nucleating agent on both growth kinetics and lamellar morphology, resulting in the possibility to relate these two aspects of the crystallization process to one and the same mechanism.

However, the physics involved in the nucleant-assisted crystallization of polymers is complex, and branches in a large number of seemingly different situations. So, the present contribution sheds only a little light on the rich phenomenology characterizing this area of research. Wishing for a more detailed description of these processes also follows from the coarse grained character of the above treatment consisting, besides the absence of microscopic modeling, on the number of oversimplifications which have been made.

On the other hand, the above treatment gives some hints for further investigations, as for instance the extent to which geometric aspects (i.e. the steric hindrance of the nucleant molecules, or the weak Γ_e vs. χ_s dependence in Fig. 20.6b) dominate. From a theoretical point of view, a first step in this direction would be to reformulate a more elaborate model where the nucleant is let free to move in the lattice, thus releasing the above item 3 [preceding (20.3)] as an hypothesis

As a last comment, the use of equilibrium thermodynamics concepts eludes the process of nucleant diffusion towards the interfacial regions which are about to form: a basic aspect deserving thorough investigations.

Acknowledgement

The authors are indebited to Prof. Roberto Triolo for performing SAXS on PEO samples at ORNL.

References

1. E.J. Donth: *Relaxation and Thermodynamics in Polymers* (Akademie, Berlin 1992)
2. M. Pieruccini, G. Di Marco, M. Lanza: J. Appl. Phys. **80** 3, 1851 (1996)
3. G. Di Marco, M. Pieruccini: Il Nuovo Cimento **20 D** (12 bis) , 2479 (1998)
4. F. Aliotta, G. Di Marco, M. Pieruccini: Physica A **298**, 266 (2001)
5. Yu.K. Godowsky, G.L. Slonimsky: J. Polym. Sci. (Phys. Ed.) **12**, 1053 (1974)
6. M. Pieruccini: J. Appl. Phys. **81** 7, 2995 (1997)
7. K. Mezghani, P.J. Phillips: 'Crystallization Kinetics of Polymers'. In: *Physical Properties of Polymers Handbook*, ed. by James E. Mark (AIP Press 1996) pp. 417–425
8. L.D. Landau, E.M. Lifsits, L.P. Pitaevskij: *Fisica Statistica (parte prima)* (Editori Riuniti, 1978; Edizioni MIR, Moscow, Russia)
9. P.J. Flory, D.Y. Yoon, K.A. Dill: Macromolecules **17**, 862 (1984)
10. S.K. Kumar, D.Y. Yoon: Macromolecules **22**, 3458 (1989)
11. F. Aliotta, M. Pieruccini: Physica A **287**, 105 (2000)
12. F. Aliotta, G. Di Marco, R. Ober and M. Pieruccini: submitted
13. G. Strobl: *The Physics of Polymers* (Springer, Berlin, Heidelberg 1996)
14. J. Schmidtke, G. Strobl, T. Thurn-Albrecht: Macromolecules **30**, 5804 (1997)

21 Outlook and Open Questions: A Personal View

Günter Reiter and Jens-Uwe Sommer

Institut de Chimie des Surfaces et Interfaces CNRS et Université de Haute Alsace
Mulhouse, 15 rue Jean Starcky, F–68057 Mulhouse, France g.reiter@uha.fr

Why has crystallization of polymers interested so many researchers for so long time and why is nonetheless our understanding of the underlying fundamenta processes still at a rather rudimentary level? This is the more surprising as the subject of interest can be defined in an extremely simple way: Polymer crystallization concerns the transition from a randomly coiled to a perfectly ordered state. Probably, the key difficulty of polymer crystallization arises from the fact that the building blocks of a polymer crystal, the monomers, cannot move independently due to their connectivity to many other monomers forming a polymer chain.

The most crucial consequence of connectivity is that the formation of a per- fectly ordered state would need extremely long times, sometimes much more than the duration of any feasible experiment. Thus, at finite crystal growth velocities only imperfect crystals consisting of folded molecules are formed. Because better ordered states always represent a lower free energy these imperfect crystals can only be metastable. The sometimes quite significant lifetimes of such metastable states represents a major complication in the study of polymer crystals. More over, such metastable states are not always well defined and their characteriza- tion is by far non trivial. Although it is possible to reproduce similar metastable states, it is clear that the thermodynamic pathway (sample history) is crucial Typically, reaching a given crystallization or annealing temperature along two different pathways may result in two distinctly different crystalline states. In addition, the representative features of these states (and polymer crystals in general) may depend on the lengthscale of observation and the experimenta techniques used. Summarizing these observations, one may conclude that there exists no single and well-defined state of a polymer crystal.

Starting at the molecular level, observations in direct or reciprocal space show that a common feature of polymer crystals is a sheet-like lamellar structure which is characterized by a typical (and rather uniform) thickness. Interestingly with the exception of comparatively short non-entangled polymers, the lamella thickness is typically much less than the length of the fully extended polymer chain. In contrast, lamellae are usually much wider (up to many thousand times than thick. In such cases, thermodynamic properties like the melting tempera ture depend on the small length scale (the lamellar thickness) while the latera extension of such lamellar crystals is not much of concern. However, if this size becomes comparable to the thickness of the lamella, it cannot be neglected. In general, the lateral extension of lamellae depends on growth rate controlled by the crystallization temperature. In addition, the lateral extension may also be different in the directions perpendicular and parallel to the growth front, yielding besides the lamellar thickness two additional lengthscales. The complexity and the superposition of lengthscales is particularly visible in spherulitic structures at larger sizes but also in the finger patterns obtained by crystallization of quasi two-dimensional thin films.

On a macroscopic level, the morphology resulting from crystalline domains embedded in amorphous regions certainly affects (mechanical) properties like toughness or elasticity. In fact, a large amount of application-oriented work is devoted to the understanding of how the organization of polymers at the microscopic level (crystalline micro-structure) controls and may help to improve mechanical properties of polymeric materials. From this point of view, it is thus necessary to relate processes at a molecular level, leading to various structures at multiple lengthscales, with the macroscopically observable mechanical and thermodynamic properties. Unfortunately, this is usually not possible in an unambiguous way. The main difficulty might be due to the fact that it is not sufficient to characterize polymeric semicrystalline materials by (static) parameters like lamellar thickness or nucleation density. In addition, besides other factors, thermal history and processing conditions in general play a significant role.

In our opinion, one has to distinguish clearly between at least three major phenomena involved in the formation of polymer crystals, as detailed below. It should be noted that in many (typical) experiments these three phenomena are strongly intermingled. In such cases, it may be difficult to unambiguously attribute the resulting structure to the basic processes.

A: How Does Polymer Crystallization Start?

As in any first order phase transition, growth will only begin after a nucleus of sufficient size has been formed. Also here, the special difficulty of polymers arises from the connectivity of the individual segments. What would be the ideal way and what is the most efficient way to order these segments in stems and to nucleate a crystal? Do (spinodal) density fluctuations in the still molten state induce alignment of the stems or does local (partial) alignment of chain segments in the undercooled melt lead to density fluctuations? While it is clear that certain substrates or "nucleating agents" may favor or enhance nucleation, a profound understanding of the fundamental processes is still missing. Several new and interesting experimental observations have stimulated research. Small angle X-ray scattering allowed to detect spinodal-like density fluctuations or transient mesophases well before the appearance of a signal in the wide angle range, which would indicate the onset of crystal growth. Alternatively, the early stages of crystallization have been viewed as a physical gelation process. In addition, promising attempts to answer the relevant questions of polymer crystal nucleation come from computer simulations and experiments on crystallization in restricted geometries like e.g. in the mesophases of block copolymers. Understanding the nucleation process may eventually allow to predetermine where crystallization should start. This, in turn, will allow to direct the growth of polymer crystals and so may provide us with possibilities of tailoring (macroscopic) material properties.

B: How Do Polymer Crystals Grow?

A key process in growing any crystal concerns the transport of molecules towards the crystal front. If this process starts to control growth one may observe also for polymers dendritic patterns similar to snowflakes as in any other material. This analogy becomes clearly visible when growth in two dimensions is studied. There, patterns may be explained by rather simple models of diffusion controlled aggregation (instability of the growth front). While the underlying fundamental process (crystallization) creates order on a molecular scale, the instability of the growth front is responsible for some kind of disorder on larger scales. Spherulites (radially grown superstructures consisting of an arrangement of lamellar structures) may be considered as objects exhibiting disorder and order at the same time. In some sense, the structure is a single crystal as growth of the initial skeleton links all parts of the spherulite in a definite way. However this initial skeleton of the spherulite does not fill space (a large amount of material was not crystallized when the growth front passed). Thus, at a later stage additional lamellae may be formed (one may call this secondary crystallization) to fill the space in between the skeleton of the spherulite. Using a focused micro X-ray beam, one can detect that the crystallinity at the growth front is far less than at some distance behind this front. Secondary crystallization is strongest directly behind the growth front. Thus, one may conclude that on large enough scales spherulites cannot be compact objects but are rather open (e.g. fractal) structures. This conclusion is consistent with direct AFM measurements.

Growth of polymeric crystals certainly depends on the conformational properties of the polymer chains. Using special ways (freeze-drying, melting of chain extended crystals) to prepare non-equilibrated amorphous states of only partially "entangled" polymers allow to show that spherulites grow faster from partially "disentangled" states. Besides this effect, for long polymers molecular weight does not affect crystal growth significantly. This may indicate that the resulting crystalline structure (e.g. the lamellar thickness, but maybe also a the lateral extension of an initially formed mesostructure) is smaller than the size of the unperturbed polymer coil in the melt.

It has to be noted that due to connectivity polymer crystals are anisotropic, at least at the level of a lamella. Thus, three dimensional growth is only possible by the superposition of lamellae (e.g originating from screw dislocations), twisting of lamellae, or stacking of (progressively tilted) lamellae. Consequently, on large scales (many microns, significantly larger than a single lamella), growth may result in superstructures like banded spherulites, caused by periodically varying orientation of the radially growing crystal entities.

C: What Happens to the Metastable Polymer Crystals AFTER They Have Been Grown?

As already noted above, polymer crystals are typically non-equilibrium structures. Thus, whenever possible, such crystals will relax leading to re-arrangements and also to morphological changes. Such changes may occur even at temperatures far below the melting point. In some cases, samples stored at room-temperature for extended times may exhibit different morphologies and melting behaviors even if the melting point is above 100°C. Naturally, relaxations are more pronounced when the melting point is approached. Consequently, it is almost impossible to vary the temperature of a polymer crystal without also changing it. In particular, when one melts a polymer crystal one never melts the same crystal which was grown initially. In our opinion, this reflects the major difficulty in characterizing thermodynamic properties of polymer crystals.

What Are the Major Issues for the Future? From what we said above, we conclude that one of the most important tasks for future research on polymer crystallization is to identify additional parameters besides lamellar thickness, growth rate, habit, or, on a more macroscopic level, crystallinity, which are necessary to characterize polymer crystals more completely. Morphology on all length scales and thermal history have to be taken into account more explicitly. The identification and a clear definition of the relevant parameters may also remove some ambiguity existing at present and arising from the use of parameters (e.g. the crystallinity of a sample) which for the most part hide the complexity of the process and the corresponding structure. Just as an example, the same crystallinity may be obtained for two completely different morphologies resulting from different thermal pathways. Similar statements may be true for the lamellar thickness.

The metastability of polymer crystals has long been realized but the consequences resulting thereof are still only partially understood. The mobility of chains in the crystalline state certainly represents a major factor which needs to be understood in more detail. The special importance of crystal (lamellar) boundaries and the relevance of regions in between the "purely" crystalline and the "purely" amorphous parts have to be considered more explicitly.

Some insight into the process of polymer crystallization may also come from model studies of how external influences and constraints can and do affect the formation of polymer crystals. Such external constraints may be caused by flow, temperature gradients, magnetic or electric fields. Confining crystallizable polymers in thin films or block copolymer mesostructures may help to reduce the number of possible metastable states and thus allow to study how and why these metastable states are selected and realized.

Certainly, additional efforts are needed to overcome experimental difficulties by improving techniques, by combining complementary techniques, or by choosing and designing the optimal samples. Here, polymer chemistry may be of significant use. One also may expect new insight into the fundamental processes coming from computer simulations. Eventually, theoretical approaches may emerge which can deal with the formation and the subsequent evolution of non-equilibrium states of complex molecules. Such approaches may go far beyond polymer systems and be relevant in the field of non-equilibrium phenomena in general. At the end, understanding the principles of how how synthetic polymers order may also provide some information on ordering processes in more complex biological systems found in Nature.

Subject Index

Druck: Strauss Offsetdruck, Mörlenbach
Verarbeitung: Schäffer, Grünstadt